Quantitative Studies in Agrarian History

Quantitative Studies in Agrarian History

EDITED BY

Morton Rothstein
Daniel Field

Iowa State University Press / Ames

Morton Rothstein is professor of history at the University of California, Davis, where he has been the editor of *Agricultural History* since 1984. He has written widely on U.S. trade in agricultural commodities and on planter-businessmen elites in the antebellum South.

Daniel Field is professor of history at Syracuse University, specializing in Russian history, on which he has written many books and articles. He served as the editor of *The Russian Review* for a full term in the 1980s and translated the articles of Russian scholars for the special issue of that publication devoted to quantification.

Authorization to photocopy items for internal or personal use, or the internal or personal use of specific clients, is granted by Iowa State University Press, provided that the base fee of $.10 per copy is paid directly to the Copyright Clearance Center, 27 Congress Street, Salem, MA 01970. For those organizations that have been granted a photocopy license by CCC, a separate system of payments has been arranged. The fee code for users of the Transactional Reporting Service is 0-8138-1673-4/93 $.10.

♾ Printed on acid-free paper in the United States of America

First edition, 1993

ISBN 0-8138-1673-4

Library of Congress Cataloging-in-Publication Data

Quantitative studies in agrarian history / edited by Morton Rothstein, Daniel Field. —1st ed.
    p.   cm.
    Includes bibliographical references.
    ISBN 0-8138-1673-4 (acid-free paper)
    1. Agriculture—Economic aspects—United States—History—Congresses. 2. Agriculture—Economic aspects—Russia—History—Congresses. 3. Agriculture—Economic aspects—History—Econometric models—Congresses. I. Rothstein, Morton. II. Field, Daniel.
  HD1755.Q36   1993
  338.1′0973—dc20   93-24729

CONTENTS

Part II. RUSSIAN SCHOLARS

INTRODUCTION

These essays were prepared for a conference held in Tallinn, Estonia, under the auspices of the Soviet Academy of Sciences, the American Council of Learned Societies, and the International Research and Exchanges Board (IREX). The Soviet papers appeared in a special issue of *The Russian Review,* 47 (October 1988) no. 4. A special issue of *Agricultural History,* 62 (Summer 1988) no. 3, edited by Alan Olmstead and Peter Lindert, comprised the American contributions.

Together they are tangible evidence of the conference's success as part of a series of meetings to share ideas and problems concerning "Quantitative Research in History." As editors of the two journals, we believe they complement each other in addressing issues of agrarian history. They demonstrate the power and the potential of relatively new and rapidly developing research techniques and technology. We also believe they deserve the wider audience that a book combining both sets of essays can reach.

On a bright morning in early June 1987, a delegation from various parts of North America who had gathered in Helsinki, Finland, sailed from that port to Tallinn, Estonia, where our Soviet confreres had gathered, along with Jeremy Atack, who had arrived in the former USSR some weeks before. Tallinn had been the site of several other conferences, under the same auspices, on quantitative techniques in historical research. Like those, ours needed much advance preparation and negotiation, beginning with a gathering in 1984 that the International Research and Exchanges Board sponsored in Montreal, Canada. The idea for a conference on this subject came from Carol Leonard, of SUNY and the Russian Research Center, Harvard University, who is deeply involved in research on eighteenth- and nineteenth-century Russian rural life. She served as an effective intermediary between the Soviet scholars, led by Academician I. D. Koval'chenko, and a small band of North American researchers, led by Alan L. Olmstead, then Director of the Agricultural History Center, University of California, Davis. Several other meetings while cooperative work was under way brought the groups closer together, including a memorable dinner at the 1986 International Congress of Economic History in Bern, Switzerland. At critical early stages in planning, George Grantham, of McGill University, Montreal, provided special help. He also was a vigorous participant in discussions during the conference itself.

What the essays cannot convey is the strong sense of camaraderie that developed among the participants, in spite of some frustrations due to language barriers. There were frank exchanges, not only at the tables during formal sessions, but on the buses that took many of us to visit an outdoor agricultural museum and kolkohz outside Tallinn, the sight-seeing walks around that beautiful city, the special tours of recently restored palaces near Leningrad, as well as of the Hermitage, and the visits to several sites in and near Moscow, including a lunch at Moscow State University. Throughout their journeys, the American delegation piled up special debts to Sergei Stankevich, who managed many of the details of negotiating with hotels, transport facilities, restaurant managers, and so forth. Since guiding us, he won election from a district in Moscow to the Congress of People's Deputies and served as Deputy Mayor of Moscow and in President El'tsin's cabinet.

The contacts have been sustained and have grown stronger. Our chief host in Tallinn, Juhan Kahk, visited California, among other places, in 1988; the gifted historian Boris Mironov, who has published several major articles in western journals, also visited Davis and other American institutions in 1989, the University of Toronto in 1991, and the Kennan Institute in 1992. Finally, Academician V. A. Takhanov, president of the Union of USSR Amalgamated (Agricultural) Cooperatives, visited campuses in California and Iowa, among others, in the spring of 1990. He is also a deputy to the Congress of People's Deputies and served as a major advisor to President Gorbachev on farm policy. We hope that from these and other meetings between scholars sharing mutual interests in farming and rural life, past and present, in both societies, more essays will emerge to inform academics and the public about each nation's accomplishments and the problems each encounters in improving their farm sectors.

The four essays by Russian scholars are, by and large, representative of the application of quantitative methods to problems of Russian history.[1] Despite the long-standing and increasing interaction between Russian and American agrarian historians, these articles emerge from a tradition distinct from that within which Parker, Wright, and their compatriots operate. Of the six authors, five are historians by training; in the United States, by contrast, economists do much of the significant work in economic history. Nor are these authors, strictly speaking, cliometricians, since cliometrics ultimately rests upon the concepts and vocabulary of the neoclassical economists. Soviet historians, like Soviet economists and policy makers, prefer physical to monetary measures and attach very little importance to such concepts as rent and interest. (The turn toward a market economy in Russia may gradually undercut this preference.) On the other hand, traditions extending back to the prerevolutionary period encourage Russian historians to employ quantitative methods. In the English-speaking world, scholarly study of history derived from law, journalism, and even literature, which to some extent provide models for historians even today. In the USSR, the natural sciences provided the corresponding model; in the Russian language, as in French and German, history is a science (*nauka*) without any modifier. Hence the rigor, no less than the power, of quantitative analysis made it congenial to Soviet historians.

The articles by Selunskaia and by Koval'chenko and Borodkin are characteristic of Soviet quantitative historical literature in their attempt to embrace Russian agriculture as a whole and to establish the trajectory of its development.[2] Also characteristic are their search for the economic determinants of the revolutions of 1917 and their attempt to establish the peasants' contribution to the agricultural economy. The concern with typology in Koval'chenko and Borodkin's article reflects a new preoccupation with pattern recognition, cluster analysis, and the like—a tendency perhaps associated with the somewhat easier and more regular access to computers that historians in Russia are beginning to enjoy. Until recently, computer time was severely rationed, which made, on the one hand, for a salutory concern for research design and, on the other, for a certain inflexibility.

Oleg Bukhovets's quantitative study of texts is one of the first of its kind undertaken by a historian of Russia, but Soviet historians of classical and medieval Europe have used computers to analyze a number of literary and narrative sources. As for the article by Milov and Garskova, the authors have blazed a trail into a period and subject matter that are new for quantitative methods but richly endowed otherwise. A tradition extending back to before 1917 confers maximum prestige on historians of medieval and early modern Russia; V. O. Kliuchevskii, the most renowned of all Russian historians, was a specialist on the sixteenth and seventeenth centuries. The remarkable novelty of Milov and Garskova's findings concerning the *locus classicus* for Russian historians testifies not only to their acumen and diligence but also to the power of quantitative methods.

The Russian Review is devoted strictly to Russian, rather than Soviet, studies; *Agricultural History* is more eclectic, publishing work on farming in any time or place, so it includes the article by Juhan Kahk on agricultural mechanization in nineteenth-century Estonia. Kahk's article drew on extensive work in local records to analyze the introduction of farm equipment into the part of the Baltic now roughly within the borders of present-day Estonia. He presented evidence about the reaction by farmers to the particular machines they bought, the relative costs of competing technologies, and speculations on the role of persistent manorial farms in impeding the adoption of these new machines. Also rather traditional in subject and methods is Rothstein's survey of farmer movements and organizations in late nineteenth-century United States, a story with some parallels in Canada, where it reached its climax well into the twentieth century. His main concern is assessing recent work on both levels of membership and effectiveness of these movements. He also explores the implicit irony that although the more radical groups were influential in pushing broad reform efforts of the time, which have attracted much scholarly interest, they were on the whole unable to get government help in improving their lot. It was the more conservative farm organizations and their leaders who got help for commercial farmers at state and national levels even as their numbers declined relative to those in other economic sectors.

The rest of the U.S. group come from economics departments and were trained in the use of theoretical models, estimation and sampling techniques, and in handling large masses of data from previously untapped sources or their surrogates in previously unimaginable ways with computer technology. These hallmarks of the "new economic history," as we called it until very recently (*cliometrics* as a descriptive term had a smaller circulation, though derived from econometrics), have enabled scholars to raise and answer new questions about the past and to test old verities. One of the pioneers in such endeavors is William N. Parker, who with Robert Gallman began (about three decades ago at the University of North Carolina) a project for intensive sampling of the manuscript census data for 1850 and 1860 for almost every county in the eleven southern states in order to find more precise answers to long-standing questions about the region's deviant agricultural development.[3] It was wholly appropriate, therefore, that he led off the free discussion of the conference participants in a windup seminar at Moscow State University. In the paper he prepared for the meeting, he assesses the work he began and that others followed in explaining differences in labor productivity for different crops. The first results of that work appeared about the same time that Conrad and Meyer used new calculations based on rent theory to put any discussion of the profitability of antebellum slavery in the United States on a new footing and that Robert W. Fogel began his efforts to measure more precisely the contributions of railroads in nineteenth-century America. The scale and techniques used in these studies, as well as the style of reporting results, were quite different, but there is no doubt that they have put the field through a virtual revolution.[4]

Jeremy Atack used logit regression techniques to explore the causes of tenancy in the North in 1860, using a data base drawn from manuscript census materials for sample counties in that region which he and Fred Bateman, Indiana University, have worked on for some time. This essay, which appraises much of the previous literature, is an extension of *To Their Own Soil,* the prize-winning book these two scholars produced in 1987, yet it moves across much of the nineteenth century to address questions of social structure and class in that region. Gavin Wright, too, extends the analysis of his recent major book on the American South toward a class analysis that examines tenancy rates, farmer debt, and urban labor markets of a region that slowly was undergoing the transformation of its farm economy by mechanization.

Peter Lindert takes his subject to a broader macroeconomic level with a study of farmland values over a century and a half, asking basic questions about periods of instability, in real terms, that he identifies at three widely separated points in time. He relies on summary data for the nation as a whole, as well as general-equilibrium reasoning, to pinpoint three periods of great instability (the most severe from 1973 to 1988) and offers some tentative interpretations of their causes. Roger Ransom and Richard Sutch use similar techniques to apply their recent work on savings rates among various social groups and the relatively weak structure of southern financial institutions to give greater analytical bite to work on the postbellum South.

Finally, Alan Olmstead and Paul Rhode use their analytical tools to produce measures about the exceptional speed with which California farmers adopted and adapted the rapidly evolving animal-powered machinery that dominated their production, especially of small grains during the late nineteenth and early twentieth centuries. They emphasize the rational market response of the landowners; the complementarity of various implements and machines in the production process; and the special nature of geography, soil, and climate that made returns on such investment unusually high. An elegant exercise in terms of substance, their essay also throws into bold relief the article on Estonian mechanization.

Indeed, there is a common thread to all the articles. They are not simply the product of further digging in the archives or private holdings. Questions come first—basic questions about important issues in agrarian history and about the need to explain more cogently the differences between nations and between regions within those nations. Then the search for better answers leads to the search for more and better evidence, to the mustering of that evidence with the insights of theory, and to the increasing ease of handling data through computer technology. In this sense, the diffuse subjects addressed in this volume should be seen as promising samples of the various ways we might all reach a deeper and firmer understanding of our subject.

Daniel Field
Morton Rothstein

Notes

1. For relatively recent overview, see L. I. Borodkin, *Mnogomernyi statisticheskii analiz v istoricheskikh issledovaniiakh,* Moscow, 1986.
2. An example available in English is I. D. Koval'chenko and N. B. Selunskaia,

"Labor Rental in the Manorial Economy of European Russia at the End of the Nineteenth and Beginning of the Twentieth Centuries," *Explorations in Economic History,* 18 (no. 1, 1981): pp. 1–20.

3. A somewhat dated, but still valuable collection of essays that demonstrate the sweep of the "new" history is Lance Davis, et al., eds., *American Economic Growth: An Economist's History of the United States* (New York: Harper and Row, 1972), which has two outstanding essays by Parker, one on natural resources, the other a model overview of agriculture, pp. 369–417. It followed by one year a collection of reprinted articles, edited by Fogel and Stanley L. Engerman, *The Reinterpretation of American Economic History* (New York: Harper and Row) with a rather more contentious introduction.

4. For some early examples of Parker's work and that of his colleagues and students, see the collection he edited for *Agricultural History,* "The Structure of the Cotton Economy of the Antebellum South" (January 1970); for Parker's overview of the literature on the region after the Civil War, see his "The South in the National Economy, 1865–1970," *Southern Economic Journal* 46 (April 1980): 1019–48. An excellent overview that includes pungent assessments of the literature is Susan Previant Lee and Peter Passell, *A New Economic View of American History* (New York: W. W. Norton, 1979).

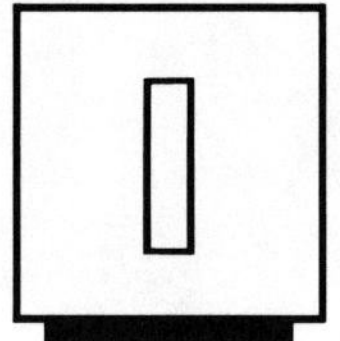

American Scholars

Tenants and Yeomen in the Nineteenth Century

JEREMY ATACK

A widely-held belief in American agricultural history has been that anyone who wanted land in the nineteenth century could have it—courtesy of the benevolent and enlightened public land policy.[1] Under such conditions tenant farming was thought unlikely. It was therefore something of a surprise that when the first statistics on tenancy were collected in 1880 they showed that nationwide a quarter of all farmers did not own their own land. This proportion seemed high and thereafter it showed a steady and marked propensity to rise.[2]

In light of the significant advantages to landownership and inferior economic and social status of the tenant farmer, it seems unlikely that large numbers would voluntarily chose tenancy. Therefore in the absence of statistical evidence to the contrary, apologists sought solace in the assumption that tenancy must have been lower—indeed, many argued, nonexistent—at some earlier time when land was relatively more abundant and there existed unsettled land on the frontier. Only one voice, that of Paul W. Gates, was raised in disagreement. He argued that the very institutions that were supposed to have promoted owner-occupancy through the progressive liberalization of terms, in fact, promoted land speculation, monopolization, and tenancy.

JEREMY ATACK is Professor of Economics, University of Illinois and Visiting Professor of Economics, Harvard University. Earlier versions of this paper were given at workshops at the University of Illinois, Indiana University, and the bilateral Soviet-American Conferences on Quantification in Agrarian History in New Orleans, 1986. The author benefited from the comments of the participants and the able research assistance of Dan Barbezat who helped program the Logit model in SAS.

1. See, for example, the extensive literature on the "Safety Valve," especially Frederick Jackson Turner, *The Frontier in American History* (New York: Henry Holt & Co. 1920), and Ray H. Billington, *Westward Expansion: A History of the American Frontier* (New York: Macmillan & Co. 1967).

2. U.S. Department of the Interior. Census Office, *Report upon the Statistics of Agriculture* (Washington DC: GPO, 1883) Volume 3 of the Tenth Census, pp. xiii–xiv.

In this paper, I review the debate over rising tenancy and present the available quantitative and qualitative evidence on the trends in the northern half of the United States during the nineteenth century.[3] I then develop estimates of tenancy rates for the northern half of the United States in 1860—a time when there was unsettled land on the frontier in Kansas and Minnesota and public policy was ostensibly directed toward promoting smallholdings. These estimates reveal levels of tenancy that are generally consistent with those in 1880 and the trend thereafter. Using information on the personal, familial, and farm characteristics of these individual tenants and yeomen in 1860 I then determine the probability that a person with particular characteristics would be a tenant or owner-occupier at that time.

The Superintendent of the 1880 Census, Charles W. Seaton, described the first official data on tenancy as of the "highest economical and sociological importance," and drew attention to the sharp regional differences in tenancy rates.[4] The tenancy rate was lowest in the Northeast, 16 percent, and highest in the South where it exceeded one-third (Table 1).[5] In the Midwest, it lay between these two extremes at just over 20 percent. The high incidence of southern tenancy might be excused as an aberration induced by emancipation and the failure to provide the freeman with 40-acres and a mule. However, it is harder to rationalize the higher rate of tenancy in the Midwest *vis á vis* the Northeast. For example, the Northeast had been settled earlier and hence those forces promoting economic concentration such as luck, superior ability, and inheritance had had longer to operate. Moreover, settlement, particularly in the Middle Atlantic states, had often taken place under adverse tenure conditions.[6] In contrast, settlement in the Midwest was much more recent and land alienation had taken place under the increasingly liberal provisions of successor land legislation to the 1785 Ordinances. Furthermore, by the time the 1880 Census was taken, western parts of the Midwest had experienced eighteen years of homesteading where 160 acres was free to those who cultivated it for five years.

3. This paper was originally commissioned by Alan Olmstead as an article detailing a typology of American agriculture but it remains far short of that much broader goal. It deals with only one typological division that between tenant and yeoman and even then only within the context of the northern half of the United States during the nineteenth century. By focusing upon the North, I am able to avoid the complex changes in southern agriculture wrought by Emancipation. By concentrating upon the nineteenth century, I can ignore the effects of the tremendous technological changes that have taken place in American agriculture since the First World War and the current high degree of agricultural dependency upon other sectors of the economy and the rest of the world.

4. U.S. Department of the Interior. Census Office, *Statistics of Agriculture,* pp. xiii–xiv. Seaton assumed the post of Superintendent of the Census in 1881 following the resignation of Francis A. Walker. He held the post until it was abolished in 1885.

5. Excluding the eight Mountain states where there were only about 25,000 farms.

6. For example, the Hudson Valley was originally settled under a manorial system reminiscent of feudal Europe and Pennsylvania through the Crown grants to William Penn.

The Census, however, did not pursue the paradox of tenancy under the supposedly favorable land settlement conditions in the Midwest and virtually no analysis of the data was conducted. Instead, the Census reprinted an optimistic article by Francis A. Walker, the former Superintendent of the Census, on American agriculture. In his view, the data showed that the land tenure system in the United States was "highly popular." He attributed this in part,

> to the existence of vast tracts of unoccupied lands "at the West" whatever that phrase may at the time have meant . . . [and also to] the liberal policy of the government relative to the public domain; partly to excellent laws for the registration of titles and the transfer of real property . . . ; and partly to the genius of our people, their readiness to buy or sell, to go east or to go west, as a profit may appear.[7]

When the inquiry was repeated at the Eleventh Census in 1890, it was found that tenancy had risen in every region except the West. Once again, though, the data were viewed with little concern and no extensive analysis was performed. Indeed, the Census officials drew an erroneous inference from the data which show an increase in the total number of farms of 555,734 from 1880 with 285,422 more owner-occupied farms. Consequently, the number of tenant farms increased by 270,312.[8] The Census Office, however, concluded, mistakenly, that these additional tenant farms came only from the population of new farms established during the decade and not by the decline of existing yeomen into tenancy.[9]

Tenancy rates rose even more sharply in most areas between 1890 and 1900 (see Table 1) and at last the data began to attract more official attention. Their analysis by the Superintendent of the Twelfth Census in 1900 was more sophisticated and extensive than in the past. Nevertheless the official view remained that the rising level of tenancy was not a cause for concern:

> It was taken for granted almost universally that the number of tenants was increasing at the expense of the number of owners and that the movement expressed by the increase of tenancy was an ill omen for the republic . . . [but] the popular conclusion overlooks some very important social facts . . . that the farms operated by owners have increased faster

7. Francis A. Walker, "American Agriculture" as reprinted from the *Princeton Gazette* with addition from the *Agricultural Review* by the Tenth Census. See U.S. Department of the Interior. Census Office, *Statistics of Agriculture,* p. xxviii.

8. That is, by the difference between the total number of farms and the number of owner-occupied farms.

9. U.S. Department of the Interior. Census Office, *Report of the Statistics of Agriculture* (Washington DC: GPO, 1895), Volume 5 of the Eleventh Census, pp. 3–6, especially p. 5.

Table 1. Percentage of All Farms Operated by Tenants, by Geographic Region, 1880–1900

Region	1880	1890	1900
New England	8.5	9.3	9.4
Middle Atlantic	19.2	22.1	25.3
East North Central	20.5	22.8	26.3
West North Central	20.5	24.0	29.6
South Atlantic	36.1	38.5	44.2
East South Central	36.8	38.3	48.1
West South Central	35.2	38.6	49.1
Mountain	7.4	7.1	12.2
Pacific	16.8	14.7	19.7

Regions are as follows:
New England: CT, ME, MA, NH, RI, VT.
Middle Atlantic: NJ, NY, PA.
East North Central: IL, IN, MI, OH, WI.
West North Central: IA, KS, MN, MO, NB, ND, SD.
South Atlantic: DE, DC, FL, GA, MD, NC, SC, VA, WV.
East South Central: AL, KY, MS, TN.
West South Central: AR, LA, OK, TX.
Mountain: AZ, CO, ID, MT, NM, NV, UT, WY.
Pacific: CA, OR, WA.

Source: E. A. Goldenweisser and Leon E. Truesdell, *Farm Tenancy in the United States,* (Washington DC: GPO, 1924), p. 23. (US Department of Commerce. Bureau of the Census, *Census Monograph IV*).

since 1850 than the agricultural population. Such an increase can only be possible providing the increase in the number of tenants has been by the elevation of former wage employees to the position of farm tenants. Such an increase in the number of tenants has been by recruits from the ranks of wage employees and not from farm owners or their children.[10]

Not until the publication of the results of the Fourteenth Census for 1920 were the data on tenancy examined seriously. They were then the subject of a separate monographic study.[11] The authors, Goldenweisser and Truesdell, described farm tenure as a two-fold problem, on the one hand dealing with the relationship between the cultivator and the land and on the other, with the distribution of wealth between those who furnish land, those who furnish capital, and those who provided the labor for farming. As economists, they argued that the economic advantage of one

10. U.S. Census Office. Twelfth Census, *Agriculture, Part I,* p. lxxvii. The view of tenancy as a way-station on the climb up an agricultural ladder is a common one in the literature and one to which I return below.

11. E. A. Goldenweisser and Leon E. Truesdell, *Farm Tenancy in the United States,* Washington, DC: GPO, 1924 (U.S. Department of Commerce. Census Bureau. *Census Monograph IV*)

status versus the other could be determined by comparing the annual cost of land possession under the two forms of tenure but they concluded that "it is accepted as a foregone conclusion that ownership is preferable . . . based in part upon conditions purely accidental and having little connection either with the net income of individual farmers or with the productivity of agriculture in general."[12] Among the forces that they identified as promoting owner-occupancy were the federal land policy of putting land directly into the hands of those who would cultivate it, rising land values and the importance of these to farm profitability, and the lack of an efficient system of leases.[13] Like earlier observers of tenancy, Goldenweisser and Truesdell thought that "when a given area was newly settled, especially so long as free land was to be had, there was little tenancy."[14]

What then gave rise to the steadily increasing rates of tenancy? Goldenweisser and Truesdell attributed it to "the fact that free land was practically exhausted by 1900, and . . . to the hard times that prevailed in the nineties and caused a large number of mortgages to be foreclosed, making it necessary for many farm operators to rent farms in order to continue farming."[15] Tenancy was therefore a very recent phenomenon, but one expected to persist.

The closing of the frontier and economic adversity were also cited as causes of rising tenancy in the *1923 USDA Yearbook.* There, however, the authors (Lewis C. Gray among them) went one step further and attributed long-term tenancy to the below average capabilities of those farmers.[16]

Not until the Report of the President's Committee on Farm Tenancy in 1937 were these views seriously challenged by a more sympathetic portrait of the tenant farmer and concern with the rising tenancy rates.[17] In the Committee's analysis, tenant farmers were trapped by the system in an economically and socially undesirable situation and had little chance of escape to the freedom of independent owner-occupancy. In part, the Committee blamed current economic conditions but they also thought that "policies for disposing of the public domain have permitted acquisition of large areas, mostly for speculative purposes, by those who have no intention of farming them."[18] They thus echoed the views of Paul Gates that were then appearing in the scholarly journals.

12. Ibid., p. 12.
13. Ibid., p. 13.
14. Ibid., p. 13 and p. 19.
15. Ibid., p. 21.
16. L. C. Gray et al., *Farm Ownership and Tenancy, USDA Yearbook of the Department of Agriculture 1923,* pp. 507–600.
17. National Resources Committee, *Farm Tenancy: Report of the President's Committee* (Washington, D.C.: GPO, February 1937).
18. Ibid., p. 6.

Tenancy, however, received scant attention from the first generation of American agricultural historians. In *the* classic study of northern agriculture, Bidwell and Falconer make virtually no mention of tenancy.[19] For them, it simply was not an issue in American agriculture before the Civil War. Where tenancy was mentioned, the traditional view was that cheap or free land promoted owner-occupancy. This is exemplified by Percy Bidwell's dismissal of tenancy in his essay on the rural New England economy at the start of the nineteenth century with the simple statement that "it is well known that almost every farmer owned his own land." Similar statements were made later for lands lying further west. Wrote one scholar of Illinois in the 1850s, "with so much land yet unoccupied, the cultivated portions could command but little rent and tenancy was not common."[20] Statistically unseen and—by argument—logically impossible, tenancy could not have existed.

Paul Gates, however, was less easily convinced.[21] Reversing the logic of the traditional argument that free or cheap land promoted ownership, Gates claimed that this very same system in fact promoted the growth of tenancy:

> The Land Ordinance of 1785 and subsequent laws had placed no restrictions upon the amount of public land that individuals or groups could acquire . . . The policy of unlimited sales and unrestricted transfer of titles made possible land monopolization by speculators, who acquired most of the choice lands in certain areas . . . This resulted in the early disappearance of cheap or free land and the emergence of tenancy.[22]

The result was, he argued, an incongruous land system that fostered tenancy at the expense of ownership. Cheap land was a double-edged sword. Not only was it more affordable for the person with limited means, but in the absence of restrictions on the size of holdings the wealthy could buy huge tracts at minimal cost. Passage of the Homestead Act in 1862 did

19. Percy W. Bidwell, "Rural Economy in New England at the Beginning of the Nineteenth Century," *Transactions of the Connecticut Academy of Arts and Sciences* 20 (April 1916): 241–399. The quote is from p. 371. Percy W. Bidwell and John I. Falconer, *History of Agriculture in the Northern United States, 1620–1860* (Washington DC: Carnegie Institution, 1925), 242.

20. Russell H. Anderson, "Agriculture in Illinois During the Civil War Period, 1850–1870" (unpublished Ph.D. thesis, University of Illinois, 1929), p. 63.

21. See Paul W. Gates, "The Homestead Law in an Incongruous Land System," *American Historical Review* 41 (1936): 652–81. Also "The Role of the Speculator in Western Development," *Pennsylvania Magazine of History and Biography* 66 (July 1942): 314–33; "Land Policy and Tenancy in the Prairie States," *Journal of Economic History* 1 (May 1941): 60–82; "Land Policy and Tenancy in the Prairie Counties of Indiana," *Indiana Magazine of History* 35 (March 1939): 1–26; "Frontier Landlords and Pioneer Tenants," *Journal of the Illinois State Historical Society* 38 (June 1945): 142–206.

22. Gates, "Land Policy and Tenancy," 3.

little to relieve this because of the ability of speculators to find dummy entrymen and take advantage of the commutation privilege. As a result, large tracts of land were acquired by speculators, land companies, and the wealthy. Moreover, the railroads were given vast acreages through federal land grants.[23]

In Gates' view, speculative landholdings and railroad land grants reduced the market supply of land and bid up the price. Since demand was inelastic, would-be farmers were therefore compelled to spend more of their income on land than would otherwise have been the case. Those who could not afford to pay the higher prices or borrow were faced with a choice of farming a smaller area or becoming tenants and some may have been excluded from the market altogether.

Gates made a persuasive case, supporting his argument with extended discussions of the operations of land speculators such as Samuel Allerton, Michael Sullivant, and William Scully, institutional landholders such as the Illinois Central Railroad, and the actions of promoters of tenancy such as Henry L. Ellsworth.[24] However, the debate could not be closed because the statistical evidence to substantiate the case for early tenancy was lacking. Moreover, Gates' economic argument is incomplete. In particular, it ignores the effect of competition between landlords for tenants which should have driven down rents thus making tenancy relatively more desirable.

Contemporary observers in the early nineteenth century are mute on the issue of tenancy. Indeed, the most insightful of them, Alexis de Tocqueville, opened *Democracy in America* with the statement "nothing struck me more forcibly than the general equality of condition among the people"—something not generally associated with landlord-tenant societies.[25] The foundations for this condition, he argued, were the legal institutions governing land ownership and inheritance and the abundance of the fundamental resource in which the "lands of the New World belong to the first occupant; they are the natural reward of the swiftest pioneer."[26]

While there is an enduring and continuing debate over the extent of Thomas Jefferson's contribution to the Land Ordinance of 1785 and the Northwest Ordinances of 1787, there seems little doubt that his views exercised an important influence on the character of the legislative debate and its outcome. Nor can there be any doubt that Jefferson abjured tenancy. His views on the subject are perhaps best reflected in a letter that he

23. Federal railroad land grants between 1850 and 1870 totaled 131 million acres and Texas granted an additional 27 million acres.

24. See, especially, Gates, "Land Policy and Tenancy," and "Frontier Landlords."

25. Alexis de Tocqueville, *Democracy in America,* (first published, Paris, 1835). Quotes and references are to the Henry Reeve text as revised by Francis Bowen with corrections by Phillips Bradley (New York, Alfred Knopf, 1946), v. 1, p. 3.

26. Ibid., 431.

wrote to James Madison from Fontainebleau, France in 1785 wherein he recounted a conversation with a poor woman whom he met along the road:

> This little *attendrissement* with the solitude of my walk, led me to a train of reflections on that unequal division of property which occasions the numberless instances of wretchedness which I have observed in this country and is to be observed all over Europe.
>
> The property of this country is absolutely concentrated in a very few hands. . . . I asked myself what could be the reason so many should be permitted to beg who are willing to work, in a country where there is a very considerable proportion of uncultivated land. . . . It should seem then that it must be because of the enormous wealth of the proprietors which places them above attention to the increase of their revenues by permitting these lands to be labored. . . . Whenever there are in any country uncultivated lands and unemployed poor, it is clear that the laws of property have been so far extended as to violate natural right. The earth is given as a common stock for man to labor and live on. If for the encouragement of industry we allow it to be appropriated, we must take care that other employment be provided to those excluded from the appropriation. If we do not, the fundamental right to labor the earth returns to the unemployed. It is too soon yet in our country to say that every man who cannot find employment, but who can find uncultivated land, shall be at liberty to cultivate it, paying a moderate rent. But it is not too soon to provide by every possible means that as few as possible shall be without a little portion of land. The small landholders are the most precious part of a state.[27]

In keeping with this philosophy, Jefferson favored giving the land to settlers, "for by selling land to them [poor settlers], you will disgust them, and cause an avulsion to them from the common union. They will settle the lands in spite of everybody," but on this point he lost.[28] The immediate revenue needs of the federal government required that the land be sold and in the short-run the terms for alienation became more restrictive as land prices rose in real terms, credit terms were abolished, and payment in specie demanded. However, in the long-run, the trend was toward relaxation. Minimum acreages were reduced and the preemption rights of squatters were recognized. Furthermore, there was continual pressure from both within and without the government to abolish sales in favor of donation. Thomas Hart Benton, for example, frequently made impassioned speeches on the floor of the Senate on this point, arguing that:

27. Thomas Jefferson to the Reverend James Madison, October 28, 1785.
28. Thomas Jefferson to Edmund Pendleton (?), August 13, 1776.

> Tenantry is unfavorable to freedom. It lays the foundation for separate orders of society, annihilates the love of country, and weakens the spirit of independence. The farming tenant has, in fact, no country, no hearth, no domestic altar, no household god. The freeholder, on the contrary, is the natural supporter of free government; and it should be the policy of republics to multiply their freeholders . . . pass the public lands cheaply and easily into the hands of the people; sell, for a reasonable price, to those who are able to pay; and give, without price, to those who are not able to pay.[29]

This dream was eventually realized in the Homestead Act of 1862, but the evidence from the 1880 Census suggests that tenantry was not eliminated by its passage. Whether or not it was successful in reducing the levels of tenancy, however, remains to be shown in this paper.

If federal land policy was to prove unsuccessful in preventing tenancy in the long-run, there is some evidence of earlier success. According to de Tocqueville, "the English laws concerning the transmission of property were abolished in almost all states at the time of the Revolution. The law of entail was so modified as not materially to interrupt the free circulation of property" with the result that, as of 1830 at least, "the families of the great landed proprietors are almost all comingled with the general mass."[30]

In recent years, a number of quantitative estimates of tenancy pre-dating 1880 have appeared. These rely upon scattered, direct and indirect, evidence. On the eve of the Revolution, for example, it is estimated that there were 6000–7000 tenant farmers in New York.[31] Further south, in the "best poor man's country"—Chester and Lancaster counties, Pennsylvania—James Lemon estimated that perhaps 30 percent of the married taxpayers were landless in the late colonial period.[32] Such rates imply that tenancy was as high (and maybe even higher) at the time of the Revolution as it was a century later. One rationalization of these statistics is that they represent outcomes under old land tenure systems that were based on large individual grants—systems that the Land Ordinances replaced as the basis for

29. See Thomas Hart Benton, *Thirty Years' View* (New York: D. Appleton & Co., 1854), 2 vols., Volume 1, 103–04.

30. de Tocqueville, *Democracy in America,* pp. 50–1.

31. John Watt, *Pennsylvania Ledger: or the Weekly Advertiser,* Oct. 29, 1777 quoted by Sung Bok Kim, *Landlord and Tenant in Colonial New York Manorial Society, 1664–1775* (Chapel Hill: University of North Carolina Press, 1978), vii. Unfortunately, the lack of data on the number of farms precludes expressing this as a tenancy rate. Based on the enumeration at the first census in 1790, however, it seems unlikely that there would have been more than about 80,000 families in New York at the time of the Revolution and that more than 70,000 could have been engaged in farming. On this basis, no more than 10 percent of farmers were tenants.

32. James T. Lemon, *The Best Poor Man's County* (Baltimore: Johns Hopkins University Press, 1972), 94.

settlement in the West. Nothing definite though was known about conditions in the Midwest beyond the speculation by traditionalists that tenancy was unlikely under the circumstances and the denial of this by Gates and his supporters.

However, as a result of the pioneering efforts of Allan C. Bogue and thanks to the diligence of a census enumerator in exceeding the scope of his instructions, we came to have some inferential statistics on tenancy at mid-century for the Midwest. Bogue discovered that in Jones County, Iowa for 1860 the Assistant Marshal had noted the tenure status of a large number of respondents who did not own real estate in that county. Based upon this, Bogue has argued that those persons named in the agricultural schedules as operating a farm and who listed their occupation as farmer but reported no real estate value on the population schedules were tenants.[33] Using this technique, he estimated the 1860 tenancy rates in three Iowa townships as ranging from 6.6 to 11.2 percent.[34] Such tenancy rates are lower than those for colonial Pennsylvania and probably about the same order of magnitude as those in Revolutionary New York State.

Tenancy in Iowa has also been the focus of two other studies by students of Bogue. Using the same methodology as Bogue, Seddie Cogswell studied the relationship between tenancy, age, and nativity in a six-county area of eastern Iowa.[35] In that area, tenancy rates increased irregularly from 17.6 percent in 1850 to 27.3 percent in 1880, though the rate fell between 1850 and 1860 to 15.1 percent, declining in all but two of his sample counties.[36] Cogswell argues that the increasing tenancy did not reflect the emergence of a class of economically distressed farmers in eastern Iowa but rather a changing age structure of the farm population.[37] Indeed, among the 126 farmers whom he can trace between 1860 and 1870, only 2 became tenants whereas the overall tenancy rate during the same period increased by four percentage points.[38] Based upon his analysis of statewide-trends in tenancy in Iowa from 1850, Donald Win-

33. Bogue also identifies a separate class of "farmers without farms" who gave occupations as farmer on the population schedules and reported a zero real estate value but for whom no farm was located in the agricultural schedules. He argues that these comprised optimistic farm laborers and recent settlers in the process of looking for a suitable farm. See Allan G. Bogue, *From Prairie to Corn Belt: Farming on the Illinois and Iowa Prairies in the Nineteenth Century* (Chigago: University of Chicago Press, 1963), 56–66, especially 63–65.

34. These are the recalculated tenancy rates from Bogue, *From Prairie to Corn Belt,* Table 9, p. 65, if "farmers without farms" are excluded from both the denominator and numerator.

35. Seddie Cogswell, Jr., *Tenure, Activity and Age as Factors in Iowa Agriculture, 1850–1880* (Ames, IA: Iowa State University Press, 1975).

36. Ibid., Table 3.1, p. 23.

37. Ibid., pp. 152–3.

38. Ibid., p. 153 and Table 3.1, p. 23.

ters concluded that the changes were "an integral part of an evolving, maturing agricultural system," placing more emphasis upon cash grain farming.[39]

All of the published quantitative studies of tenancy pre-1880 focus on narrow geographic areas—two counties in the case of work by Bogue and Lemon to as large an area as an entire state.[40] The new evidence on early tenancy developed in this study, on the other hand, covers the entire North. It is derived from the large-scale, matched sample of agricultural and demographic data collected by Fred Bateman and James D. Foust. This sample was selected from a population of twenty states from Maryland north and from the East Coast westward to the frontier of settlement. It was stratified to be representative of both the Northeast and Midwest. The data were actually drawn from 102 randomly selected rural townships from the pool of all non-urban counties and townships and sixteen northern states are represented.[41] Statistical tests indicate that the sample is not only a reasonable approximation of the entire North and the two subregions, the Northeast and Midwest, but also that the township data for particular states correspond closely to that for the state as a whole.[42] Generalizations at the state level, however, may be somewhat suspect, especially with respect to a single statistic such as tenancy.

There are 11,943 farms in the sample, 10,288 of which met my criteria for inclusion in this study. To be included, acreage, crops, and farm, implements, and livestock values had to be recorded for each farm. It also had to be matched with a household in which at least one member reported an occupation of farmer, tenant, agriculturalist, or part-time farmer. These criteria are somewhat more stringent than those used by Atack and Bateman and led to the exclusion of a somewhat higher percentage of poten-

39. Donald L. Winters, "Tenancy as an Economic Institution: The Growth and Distribution of Agricultural Tenancy in Iowa, 1850–1900," *Journal of Economic History* 37/2 (June 1977): 382–408. Quote is from p. 406.

40. The statement is qualified because since starting work on the project I have become aware of a chapter in Donghyu Yang's doctoral dissertation from the Harvard University that uses the same data set in a manner similar to that employed here. See Donghyu Yang, "Aspects of United States Agriculture, Circa 1860" (unpublished Ph.D. dissertation, Harvard University, 1985), especially Chapter 3, "Farm Tenancy in the Northern United States, 1860." Yang touches upon some of the issues such as the differences in the demographic and socio-economic characteristics of tenants and yeomen discussed here, in addition, he develops a economic model of the rental market and investigates the productivity differential between tenant farmers and owner-occupiers. The data that I ultimately use for my analysis also differs because of variations in selection criteria and my inclusion of Maryland and Missouri.

41. Fred Bateman and James D. Foust, "A Sample of Rural Households Selected from the 1860 Manuscript Censuses," *Agricultural History* 48 (1974) 1:75–93.

42. Jeremy Atack and Fred Bateman, *To Their Own Soil* (Ames: Iowa State University Press, 1987), Chapter 7 and unpublished work.

tial tenant farmers than yeomanry.[43] As a result the tenancy rates reported here are generally fractionally lower.

Like Bogue, I classify as tenants all those farmers who reported owning no real estate. The lack of written rules for enumerators concerning the treatment of tenant farmers prior to 1880, however, seems to have led to a variety of other coding patterns for tenants elsewhere. In the South, Frederick Bode and Donald Ginter similarly found that enumerators deliberately listed persons who called themselves farmers but who were really tenants as having no real estate, but the convention among southern enumerators was also to list no farm value or acreage on the agricultural schedules. Production data would, however, be recorded. Sometimes there were other variations on this theme such as recording acreage and farm value in whole or in part.[44] Such practices were clearly contrary to the enumerator's instructions and do not seem to have been widespread. Otherwise tenancy rates were extremely low in the northern half of the country.[45] The partial recording of data in particular would confound analysis. To the extent that tenants were poorer than owner-occupiers, we would expect that, other things equal, they would have smaller farms with fewer improvements. It would therefore be difficult if not impossible to distinguish this relative poverty hypothesis from the measurement error implied by Bode and Ginter's alternative hypothesis of the partial recording of relevant values.

There is no evidence of deliberate partial recording of information in the North and there is broad agreement among researchers that farmers without real estate were tenants. Contrary to the practice of others, however, I do not accept the simple dichotomy of farmers into tenants and yeomen. There was, I will argue, a third group, namely those who listed their occupation as farmer or agriculturalist or who claimed to follow two or more occupations one of which was that of farmer and for whom a farm was located in the agricultural schedules but who probably owned only a fraction of the land that they farmed. They are identified by a value of real estate on the population schedules greater than zero but less than the

43. See Atack and Bateman, *To Their Own Soil,* Chapter 7. The tenancy rates reported in *To Their Own Soil* are based only upon those farms that had a farmer. For inclusion here, however, no missing values were tolerated. For example, the tenancy rate reported in Table 7-1 of *To Their Own Soil* for Indiana is 210 per thousand compared with 190 per thousand reported in Table 2 below. Most were closer. The effect of these more stringent criteria, however, was dramatic for the estimate of the tenancy rate in Missouri, reducing it from 205 to 120 per thousand.

44. Frederick A. Bode and Donald E. Ginter, "A Critique of Landholding Variables in the 1860 Census and the Parker-Gallman Sample," *Journal of Interdisciplinary History* 15 (1984): 277–95, esp. pp. 278–80.

45. See U.S. Census Office. Eighth Census. *Instructions to U.S. Marshals. Instructions to Assistants* (Washington DC: G. W. Bowman, 1860.)

value of the farm which they operated. I shall refer to this group as "part-owners."[46] Such farmers are common today and they also proved numerous in the nineteenth century. They constituted about a third of all tenants and have been ignored by other researchers during the nineteenth century.[47] In much of the analysis that follows, I too will ignore them.

Using this classification scheme, the tenancy rates from the Bateman-Foust sample by state, for the sub-regions, and the region as a whole are shown in Table 2.[48] Rates for part-owners and tenants are reported separately so that the data can be compared directly with the estimates by Bogue, Winters, and others. For example, my estimate of 130 tenants per thousand farmers in Iowa is somewhat lower than that reported by Cogswell for eastern Iowa as a whole and for most of his individual sample counties, but is in the upper half of the county estimates given by Winters.[49] Although the part-owner rate is quite strongly positively correlated with the rate for pure tenants, in a number of states more than half of our overall tenancy rate is accounted for by this group whom others would exclude.[50] Indeed, in Michigan, New Hampshire, and Ohio more than 70 percent of the tenants owned some real estate. Rates for part-owners range from 30 to 100 per thousand farms across the sample and averaged 60 per thousand in the sub-regions and for the entire North. Rates for those farm operators owning no real estate whatsoever were much more diverse, ranging from as few as 20 per thousand in Michigan to as many as 450 per thousand in Maryland. Notwithstanding these two extremes, however, the rates were generally higher in the Midwest than the Northeast.

If we count part-owners as tenants, then less than half of the farms in Maryland were operated by yeomen who could have owned their farms outright. At the other extreme, only about 7 percent of Michigan farmers and 8 percent of farmers in Connecticut and New York were tenants.[51] The finding that, in general, tenancy was more prevalent in the newer regions of settlement lends superficial support to Paul Gates' assertions about the

46. This represents a change in terminology (though not group) from that Atack and Bateman used in *To Their Own Soil* where they are referred to as "tenants in part." This name change reflects the results in Table 2 below and is consistent with the terminology of Gray et al., *Farm Ownership* and Goldenweisser and Truesdell, *Farm Tenancy*.

47. The instructions to census enumerators for 1910 and 1920 were unambiguous that farmers renting only a part of their land were to be classified as owners. However, the instructions for earlier censuses did not state how such cases were to be treated by enumerators.

48. To avoid the appearance of unwarranted precision in the estimates, they are reported as rates per thousand farms rounded to the nearest ten.

49. See Cogswell, *Tenure*, Table 2-1, p. 23; and Winters, "Tenancy," 384.

50. Unless noted otherwise, the tenancy rates for 1860 reported throughout this paper are those including part-owners as tenants.

51. The overall figure for New York compares favorably with that by Kim for the Revolutionary period. If part-owners are excluded then the rate is much lower. See Kim, *Landlord and Tenant*.

Table 2.   Estimated Tenancy Rates per Thousand Farms from the Bateman-Foust Sample for 1859–60, by State

State/Region	Overall Tenancy Rate[a]	Tenancy Rate[b]	Part-Ownership Rate[c]
Illinois	170	130	40
Indiana	190	130	60
Iowa	190	130	60
Kansas	330	220	110
Michigan	70	20	50
Minnesota	360	310	50
Missouri	120	60	60
Ohio	130	40	90
Wisconsin	110	50	70
Midwest	170	110	60
Connecticut	80	50	30
Maryland	550	450	100
New Hampshire	100	30	70
New Jersey	150	70	80
New York	80	40	40
Pennsylvania	190	130	60
Vermont	160	50	100
Northeast	150	90	60
THE NORTH	160	100	60

Notes:
   [a]Overall Tenancy rates = Tenancy Rate + Part-Owners Rate
   [b]Those farm operators reporting no real estate value.
   [c]Part-owners owned real estate valued at less than the value of the farm that they operated.
Totals may not sum across the rows because of rounding.

failure of public land law to promote smallholders as Jefferson had desired. In particular, tenancy was especially high on the Kansas and Minnesota frontiers, where the rate exceeded one-third, despite the abundance of cheap land in the immediate vicinity.[52] This directly contradicts those in the 1920s who argued that tenancy could not have existed under such circumstances of land availability.[53] Nevertheless, the relationship seems

52. Even if part-owners are excluded, the tenancy rates in Minnesota and Kansas remain among the highest rate for the sample states. See also Atack and Bateman, *To Their Own Soil,* Chapter 4.

53. It has been suggested that a high percentage of frontier farmers reported no real estate value because they did not yet have property rights in the land that they farmed since they were illegal squatters in the process of securing preemption. I cannot confirm or refute this hypothesis with the data at my disposal.

to have been spurious. Using the Atack and Bateman estimates of land availability for the 102 sample townships and correlating these with the tenancy rate show little correlation between them.[54]

Despite the caveat regarding the use of the Bateman-Foust data at the state level, my estimates of tenancy rates appear consistent with the levels at the Tenth Census in 1880 and with their trend thereafter (Figure 1). In 1937, the *Presidential Committee* surveying the published statistics on tenancy remarked that "for the past 55 years, the entire period for which we have statistics on land tenure, there has been a continuous and marked decrease in the proportion of operating owners and an accompanying increase in the proportion of tenants."[55] Based upon my estimates for 1860 this trend was established much earlier. Rising tenancy in the northern half of the United States thus appears part of a long-term evolutionary process rather than the result of any revolutionary change or sudden crisis. Only the rates for Kansas, Maryland, and Minnesota appear far out of line with what one might have expected based on the post-1880 data and these can be explained. The recording of squatters as tenants on the frontier may account for the high tenancy rates in Kansas and Minnesota. Maryland is even more easily explained by the peculiar atypical nature of the sample observation of Costin District in Worcester County. If the post-1880 data for Maryland are also restricted to this same county, the tenancy rate for 1860 no longer appears that unusual.[56]

The estimates of the tenancy rates in 1860 also show substantial intrastate variation among the sample townships. This is consistent with the findings of Cogswell and Winters using Iowa data.[57] Rates, by broad range, for the individual townships in the Bateman-Foust sample are shown in Figure 2. In Illinois, for example, rates varied from as low as 43 per thousand in Bureau and Whiteside counties to 406 per thousand in Knox county. Although at the township level the number of tenant farmers tended to be small, the data exhibit spatial regularity in that tenancy rates in adjacent townships were generally more similar than those in townships that were more widely removed. There is also a geographic regularity apparent in the data. Tenancy was generally very low (under 100 per

54. For the methodology for estimating land availability, see Donald Leet, "Human Fertility and Agricultural Opportunities in Ohio Counties: From Frontier to Maturity, 1810–60," in D. Klingaman and R. Vedder, eds., *Essays in Nineteenth Century Economic History* (Athens: Ohio University Press, 1975), 138–58. The measure is an index of excess demand for land. The correlation coefficient was estimated at −0.071.

55. National Resources Committee, *Farm Tenancy*, 3.

56. In 1880 the tenancy rate in Worcester County was 48 percent; by 1890, 49 percent; and in 1900, 52 percent. In Costin District, the rates may have been even higher.

57. See Cogswell, *Tenure*, Table 2–1, 23 and Winters, "Tenancy," 384 and Maps 1–3, pp. 401–3.

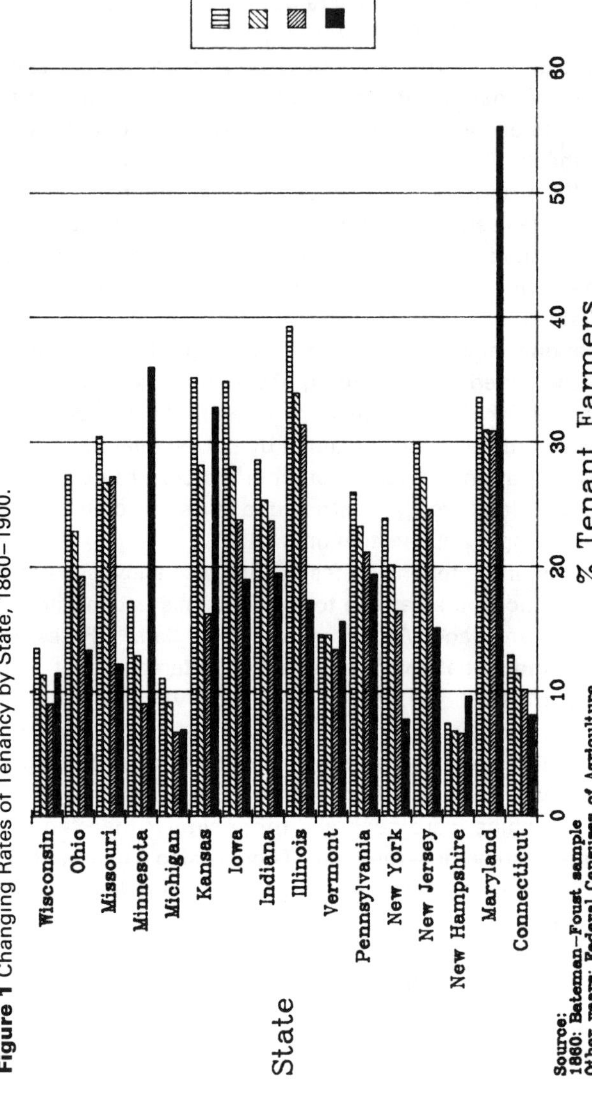

Figure 1 Changing Rates of Tenancy by State, 1860–1900.

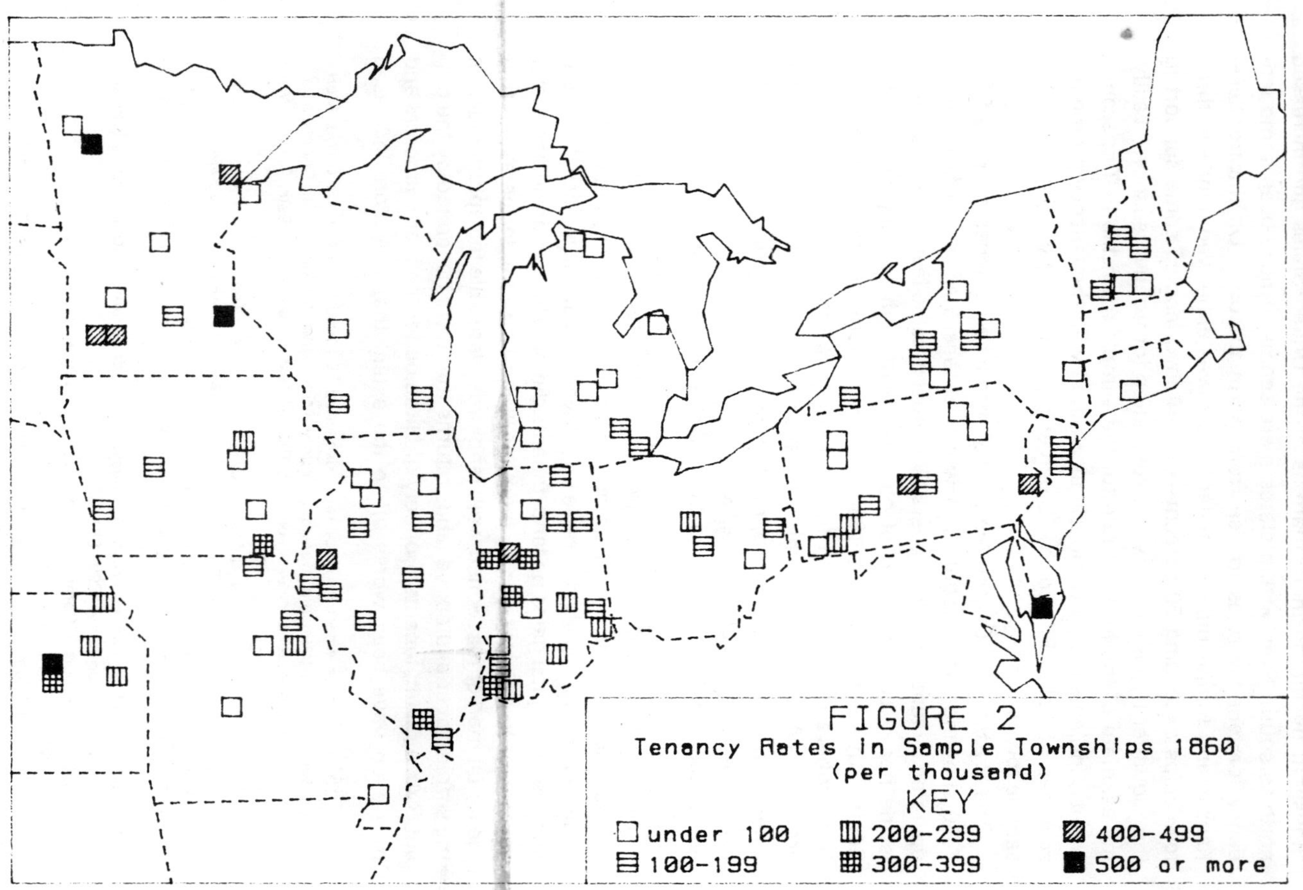
FIGURE 2
Tenancy Rates in Sample Townships 1860
(per thousand)
KEY
under 100
100-199
200-299
300-399
400-499
500 or more

thousand) on the northern margins of the United States and increased north to south. Even within states there seems to have been some tendency for tenancy rates to rise from north to south. On the frontier—Kansas and Minnesota—the tenancy rate was either very high—in three townships exceeding 50 percent—or very low, lending some support to the "squatter hypothesis" as an explanation for frontier tenancy.[58] Lastly, my estimate of the tenancy rate for the sample township in Tippecanoe county, Indiana, a county closely associated with the tenancy promotion schemes of Henry Ellsworth, 42 percent, was among the highest in our sample townships.[59]

The *Presidential Committee* in 1937 produced a series of maps from 1880 showing those counties in which at least half of the farms were operated by tenants and croppers. The most notable feature in those maps is the heavy concentration of counties with high tenancy rates in the South. But even as early as 1880 there were some entire counties in the North where more than half the farmers were tenants. Moreover, those areas for which I estimate high tenancy rates in 1860 were also those where it was high later. The spread of high tenancy rates in the northern half of the United States is westward and northwards from central Illinois and western Indiana back into the Plains states from at least as early as 1860.[60]

Notwithstanding the secular growth of tenant farming, it seems unlikely that tenancy was the unconstrained first choice of the farmer. It was certainly less economically desirable than owner-occupancy. Since they were deprived of capital gains on land, tenants did not share in one of the major sources of farm profit during this time.[61] Furthermore, to the extent that the rental market functioned well, landlords were able to capture much of the surplus over and above labor returns that were generated by the tenant.[62] As a result, tenants probably did little better than farm laborers and may even have been worse off to the extent that they bore increased

58. The townships where tenancy rates exceeded 500/1000 were Council Grove, Morris County, Kansas; Vasa township, Goodhue County, Minnesota and Mahnomen County also in Minnesota. In addition, tenancy exceeded 500/1000 in Costin District, Worcester County, Maryland.

59. See Gates, "Land Policy and Tenancy in . . . Indiana." Tenancy was also particularly high in one Illinois county, Knox. This is not one of those mentioned by Gates where the largest and most famous speculators made investments but a county history lists two early residents of the sample township, George Stevens and John Wyman, who owned hundreds of acres that they leased. These two men could have accounted for between a third to a half of the tenants in 1860. See Newton Bateman, ed., *Historical Encyclopedia of Illinois and Knox County* (Chicago: Munsell Publishing Company, 1899), 881–82.

60. National Resources Committee, *Farm Tenancy,* Figure 2, p. 40.

61. Gates, *The Farmer's Age,* 399, 403.

62. Atack and Bateman, *To Their Own Soil,* especially chapter 13 and Figures 13-3, 13-4, and 13-5.

income risk without commensurate compensation because of their choice of crop mix.[63]

Why then did people become tenants instead of yeomen? It is clear from the data that tenants were not the same as yeomen. In part, they were at different stages in their life cycle, but there were also other, more immutable differences between them. Do these differences "explain" that choice? What evidence can be gleaned from the data for 1860?

The simple model used here expresses the choice of tenant or yeoman as a function of personal, household, and farm characteristics.[64] However, the issue is somewhat confused because cause and effect are intertwined. For example, personal characteristics such as age and literacy are more likely to be causal factors of tenancy, while characteristics of the farm such as a high degree of crop specialization probably result from it. In the case of a variable such as wealth, the argument is less clearcut. Owner-occupiers tended to be wealthier than tenants in part because their status as yeomen depended upon their wealth. Moreover, ownership of a farm also gave yeomen a superior income stream to that of tenants because they avoided paying rent and because they captured capital gains on land and the benefits of farm-making. *Ceteris paribus,* they could therefore accumulate wealth more rapidly.

Despite these problems, a single equation binary choice model is capable of throwing light on the factors that influence why one person became a tenant and another became a yeoman farmer. The variables that I hypothesize determine tenancy choice or are determined by it are shown in Table 3. Some of these are dummy variables, taking values of zero or one; others are continuous. The dependent variable itself is dichotomous and assumes a value of zero if the farmer was a yeoman and one if he was a tenant.

The model then uses the proportion of farmers who become tenants as the dependent variable. The odds in favor of becoming a tenant farmer may be expressed as:

$$p/(1 - p)$$

where p is the probability of tenancy. This transformation forms the basis of my logit model of farm tenancy:

$$\log [p/(1 - p)] = b_0 + b_i \{\log X_i\}, i = 1, \ldots, n$$

63. The less diversified crop, livestock, and land portfolio of tenant farmers increases the variance of tenant income. The same was true of tenants in the early twentieth century. See Gray, *Farm Ownership.* In terms of pure labor income in 1920, however, tenants did better than yeomen.

64. Part-owners have been excluded from the analysis here.

Table 3. Variables Determining or Determined by the Choice of Tenure

Variable	Interpretation
PERSONAL CHARACTERISTICS:	
Age (and its square)	years
Literacy	1 = literate
	0 = not literate
Race	1 = White
	0 = other
Migrant	1 = yes
	0 = no
Immigrant from England	1 = yes
	0 = no
Immigrant from Ireland	1 = yes
	0 = no
Immigrant from Germany	1 = yes
	0 = no
Immigrant from elsewhere	1 = yes
	0 = no
Years residency in the state	years
Wealth	$
HOUSEHOLD CHARACTERISTICS	
Size of household	number of people
Number of children	number aged under 15
Adult male equivalent workers in the household	number
FARM CHARACTERISTICS	
Proportion of improved acres	improved/total acres
Hired non-household workers	number
Farm gross revenues from wheat	Wheat $/gross revenue
Farm gross revenues from corn	Corn $/gross revenue
Farm gross revenues from oats	Oats $/gross revenue
Farm gross revenues from dairy	Dairy $/gross revenue
Farm gross revenues from home manufactures	home mfg. $/gross revenue
Cattle per acre	cattle/total acres

where X_i is the vector of independent variables representing the personal, household, and farm characteristics of the subject.[65] Because there is only one observation per decision-maker, this equation was estimated by an iterative Newton-Raphson Maximum Likelihood method.[66]

The statistical results for the Northeast and Midwest are summarized in Table 4. The equation for the Midwest converged rapidly to its optimum level; that for the Northeast, somewhat more slowly.[67] Both estimates were statistically significant. Because of the logit transformation, interpretation of the regression coefficients is more complex than usual. Increases in the value of a variable whose estimated coefficient was negative reduces the probability that the person will be a tenant. Thus, for example, being a German immigrant tended to reduce the probability that the person was a tenant. Similarly, decreases in the value of a variable whose estimated coefficient was positive also reduces the probability that the person would be a tenant. Thus, being black in the Midwest reduced the probability of tenancy, whereas in the Northeast it increased it.

In both equations, wealth was, by far, the most significant explanatory variable for tenancy.[68] It drove the equations. Since, by definition, people with zero wealth were tenants, this is hardly surprising. Both equations also predict that people without wealth would be tenants with a high degree of certainty ($p > 0.96$ in the Midwest, $p > 0.85$ in the Northeast).[69] What proved surprising, however, is that wealth did not have to be very great before the probability of tenancy fell sharply (Figure 3). For someone with the average regional characteristics, wealth in excess of about $565 in the Midwest and $700 in the Northeast was sufficient to tip the probability in favor of owner-occupancy rather than tenancy. The switch-over to owner-occupancy with increasing wealth was quite dramatic and swift. In the Midwest, $1,000 was more than sufficient to make it at least 90 percent certain that a farmer would be a yeoman.[70] To achieve the same degree of

65. The methodology is spelled out in detail in any standard econometric text. See, for example, George G. Judge, R. Carter Hill, et al., *Introduction to the Theory and Practice of Econometrics* (New York: John Wiley, 1982), esp. 521–25.

66. Ibid., p. 522. The algorithm used was the LOGIST PROC in SAS.

67. The equation for the Midwest converged in 8 iterations, that for the Northeast in 20.

68. Substituting personal wealth for total wealth dramatically reduces the explanatory power of the equations. The pseudo R^2 for the Midwest, for example, falls from 0.839 to 0.465. This reflects the loss of statistical significance not just for the wealth variable but for all variables, except those reflecting characteristics of the farm.

69. These are the probabilities with other characteristics set at their approximate mean or most likely values. See Table 5. With other characteristics more appropriate for a tenant, the probability of tenancy estimated from the equations was a virtual certainty ($p > 0.99$).

70. This figure is the same as that given by Clarence Danhof as the sum needed for independent farm-making in the Midwest during the 1850s. See Clarence C. Danhof, "Farm Making Costs and the Safety Valve," *Journal of Political Economy* 49 (1941): 317–59 and Jeremy Atack, "Farm and Farm-Making Costs Revisited," *Agricultural History* 56 (October 1982): 663–76.

Table 4. Logit Regression Results for Farm Tenancy in the Northeast and Midwest, 1859–60 (Dependent Variable = 0 if Owner-Occupier = 1 if Tenant)

Characteristic	Northeast	Midwest
Intercept	−1.2954	−4.2641***
Age	0.0411	0.1062**
Age squared	−0.0008	−0.0013**
Literacy	−0.1232	0.3942*
Race	−1.1666	2.4529***
Migrant	−0.5142	0.1349
English Immigrant	−0.6793	−1.0086**
Irish Immigrant	0.2107	−1.3371***
German Immigrant	−1.2249	−0.8861**
Other Immigrant	0.2412	−0.9310**
Years of Residency	−0.0021	−0.0058
Wealth	−0.0026***	−0.0058***
Size of household	0.1712***	0.0956**
Number of children	−0.0453	−0.0496
Household labor force	0.8968***	0.4198**
% improved acres	1.2741***	1.6549***
Wage labor	1.3227***	0.5115***
% revenue from wheat	9.5628***	2.2656***
% revenue from corn	3.0177***	3.3461***
% revenue from oats	10.2621***	6.8615***
% revenue from dairy	0.5044	−2.0789*
% revenue from home manufactures	2.9874	−1.1167
Cattle/acre	−0.2866	−1.9337**
Pseudo R^2	0.790	0.839
Log likelihood	−806.13	−1280.67

*** Significant at better than 0.01
** Significant at better than 0.05
* Significant at better than 0.10

certainity about tenure status in the Northeast would have required perhaps 75 percent more wealth.

The dominance of wealth as the underlying cause of tenancy is also what distinguishes American tenancy from that on the European continent. In America, accumulation over time, luck, and improved capital markets provided a remedy. In Eastern Europe especially, things were different. There, people could not buy land regardless of their financial means. As a result, they could remain tenants forever.

Although wealth or the lack thereof dominated the logit regressions,

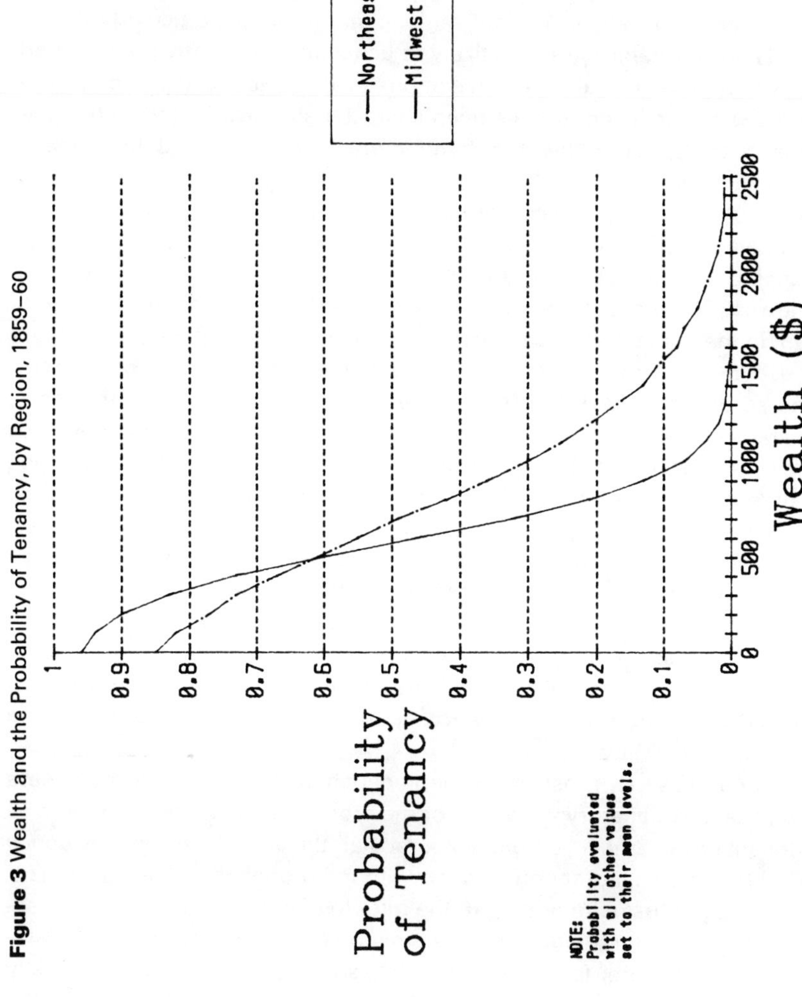

Figure 3 Wealth and the Probability of Tenancy, by Region, 1859–60

knowledge of other personal and household characteristics would also influence the probability of correctly predicting whether a given farmer was a tenant or owner. In both regions, immigrants, particularly the English and Germans, were less likely to be tenants than native-born Americans.[71] More family labor, the use of hired labor, and being part of a larger household also seem to have been characteristics associated with higher tenancy rates, but on the other hand, more children reduced the probability of tenancy.

Despite these inter-regional similarities, however, it is clear from the estimated coefficients in Table 4 that tenancy in the Northeast was determined by a different set factors from those that influenced tenancy in the Midwest. None of the personal characteristics, except wealth, was significant in the equation for the Northeast. The variables that we argued were causal factors in tenancy—personal characteristics such as age, literacy, and so on—simply do not explain tenancy in that region. Furthermore, characteristics such as illiteracy that increase the probability of tenancy in the Northeast, reduce it in the Midwest. On the other hand, those variables that reflect tenancy—particularly the degree of crop specialization, emphasis upon cash grains, preference for farms with a high ratio of improved, that is cultivable, land to total acreage, and so forth—were highly significant. In particular, the failure of the age variables to explain tenancy in the Northeast suggests that the agricultural ladder from farm laborer to tenant to yeoman was not working very well in that region. One possible explanation for this is that the successful Northeastern tenants—those moving up the ladder—were drawn westward by the lure of cheap land to become yeomen in the Midwest.

In the Midwest, most of the personal characteristics of farmers were significant explanatory variables for the probability that a person would be a tenant rather than a yeoman. In particular, the data show some evidence of a life-cycle phenomenon at work, with the probability of being a tenant increasing at first with age up to the mid- to late-forties and then decreasing. The agricultural ladder thus seems to have been working in the Midwest. Furthermore, farm characteristics such as specialization generally operated in the same way in the Midwest as in the Northeast. In the Midwest, however, tenant farmers were much less likely to be heavily involved in dairy operations, nor were they likely to operate extensive cattle ranches.

The probability that someone was a tenant rather than a yeoman depends upon the vector of their personal, household, and farm characteris-

---

71. The Irish and other unspecified immigrant groups were, however, more likely to be tenants in the Northeast.

Table 5. Probability of Tenancy for Specific Changes in Variables, Given an Initial Vector of Characteristics such that P(Tenancy) = 0.50.[a]

| | Probability of Tenancy | |
Change in Characteristic	Northeast	Midwest
PERSONAL CHARACTERISTICS		
Age + 10 years	.40	.49
Age − 10 years	.57	.46
Illiterate	.53	.41
Non-Migrant	—	.47
Migrant	.38	—
English born	.34	.24[b]
Irish born	.56	.19[b]
German born	.23	.27[b]
Born elsewhere overseas	.56	.26[b]
Wealth + $100	.44	.36
Wealth − $100	.58	.64
Years of residency + 5	—	.50
Years of residency − 5	—	.51
Years of residency − 10	.51	—
Years of residency − 20	.51	—
Black	.76	.08
HOUSEHOLD CHARACTERISTICS		
Household size +	.55	.53
Household size − 1	.46	.48
Children + 1	.49	.49
Children −1	.51	.52
Household labor force + 1	.71	.61
FARM CHARACTERISTICS		
⅔ acreage improved	—	.57
⅞ acreage improved	.54	—
One wage laborer	.79	.63
Double wheat revenue	.60	.56
Double corn revenue	.58	.67
Double oats revenue	.63	.54
Double dairy revenue	.53	.45
Double home manufacturing	.50	.51
Double cattle/acre	.50	.48

[a]The initial vector of characteristics for the Midwest was:

Age	40	Literacy	1	Race	1
Migrant	1	English immigrant	0	Irish immigrant	0
German immigrant	0	Other immigrant	0	Years of Residency	15
Wealth	$565	Household size	6	Number of children	3
Household labor	1	Wage labor	0	Proportion improved acres	.5
Wheat revenue	.1	Corn revenue	.2	Oats revenue	.02
Dairy revenue	.1	Home mfg.	.01	Cattle/acre	.04

28
JEREMY ATACK

The initial vector of characteristics for the Northeast was:

Age	45	Literacy	1	Race	1
Migrant	0	English immigrant	0	Irish immigrant	0
German immigrant	0	Other immigrant	0	Years of Residency	40
Wealth	$550	Household size	5	Number of children	2
Household labor	1	Wage labor	0	Proportion improved acres	.75
Wheat revenue	.04	Corn revenue	.1	Oats revenue	.05
Dairy revenue	.2	Home mfg.	.01	Cattle/acre	.04

[b]Migrant dummy set to zero for these estimates. Base probability of tenancy was therefore 0.47 instead of 0.50

tics. For any given set of characteristics, the probability of tenancy, given the set of characteristics X_i^*, is defined by:

$$p = 1/[1 + e^{(-b0 - bi X_i^*)}].$$

The possibilities are infinite. Therefore in Table 5, I present some estimates of the change in the probability of tenancy in response to changes in individual explanatory variables holding other variables fixed at specific levels. In most cases, those levels are close to their mean values for continuous variables or at their most likely value (0 or 1) for dummy variables, such as literacy and migrant.

Consider, the case of the 40-year-old, white, literate midwestern farmer with $565 wealth.[72] The probability that this person was a tenant and not a yeoman was almost 50/50. If this farmer somehow obtained an extra $100, say from the sale of a calf and a foal, then the probability that he would be a tenant falls to 0.36. If the farmer was illiterate, the probability of tenancy would be 0.41; if he was black, the probability would be only 0.08 that he was a tenant. If the farmer was Irish, the probability that he would be a tenant was only 0.19. Similarly, based upon the farm characteristics, if we knew that this same farmer derived, say, 40 percent of his gross revenues from corn, then the probability that he would be a tenant is 0.67.

Changes in the probability of tenancy in the Midwest were particularly sharp for farmers who were black, or who had immigrated to this country, particularly from Ireland. Farmers with such characteristics, other things equal, were much more likely to be yeomen than tenants. In the Northeast, the division between tenant and yeoman was much less clear-cut. Among the most important factors affecting the probability of tenancy were again those related to race and nativity. The German and English immigrants who settled in the Northeast were unlikely to be tenants, but the probability of tenancy for farmers belonging to other immigrant groups was little changed. Blacks, on the other hand, were much more likely to be tenants,

72. See the footnote to Table 5 for the values of the other personal, household, and farm characteristics.

though if the slave-state of Maryland is excluded from the data set, this characteristic loses its explanatory power. Two other factors provide important clues regarding the tenant status of northeastern farmers: those who had larger potential household labor forces than average and those who hired nonhousehold wage laborers were generally much more likely to be tenant farmers.

Seen against the backdrop of history, rising tenancy in the northern half of the United States during the nineteenth and early twentieth centuries seems part of an evolutionary, rather than revolutionary, process. Tenancy was probably an ever-present condition throughout the nineteenth century. It was certainly quite pervasive in 1860. In that year, lack of capital seems to have been the most likely cause of tenancy, just as it was sixty years later, and rising farm values are thus the most likely explanation of increasing tenancy in the North.[73] If this is the case then federal land policy, to the extent that it minimized entry costs, reduced the incidence of tenancy. Certainly, contrary to Gates's claim, it is unlikely to have promoted it.

73. See Gray et al., *Farm Ownership*.

The Spread of Agricultural Machines in Estonia from 1860 to 1880

JUHAN KAHK

In the middle of the nineteenth century the feeling that the world was an economic and intellectual unity was getting stronger all the time. Although the American War of Independence only marginally affected the European economy, the U.S. Civil War had a strong impact. In the Russian empire, in Latvia and Estonia, both industry (textiles) and agriculture responded to international markets. This forms the background for the introduction of agricultural machinery into the Baltic area.

Large landowners of the Baltic provinces, mostly barons of German origin, managed their estates by exacting labor services from Estonian and Latvian peasants, who were in some ways in a comparable position to that of slaves on U.S. plantations. By this time, the development of agricultural technology had already clearly demonstrated elsewhere that the use of steam engine-driven threshing machines could increase the output of grain and that machines were economically profitable on large estates. However, there were reasons why the manorial economy did not develop an advanced agricultural technology.

Of the two economies coexisting in the Baltic region, the manorial economy had for centuries been based on labor services provided by peasants who also supplied practically all the implements needed in agriculture. For many centuries the Baltic landlords had not been concerned with the technical side of agriculture; towards the end of the eighteenth century, they began to cultivate an interest. But, because they had cheap labor at their disposal—peasants who paid with their work for the land—they had neither the experience nor the economic incentive to learn from

JUHAN KAHK is Executive Secretary of the Estonian Academy of Sciences. This paper has benefited from the editorial revisions of Carol Leonard and Morton Rothstein.

the capitalist system how to introduce machinery into agriculture. In the middle of the nineteenth century, they began to sell their land to the peasants.

The second economy, small-scale peasant farming, had already introduced labor-saving devices and rational principles into agriculture during the centuries of feudalism. When serfdom was abolished, these principles expanded. For example, although on the manorial estates, for fear of loss of grain due to careless labor, the scythe was not used in the harvest, on peasant farms, where the farmer himself supervised the work, the scythe was used because it was efficient. Small-scale peasant farmers could not afford such expensive machinery as threshing-machines until later, after the abolition of serfdom, when, economically stronger, they became interested in (and capable of) organizing agricultural cooperatives.

From C. E. Muller's *Praktisches Handbuch der Landwirtschaft,* published in Tallinn (1850), one can get a clear picture of the level of technological knowledge of the Estonian and Latvian barons. The manual has data on the chemical composition of soil and plants; therefore they must have known something of Liebig's theories regarding soil chemistry. But the advice offered has mostly to do with practical matters. An indication of the relative backwardness of the readership, the *Handbuch* does not contain information about new agricultural implements available in western Europe, and it mentions only at the end a wooden cylinder used for threshing and an uprooting machine used in clearing land.[1]

This cannot be considered as evidence of very serious technical ignorance. After all, a census report in such a land of advanced agricultural enterprise as the United States indicated that "The year 1850 practically marks the close of the period in which the only farm implements and machinery other than the wagon-cart and cotton gin were those which, for want of a better designation, may be called implements of hand production."[2]

In mid-nineteenth century Europe there were two very different, contrasting regions—one marked by use of labor services, the other a region of free enterprise. Both regions were on roughly the same level of agricultural technology. But the introduction of new machinery developed in very different conditions, on one side, free enterprise, and on the other, outmoded feudal system.

A good example is given in the 1863 session of the Livonian Public Benefit and Economic Society where the same uprooting machine that Müller mentions was discussed. On some manors, it was reported, peas-

---

1. C.E. Müller, *Praktisches Handbuch der Landwirtschaft vorzugsweise für die Ostseelander Russlands bearbeitet* (Reval: 1850): 252–267.

2. J.B. Davidson, "The Evolution of Farm Machines," *Palimpsest,* XX (March 1950): 86.

ants used these machines for clearing land on the demesne, but only so long as the squire kept to the time work system. As soon as the squire switched to a piece-work system, the peasants gave up these clumsy machines and went back to the manual work, which they knew from experience was much more productive.[3] With unfree labor, lords could afford inefficient technology. But with wage labor, more effective methods spread.

Until now, Soviet historians have explained the retardation of agricultural development as originating on the one hand from the peasant adherence to tradition and distrust of innovation, and on the other hand, by the reluctance of a parasitic landlord class to invest in machines. It is important to note, however, that until at least the middle of the nineteenth century the spread of new agricultural technology was also checked by the imperfections of available machines, and the backwardness of technical science.

Also, a shortage of skilled labor slowed the spread of agricultural machines. According to the compilers of the Livonian Public Benefit and Economic Society Review in 1864, "we have procured a number of machines and improved agricultural implements and we have tried them out, but still there is no progress because we are still short of skilled labor. We may hope to make headway only after the final abolition of the corvée." The only new implements in use on Livonian manors as late as 1864 were the swing plow used on clover fields and a harrow with an iron frame and iron teeth for planting winter grain.[4]

For centuries the need to make threshing easier in the Baltic lands had been obvious. The English inventor Andrew Meikle made an improved thresher with barrels in 1786, and as early as 1797 some of those machines were brought to Riga.[5] (The United States did not import Meikle's machines until the 1820s.)[6] A variety of different threshing machines were developed locally, including one by J. G. Parrot, who would be the first rector of Tartu University. As early as 1863 a session of the Economic Society supported the idea of using steam power for threshing. A few years later the squires who gathered for another session of the Society agreed that ". . . the transportable threshers and practical grain cleaners are in great demand."[7]

---

3. *Livländische Jahrbücher der Landwirtschaft* (hereafter LJL) vol. 16, Part. 3 (Dorpat, 1863) 190.

4. State Archive of History of Estonian SSR (hereafter SAHE), collection (c.) 1185, roll (r.) 1, Nº391, lists (l.) 13–14.

5. Strods,H. "No kulmasinu konstrucasanas un ieviesanas vestures Latvija XVIII gs. Beigas un XIX gs. pirmaja puse," *Vestures problemas,* (Riga) I (1958) 141: (LJL) vol.16, Part 3 (Dorpat 1863), 198.

6. J.W. Oliver, *The History of American Technology* (New York: 1956) 227.

7. "Baltische Wochenschrift" (hereafter BW), Feb. 16, 1866.

However, there were obstacles to the widespread use of machines and improved implements. The author of a September 1860 newspaper article complained that the manufacturers of agricultural machines seemed unaware of the extremely difficult conditions under which these machines had to work. They were built "too light," he observed, "and their construction often proves to be badly suited to their purpose, as if they were meant to be used by some lady in her boudoir."[8] He also denounced rival manufacturers of agricultural machines for using dirty tricks in selling them. The basic construction of a machine would remain unchanged. Yet their makers claimed that some unimportant change in detail would work miracles. The reporter described the sturdier implements as also having serious defects, charging that "the machines mainly made of wood are very clumsy and require much strength, while the cast iron ones break easily, and both types are very expensive to obtain and use."[9] Moreover, in the mid-nineteenth century the Alutaguse harvest workers were made to switch from using scythes back to sickles, because the loss of grain was less when the rye crop was cut with sickles.[10]

Until the 1860s, the only possible alternative to manpower for threshing was horsepower. The efficiency gain from mechanization was not so great that it challenged the predominance of hand production. Changes came in threshing only when steam power was introduced. N. Wilcken, the owner of Voka manor, published in 1864 his precise calculations of receipts and expenses for threshing. The results showed that if the threshing was done by 28 peasants with flails, one bushel of rye was produced by manpower that cost the equivalent of 7 rubles 47 kopecks; if horses were used in threshing the cost was 7 rubles 97 kopecks; and if a threshing machine run by 4 horses was used (which also needed 8 men), the bushel of rye cost 7 rubles 20 kopecks.[11] If these estimates are to be believed, the savings obtained from a threshing machine was negligible. This conclusion is a result confirmed by similar calculations presented at a meeting of the Parnu-Viljandi Agricultural Association 1874, which showed that grain threshed by machine was only 3.4 kopecks (10 percent) cheaper than threshing by hand.[12] To be sure, a horse-powered threshing machine could "... considerably hasten the process of threshing, but since almost as many people had to work with the machine as when the work was done by hand ... it was not really economical," as reported to the Economic Society in January 1861.[13] A threshing machine run by steam power

8. *Revalsche Zeitung,* Sept.3, 1860.
9. LJL, Vol.13, Part 4, 25.
10. LJL., Vol. 13, Part 2, 10.
11. LJL, vol. 17, Parts 1–2, Dorpat, 1864, 30–40.
12. BW , October 31, 1874.
13. SAHE c. 1165, r. 1, N° 344,1.5.

would have required fewer workers, but such machines were much more expensive and, according to the Society, at the beginning of the 1860s there were not more than 12 in the whole of Livonia.[14]

There was great interest in Livonia, however, in the first attempts of the neighboring countries to use steam engines. A survey of the 1863 Hamburg Agricultural Exhibition published in the *Yearbook of Livonian Agriculture* gave much attention to the "steam-agriculture."[15] The survey indicated that German manufacturers were gaining on the English makers of agricultural machines. A demonstration of plowing with the help of steam engines provoked excited remarks: two Richardson-Darley traction engines named "Cupido," with a plow of the Fowler manufactory attached to them by two long steel cables, moved up and down a sod-covered field. They plowed three acres in one hour; had the work been done by horses, it would have taken almost three days. "The improvement of traction engines has brought steam usage in agriculture to a new, higher level . . ." observed the survey in the Hamburg Exhibition.[16] The public was also greatly impressed by the spectacle of the "Cupido" steam itself driving onto the field, and one could imagine ". . . a plow directly attached to the steam engine."[17]

At the Riga Agricultural Exhibition of 1865 four foreign firms displayed steam-operated threshing machines, while nine firms still demonstrated threshing machines turned by horses.[18] Yet as early as 1866 it was reported to the Economic Society that a ". . . big threshing machine with steam engine" was working at Reola manor.[19] In the 1865 meeting also, the Heimtal squire Sievers informed the landlords present at a meeting of the Economic Society that in the Viljandi district " . . . several entrepreneurs are going from manor to manor with their tractors, doing all the threshing for piece wages, which has proved very economical especially for the smaller manors."[20] According to a survey of the years 1860–1869 compiled by the Economic Society, the Oisu, Pöögle, Lisuma, Druvieni and Burtnieksi manors were using threshing machines with steam engines.[21] Furthermore, in September 1874 a competition between threshing machines with tractors of the two American firms, Buckeye and Dyke (Champion), was organized at the Raadi and Ropka manors.[22]

But the application of threshing machines with steam engines was not

14. *Ibid.*
15. LJL, vol. 16, Part 3, 248–49
16. LJL, vol. 16, Part 3. 252.
17. Ibid.
18. LJL, vol. 18, Dorpat 1865, 202–03.
19. SAHE, c. 1185, r. 1, N°436, l. 1.
20. LJL, vol. 18, Dorpat, 1865, 305.
21. SAHE, c.1185, r. 1, N° 436, l. 44.
22. BW, Sept.12, 1874.

without its problems. N. Wilcken of Voka manor, who bought a threshing machine for 910 rubles with a steam engine that cost 1666 rubles in Leiden at the beginning of the 1860s, vividly described some of the problems. Transporting the machine from Leiden to Estonia proved to be extremely cumbersome and expensive. Finding a man who could install the machine took further time and effort. When he finally started the machine he discovered that the steam engine could not work at the speed necessary to run the thresher. It took great exertion to set this right. On the basis of data he collected and comparison tests, Wilcken calculated the "prime cost" of threshing conducted in various ways (taking into account the cost of labor as well as the machines). Threshing 2,000 cartloads of grain, which would make 9,000 Estonian bushels of grain (1 bushel being approximately 25 kilograms), cost 698 rubles when the work was done by hand, 672 rubles when the grain was threshed by having horses tread on it, 342 rubles when a threshing machine run by horses was used, and 234 rubles when a threshing machine with a steam engine was used.[23]

As early as the late 1860s, Baltic landlords also could buy a variety of other manufactured agricultural implements that were becoming available, and several Livonian squires participated actively in the business of manufacturing them. For example, landlords of the manors situated near Viljandi and Tartu helped finance the factories producing implements at Kabina and Kopu.[24] At meetings of the Estonian Agricultural Association there were discussions about the merits and shortcomings of several new American and German plows.[25] In 1870 a St. Petersburg department store advertised new English, American and Finnish plows for Baltic farmers, along with new types of harrows, field rollers, plows for hilling potatoes and some special harrows called extirpators and scarificators.[26] A multiple-unit "Champion" mowing machine was commended at an 1873 meeting of the Estonian Agricultural Association.[27] We also know that in 1874 ". . . several manors in the neighborhood of Tartu were equipped with mowing machines which were actively used . . ."[28] Finally, in 1879 a competitive demonstration of the Johnston, Buckeye, Wood and Adrianne mowing machines was organized at Illuka manor, evidence of greater sophistication among potential buyers.[29]

23. LJL, vol.17, Part 1–2, Dorpat 1864, 45–59. In this calculation Wilcken took into account only the salaries of the workers, rather than all the expenses connected with the use of labor.
24. SAHE, c.1185, r.1, N°380, l. 4.
25. BW, August 21,1868.
26. BW, May 20, 1870.
27. BW, Jan.10, 1874.
28. BW, Sept.9, 1874.
29. BW, Nov.1, 1879.

At the outset, efforts by Estonians themselves to improve equipment for basic tasks proved disappointing. The first reference to a potato spinner dates from 1869 with regard to a machine produced at the Eckert factory in Berlin.[30] But within five years there were several reports that expensive efforts to construct a potato spinner in the Baltic had gone wrong.[31] The descriptions of such machines indicate that they had square frames on three wheels, with a cutter in front and a disc-shaped appliance on top for separating potatoes from the soil.[32] The Voeru master named Blomerius constructed a machine as early as the mid-1850s for swingling and breaking flax.[33] In 1864 another inventor named Friedlander also demonstrated machines for processing flax.[34] At first, farmers were far from satisfied with the quality of the machines. "What is said of the swingles is also true of the flax brakes," claimed a writer in 1866. He went on to indicate that "several experiments have proved that in the case of the Roman's as well as Friedlander's and other machines the losses suffered through the degradation of quality exceed the advantages of the time and labor saved."[35]

But Baltic farmers soon began to appreciate the locally produced machines. By 1870 a swingle (of unknown make) had become popular even with peasant farmers in the vicinity of Roneburg.[36] The swingle that Blomerius constructed was quite favorably received at the 1872 agricultural exhibition organized in Tartu.[37] When a sowing machine had been put on display at a Tartu exhibition in October 1857, "older peasants had to admit that they were surprised and impressed in spite of having at first claimed that a machine could not really compete with an experienced sower."[38] A squire named Sivers from Oisu told the 1865 meeting of the Economic Society that agricultural machines and improved implements "have been used in the neighborhood for a number of years already, and also some peasants and farm workers have acquired steel plows worked with a team of two horses and used wooden cylinders for threshing, while almost every peasant farm is equipped with a flax brake turned by hand."[39] All this was in contrast to the 1850s observation that scythes were not used for cutting grain on manor fields because landlords had no confidence in the quality of work done by corvée peasants (feeling it was safer

30. BW, June 11, 1869.
31. BW, Oct.10, 1874.
32. BW, Nov.20, 1875.
33. LJL, vol.13, Part 2, p.66.
34. BW, July 21, 1864.
35. BW, June 22, 1866.
36. BW, April 8, 1870.
37. "Eesti Postimees," July 26, 1872.
38. LJL, vol.13, Part 4, Dorpat 1858, p.23.
39. SAHE, c.1185, n.1, No. 421, l. 13.

to keep to the sickle), yet it was widely used on Northern Estonian peasant farms where the farm master who worked with the other harvesters could keep an eye on quality.[40]

The use of the improved flax brake spread quickly among peasants. An 1866 report from Rapina to the Economic Society indicated that "The flax brake turned by horses is winning popularity among the peasants. The owners of such machines lend them for use—24 hours cost 1 ruble."[41] Two years later Palamuse informed the Society of the growing use among peasants of a flax brake consisting of a roller weighted with a stone box moving on grooves 2 to 2-½ inches deep and drawn by a horse. A number of such machines were made by the peasants themselves and rented at 30 kopecks per 1000 handfuls of flax.[42]

Possibly the availability of draft animals allowed peasants to adopt innovations quickly. According to statistics for 1863 and 1867, Northern Estonian peasant farms had more draft animals than did the manorial estates. Peasant farms in three quarters of the parishes averaged one draft horse per 3–4 desiatiny (equal to 2.7 acres) of arable land, while the manors had an average of one horse per 10–20 desiatiny. Differences were less marked for yokes of oxen, with peasant farms averaging 1 yoked ox per 3–10 desiatiny of arable land, and the ratio for manorial estates at one per 4–20 desiatiny. Consequently, peasant farms were in a better position to adapt machinery with more effective draft animals.

The years from 1860 to 1880 were a time of transition from manorial agriculture to large scale junker farms. For a comparatively short time conditions favored the wealthy peasants.

Conditions also favored exports of flax instead of grain, and peasant farms were better suited for growing this labor intensive, but easily home-grown commodity. The traditional line of manorial output—distilling spirits—was also undergoing reorganization and concentration, which was best suited to the new style of junker farming. However, it also stimulated demand for the raw material, potatoes grown on peasant fields, which persisted and even grew.

Because they were more efficient than the manors, peasant farms were competitive in yields as well as labor costs. When we compare the yields of winter grain (rye) and potatoes on manorial and peasant fields of Southern Estonian provinces, we get the following results (differences under .05 were not taken into account):

<hr>

40. LJL, vol.13, Part 2, 10.
41. SAHE, c.1185, n.1, No. 436, l. 8.
42. SAHE, c.1185, n.1, No. 464, l. 47.

Table 1.

In the neighborhood of	Tartu	Pärnu	Viljandi	Number of observations
1860–1869				
The rye yields of manorial fields were				
higher than	1	3	3	7
the same as	7	4	6	17
lower than	1	2	0	3
the rye yields of peasant fields				
The potato yields of manorial fields were				
higher than	2	0	1	3
the same as	7	6	8	21
lower than	0	3	0	3
the potato yields of peasant fields				
1872–1880				
The rye yields of manorial fields were				
higher than	9	9	9	27
the same as	0	0	0	0
lower than	0	0	0	0
the rye yields of peasant fields				
The potato yields of manorial fields were				
higher than	9	5	9	23
the same as	0	2	0	2
lower than	0	2	0	2
the potato yields of peasant fields				

Note: Data for the year 1863 are missing.

The following data of the average yields of the periods clearly speak of the improvement of productiveness and of the superiority of the Viljandi province.

Table 2. The Average Yields of Winter Grain (Rye in bushels)

| | | In the neighborhood of | | |
	Tartu	Pärno	Viljandi	Saaremaa
1860–1869				
of the demesne	511	499	464	415
of peasant fields	515	486	462	416
1872–1880				
of the demesne	655	541	672	424
of peasant fields	438	474	525	397

Table 3. Distribution of Farm Machinery

| | | | | | | | Threshing machines with | | | |
Province	Sowing machines		Reaping machines		Mowing machines		horse power or water power		steam engine	
	1863	1867	1863	1867	1863	1867	1863	1867	1863	1867
Harju	105	131	13	7	12	10	106	124	5	7
Viru	127	146	3	5	28	29	104	116	6	4
Järva	89	107	5	2	9	6	84	95	1	1
Lääne	64	67	6	7	9	8	65	76	16	19
Machines	385	451	27	21	58	53	359	373	28	31
per one manor	0,77	0,90	0,05	0,04	0,12	0,11	0,72	0,75	0,06	0,06

In the 1860s, peasant farms competed in yields with manors, but in the subsequent decade peasant yields relatively declined. Over-all improvement was lacking. In comparison to 1860, rye yields of 1880 hardly grew. Potato yields in some districts grew by one-fourth and in others by one-half, but mainly on the manor.

The manor retrieved its lead thanks to the reorganization of work and the limited application of new technologies. But new technologies were not yet widely used; vestiges of manorial exploitation slowed down improvements and caused misery among the peasants.

A closer look at manors where there were threshing machines empowered by steam engines reveal that by 1864 manors were under the influence of towns: in the Parnu-Jaagupi, Parnu and Saarde parishes near Parnu; in the Viljandi and Helme parishes south of Viljande; in the Tartu, Noo, Kambja, Rouge, Sangaste, Otepae, Raepina, Kanepi, Urvaste and

Table 4.   Mechanization of Threshing, 1867–1881

		In the neighborhood of		
	Pärnu	*Viljandi*	*Tartu*	*Võru*
Initially there was no threshing machine; then a threshing machine turned by horses was taken into use	4	5	5	7
. . . . then a threshing machine worked by steam power was taken into use	9	13	35	14
The initially used threshing machine turned by horses remained in use	5	3	4	21
Initially a threshing machine turned by horses was used, then a threshing machine worked by steam power was taken into use	8	8	20	14
The initially used threshing machine worked by steam power remained in use	1	4	5	2
The initially used threshing machine turned by horses was abandoned, no threshing machine worked by steam power was procured in its place	4	9	13	9
The number of manorial estates where no threshing machine was used at all	36(51%)	69(73%)	88(57%)	48(55%)

Notes: The threshing machine worked by steam power was exchanged for one turned by horses in one manor; in one manor the threshing machine worked by steam power was put out of use.

Veru parishes in the vicinity of Tartu and Voeru. In parishes where changes in threshing technology were under way during the 1860s and 1870s between 50 and 65 percent of landlords were buying their first threshing machines, predominantly the type with steam engines. There was an exception in the hinterland of Parnu, where horse-drawn machines were still preferred. In about one-third of the cases the manors switched from horse-drawn machines to those with steam engines. Still, the number of manors that abandoned the unprofitable horse-powered threshing machines and did not replace them with steam engines was also quite large, representing a quarter to one-third of the total.

In conclusion, technical progress, a shift from manual labor to machines and from horse- to steam-power, made headway intermittently, with manors sometimes abandoning horse-drawn threshing machines and reverting to hand labor for some time. Even 5 to 10 years after large-scale production of steam-powered threshing machines was well under way, only 20 to 50 percent of Southern Estonian manors had them, although their merits must have been understood by all great landowners. Machines were presumably less profitable on small scale peasant farms, but there are no statistics available on technical devices on such farms. The few surviving bits of data suggest that peasants expressed a great interest in farm machines and attempted to use their limited supply of equipment collectively.

The rate of development depends to some extent on the preceding period. Productive forces "... are thus a result of energy generated by people, but the energy itself has its source in the living conditions of the people, the productive forces of the past years, the form of society that has existed before these people were born and which was not created by them but by previous generations," wrote Karl Marx in a letter to P. Annenkov on December 28, 1846.[43]

Although the application of new technology in agriculture was an arduous process, because of new social demand, the period 1860 to 1880 witnessed accelerated growth. Socio-economic change was not automatically triggered by technological progress. Historical materialism requires a more complex explanation. Until social demand and social relations in production underwent major change, agricultural technology lagged behind. The 1860s and 1870s were a turning point in this sense, and out of a greater need for new machinery, the threshing machine, with its almost century-long history, was at last put into use.

---

43. K. Marx and F. Engels, *F. Valitud kirjad* (Tallinn, 1977), 27.

Long-run Trends in American Farmland Values

PETER H. LINDERT

The scarcity of agricultural land richly deserves its own history. It is an actor in many of our histories of social structure and economic growth. The rent and price of agricultural land have a two-fold social meaning. One is that the return to land is largely a site rent, the controversial "unearned increment" that allows owners to "grow rich in their sleep," in Mill's phrase. The other is that agricultural land is largely owned by one or another special social class. In many countries, especially in England, it has been a rich man's asset, concentrated into the hands of persons having high overall income and wealth, even apart from their agricultural land holdings.[1] In others, such as the United States, agricultural land has been owned largely by farm operators, so that social tensions over the position of the farmer are tied to the price of farm land. Periods of farm crisis are ones in which farmers fear that the value of their land will plummet. So it was in the famous U.S. farm protests of the late nineteenth century, the farm crises of the early 1920s and the 1930s, and the severe farm crisis of the 1980s.

Agricultural land scarcity has often had a prominent role in theories of economic growth. Yet when we turn to the task of explaining land price movements across modern history, we find an incomplete story filled with

PETER H. LINDERT is Professor of Economics and Director of the Agricultural History Center at the University of California, Davis. The author wishes to thank participants at the June 1987 Tallinn conference, particularly Alan L. Olmstead, Andrey Poletayev, and Paul Rhode, for their helpful comments on the preliminary conference draft. Maite Cabeza-Gutes and Rob Mc-Clelland are also thanked for their research assistance, and Chris Grandy and Thomas Mayer for additional comments. A much longer version of this paper under the same title is available upon request at the Agricultural History Center, University of California-Davis, Davis, CA 95616 USA. (Hereafter cited as "WP.")

1. Peter Lindert, "Unequal English Wealth since 1670," *Journal of Political Economy* 94 (1986) 6: 1127–62; Lindert, "Who Owned Victorian England? The Debate over Landed Wealth and Inequality," *Agricultural History* 61 (1987) 4: 25–51.

puzzles. Our theoretical tradition is much richer than our empirical history of land values.

Traditional theories about land and growth build a strong presumption that growth of population and the economy should cause the price of land to go on rising, as long as land scarcity does not choke off that growth. In the pessimism of Malthus's essays on population, increasing population pressure entailed a rising price of land as an implicit mirror image of the declining ratios of land and food per person. What Malthus left implicit was made explicit by Ricardo's famous theory of rent. That growth should push up the price of land indefinitely was a proposition on which John Stuart Mill, Henry George and Alfred Marshall agreed, though George and Marshall agreed on little else. Rising land scarcity is also predicted by postwar mathematical models of the growth process.[2] The key to the argument is obvious enough: the supply of land is harder to expand than the supplies of human beings, capital goods and other productive inputs.

Yet the tradition of predicting a monotonic rise in the price of land, both in and out of agriculture, clashes with some of the facts. The most familiar counter-tendency is the declining share of agricultural land and its rents as shares of wealth and income. We find it only natural that its share should fall, just as agricultural output falls as a share of national income. Theodore W. Schultz developed this point in a series of anti-Malthusian essays on the declining economic importance of land.[3]

The fall in land's share of national income and wealth can be reconciled with a persistent rise in land's price. Its price can rise while its share of the economy declines as long as other productive forces grow much faster than the supply of land.

More troublesome for the presumption that land values should rise is the fact, confirmed here, that there have been prolonged periods of falling land values in America. There have also been sharp short-run crises in the farm real estate market, and these crises demand special explanation. The American case also raises special issues relating to the role of the frontier in wealth accumulation and the character of economic growth. The history of the price of land is rich and varied, and it challenges us to develop a more careful explanation of the movements in this key social barometer.

Three perspectives on the long-run history of American farmland values are offered in the remainder of this paper. The first section gathers the

2. Eg., Donald A. Nichols, "Land and Economic Growth," *American Economic Review* 60 (1970) 3: 332–40.

3. Theodore W. Schultz, "The Declining Economic Importance of Land," *Economic Journal* 61 (1951) 4: 725–40; Schultz, "Land in Economic Growth," in Harold C. Harlow, et al., eds., *Modern Land Policy* (Urbana, Illinois: University of Illinois Press, 1963); Schultz, *Economic Growth and Agriculture* (New York: McGraw-Hill, 1968).

abundant, but still underused, data on farmland values since colonial times. Their curious history is followed for the nation and for nine regions, with adjustments for changes in land quality. The implied capital gains on farm land are compared with other types of asset return and with wage rates. The second section points out two important implications of the capital gains on farm land: implications for our perspectives on the rise of U.S. "saving" rates in the nineteenth century, and implications for the long-standing debate over the effects of land abundance for American industrial development. The third section takes some tentative steps toward the goal of weighing the determinants of movements of farmland values.

Movements in American Farmland Prices and Rents

The official United States statistics offer a fairly accurate view of the average price of farm land by state and county since 1850, at first in decennial censuses and then, since 1910, in annual surveys. Before 1850, we have only a few scraps of local data, as we shall see. One peculiarity of the American data is that the purchase price of land is much more available than the annual rental price, perhaps because American land abundance and land policy made it easier than in most countries for a farmer to buy his own land instead of renting it from a landlord.

Land Price vs. Land Quality. Any attempt to measure movements in the "price" of land must grapple with its varied quality. Each parcel of land is a unique asset. Agricultural land values vary greatly with climate, soil conditions and other dimensions of nature. They also vary because of human investment of money and effort into the land itself, such as clearing forest, fencing, fertilizing, and laying drains. And they depend on the whole economic environment—the macroeconomy, transportation costs, taxes, liens, and such property entitlements as water rights, local schooling and farm income support programs. Yet we know that the values of millions of separate and immobile acres respond to common influences over time, because they are linked by farm-product markets, asset markets, other input markets and common governmental policies.

The social and economic meaning of a change in farmland values depends critically on whether or not it is a change in price, apart from any change in the owner's investments into the land. In this paper all changes in land value caused by changes in the economic environment (macroeconomic swings, transport improvements, etc.) are viewed as changes in price. Price changes are distinguished from quality changes, which result from those direct investments in the land by farm owners and farm operators.[4] The crucial price-quality separation is performed by removing known sources of quality change from the historical measures of land value, to distill as pure as possible a measure of those price changes, those capital gains and losses that owners experienced on their farm lands. In particular, the data allow removal of three quality-change components: changes in the value of farm buildings, changes in aggregate land value caused by geographic drift of America's farms between states with differing land quality, and farmers' conversion of land between the "unimproved" and "improved" categories.

It so happens that in American experience, the movements of farmland values are dominated by price movements. Removing all the quality changes does not greatly alter the view of price trends one would have gained from the raw per-acre values themselves. In fact, the long pre-1915 rise in true farmland prices was even greater than the raw value figures imply, while the price movements since 1915 were very close to those suggested by the unrefined figures. Since similar trends emerge from unrefined value-change figures or from refined price-change figures, we can economize on tables and figures here by looking only at unrefined national values (Table 1 and Figure 1), including annual values since 1900, and at refined regional and national prices for 17 benchmark dates since 1850 (Table 2 and Figure 2).[5]

Average U.S. Farmland Values since 1850. To survey the long movements, it is best to begin the main chronology in 1850, with the start of the census/USDA series, and to defer discussion of trends before 1850 to the next section, when the trends and the deficiencies of the early data can be weighed together.

Between 1850 and 1915, real farmland values advanced 2.08 percent a year, nearly quadrupling over the whole 65 years (Table 1 and Figure 1).[6]

4. Changes in nature, such as the Dust Bowl of the 1930s, could be viewed as changing either the price or the quality of farm land, depending on one's immediate purpose. The semantic choice matters little, since such occurrences play a negligible role in accounting for historical movements in farmland values.

5. For fuller development of the price-change estimates by state and region, see WP, Appendices B–D. WP, Appendix A deals with the prior question of whether the abundant official land valuations are as reliable as scarcer market sales-price data, tentatively answering in the affirmative.

6. In the following discussion of national movements, all "real" magnitudes refer to dollar units of the 1960 U..S. consumer bundle. Regional "real" land values are derived using regional cost-of-living deflators, standardized in dollar units of the 1960 New England urban workers' consumer bundle. See WP, Appendix B for a discussion of the cost-of-living deflators and the similarity of their movements to those in farm product prices.

Some of the broad trends noted in this section have been noted before. Postwar trends are heavily covered in USDA publications and scholarly articles. Few accounts span the whole period since 1850. Two convenient brief overviews are Peter H. Lindert, "Land Scarcity and American Growth," *Journal of Economic History* 34 (1974) 4: 851–84; and Folke Dovring, *Land Economics* (Boston: Breton Publishers, 1987), 282–90.

Table 1. The Purchase-value and Rent on U.S. Farmland, 1805–1986

Year	Land value per acre (nominal)	Land value per acre (1960 $/a)	Realty/acre	1960 US$, cash farm rent/acre
1986	516.71	139.56	160.98	19.29
1985	589.52	162.29	186.93	22.07
1984	678.53	193.46	222.96	25.05
1983	682.78	202.96	234.23	25.81
1982	712.19	218.51	252.51	26.76
1981	706.97	230.21	266.69	26.85
1980	635.32	228.33	264.88	27.69
1979	522.21	213.06	256.23	29.18
1978	404.97	183.83	221.52	29.64
1977	371.36	181.49	218.94	30.16
1976	320.81	166.90	201.33	27.76
1975	280.59	154.39	186.53	25.87
1974	249.45	149.80	181.36	23.91
1973	202.71	135.09	163.94	20.75
1972	177.39	125.57	155.03	
1971	164.20	120.07	148.44	
1970	157.94	120.46	149.49	20.86
1969	151.52	122.40	151.87	21.41
1968	144.52	123.01	152.37	
1967	134.77	119.54	149.02	21.07
1966	126.77	115.68	144.18	19.89
1965	116.07	108.95	137.04	18.76
1964	108.95	104.02	131.76	17.08
1963	102.01	98.68	125.75	16.71
1962	96.14	94.13	121.40	15.99
1961	91.18	90.27	116.81	
1960	89.23	89.23	116.00	
1959	84.15	85.47	112.75	
1958	77.25	79.09	105.45	
1957	71.43	75.14	102.05	
1956	67.38	73.35	99.07	
1955	61.58	68.05	93.93	
1954	58.69	64.65	90.32	9.28
1953	59.60	66.26	91.82	
1952	58.69	65.42	91.40	
1951	53.45	60.89	85.44	
1950	45.93	56.51	79.97	8.49
1949	47.14	58.56	81.98	
1948	45.84	56.39	78.74	
1947	43.24	57.30	79.51	
1946	38.23	57.96	80.36	

(Table 1 continued)

1945	33.94	55.82	77.29	
1944	30.46	51.23	72.32	
1943	27.14	46.41	64.97	
1942	24.16	43.85	61.71	
1941	22.59	45.40	64.31	
1940	22.59	47.72	67.61	8.03
1939	21.65	46.11	68.17	
1938	22.63	47.52	69.29	
1937	22.63	46.66	68.05	
1936	21.65	46.21	68.31	
1935	22.30	48.11	69.02	
1934	21.31	47.15	68.59	
1933	21.29	48.67	68.58	
1932	26.00	56.32	80.14	
1931	32.00	62.25	85.59	
1930	35.73	63.29	86.80	9.19
1929	35.98	62.14	84.62	
1928	36.24	62.58	84.62	
1927	37.24	63.46	85.21	
1926	39.00	65.27	87.03	
1925	40.78	68.80	91.12	
1924	41.62	72.00	93.41	
1923	44.04	76.44	97.20	
1922	45.72	80.71	100.63	
1921	63.15	87.96	107.57	
1920	57.50	84.93	101.92	9.31
1919	48.31	82.59	99.17	9.87
1918	44.12	86.80	104.28	10.61
1917	40.76	94.02	113.02	11.60
1916	38.25	103.77	124.81	13.01
1915	35.73	104.06	125.23	13.45
1914	36.54	107.64	129.61	13.12
1913	35.69	106.66	128.50	12.83
1912	34.84	105.65	127.36	12.71
1911	33.99	106.20	128.09	12.75
1910	33.14	103.08	124.97	12.44
1909				12.51
1908				12.12
1907				11.47
1906				11.52
1905	21.11	70.20	87.84	11.66
1904				11.57
1903				11.21
1902				11.46

(Table 1 continued)

1901				11.81
1900	15.30	54.39	70.32	11.74
1890	16.20	52.19	70.01	
1880	13.98	42.39	57.65	
1870	13.14	30.79	42.79	
1860	11.67	38.81	54.28	
1850	7.68	27.31	39.63	
1805	4.02	9.52		

Sources and Notes for Table 1 and Figure 1: Farm real estate (land and buildings) *per acre, agricultural census years, 1850–1910* and *annually, 1910–1986:* the official U.S. Department of Agriculture averages for the 48 states, beginning with the first available data or 1850. The figures for 1850–1970 are from Clifton and Crowley, *Farm Real Estate Historical Series.* The estimates for 1905, however, are taken from George K. Holmes, *Changes in Farm Values, 1900–1905,* USDA Bureau of Statistics-Bulletin no. 43 (Washington, D.C.: GPO, 1906),11 (I am indebted to Paul Rhode for this reference), spliced to the census values for 1900 using the 1905/1900 ratios from Holmes's estimates. Figures for years since 1970 are from regular publications of the USDA Economic Research Service, with the title at first being *Farm Real Estate Mortgage Market Developments* and later becoming *Agricultural Resources: Agricultural Land Values and Markets. . . . .*

Farm land value per acre, 1805: Blodget, *Economica,* 60, 196. The land area is Blodget's 39 million "cultivated" and 150 million "adjoining" acres, at his valuations. See WP, Appendix D for alternative interpretations of Blodget's figures.

Farm land value per acre, 1850–1986: same sources as for farm real estate. The shares of land value in total real estate value from 1900 on are given by the 1920 Census of Agriculture and by Clifton and Crowley, *Farm Real Estate Historical Series.* Those for 1850–1890 were assumed to equal the same state-level shares for 1900. See WP, Appendix B for further details.

Cash rents for farmland, five Midwestern states, 1900–1986: The annual data for 1900–1920 are from Chambers, *Relation of Land Income to Land Value.* Those for 1920, 1930, 1940, 1950, and 1954 are from the agricultural censuses for these years. The annual figures for 1962–1986 are from USDA Economic Research Service, *Farm Real Estate Mortgage Market Developments* and its successors.

The cash rent data are farm-area-weighted averages of the average values for five states sampled by Chambers: Ohio, Illinois, Iowa, Wisconsin, and Minnesota. For 1900–1915, his thinner-sample series were spliced onto the data rich series for 1916–1920. The annual series since 1962 refers to gross cash rents per acre of wholly-rented farms, presumably including buildings, instead of the separate series for cropland and pasture. See WP, Appendix E for a fuller discussion.

Consumer price index: this is a spliced series set at 1960=1.000. For 1850–1913 I spliced together the Adams, Hoover, Long and Rees cost-of-living indices cited in US Department of Labor *Handbook of Labor Statistics, 1970* (Washington, D.C.: GPO, 1970), 285. To extend back from 1850 to 1805, I spliced on the cost-of-living series by Paul A. David and Peter Solar, "A Bicentenary Contribution to the History of the Cost of Living in America," *Research in Economic History* 2 (1977): 1–80.

What share of the value gain was a true price gain, a capital gain on farm land? More than one hundred percent. The real price of fixed-quality farm land (Table 2 and Figure 2) more than quadrupled, rising 2.18 percent a year. The price gain exceeded the overall value gain because of the westward drift of America's farm land into lands that have always been less productive. The westward drift dragged down the average quality of American farm land faster than the 0.29 percent annual quality gain from improvements on fixed sites.

Table 2. Real Value per Improved Farm Acre, Adjusted for Interstate Drift, by Region, 1850–1986 (in 1960 New England consumer dollars per acre)

	1850	1860	1870	1880	1890	1900
N. England	41.26	46.22	33.54	49.07	54.04	71.04
M. Atl.	78.00	98.61	82.23	102.41	112.32	113.83
E.N.C.	57.37	92.66	77.96	99.44	121.53	141.64
W.N.C.	19.03	31.67	30.39	34.01	51.29	67.50
S. Atl.	20.23	29.28	14.98	26.34	41.35	45.30
E.S.C.	32.47	58.85	26.31	30.93	37.96	41.44
W.S.C.	21.73	26.89	16.75	23.43	35.39	38.90
Mountain						
Pacific						
NATIONAL	34.06	49.08	34.78	47.43	59.82	70.88
US % real capital gain/yr.:		3.72	−3.38	3.15	2.35	1.71
US % nominal capital gain/yr:		4.42	0.06	0.52	1.73	0.71

	1910	1915	1920	1930	1940	1945	1950
N. England	91.57	90.39	74.90	72.19	65.54	83.29	72.92
M. Atl.	126.66	115.94	86.81	91.46	79.54	93.84	97.49
E.N.C.	233.35	232.29	205.20	116.37	104.54	129.01	137.77
W.N.C.	136.22	147.60	146.63	89.71	57.64	67.78	75.24
S. Atl.	87.31	91.24	109.00	87.51	75.02	90.97	87.35
E.S.C.	65.00	70.49	83.04	73.74	67.71	78.01	77.08
W.S.C.	86.12	87.53	89.19	83.91	68.83	87.28	93.55
Mountain		69.22	54.11	43.09	34.67	48.36	49.90
Pacific		241.57	181.95	213.04	148.60	226.44	202.54
NATIONAL	128.19	137.95	114.35	86.98	67.19	83.86	86.26
	6.10	1.48	−3.68	−2.70	−2.55	4.53	0.57
	7.48	2.93	10.31	−4.45	−4.25	9.91	6.58

The overall price increase occurred despite a 29 percent national drop in farmland prices in the Civil War decade.[7] The era famous for farm

7. Two major qualifications must be noted about the Civil War decade. First, the real drop from 1860 to 1870 may have been less severe than the 29 percent cited here. The Census Bureau marked down its farmland valuations for 1870 by 20 percent to convert from "currency values" to "gold values" [see the *1920 Census of Agriculture*, vol. V (Washington: GPO, 1922), pp. 50–55]. Yet the cost-of-living deflator taken from the Bureau of Labor Statistics is based on work (by Ethel D. Hoover for 1851–1890) that does not mention the issue of greenbacks versus gold dollars. If the Hoover-BLS series is based on greenback prices, the national decline should be only about 9 percent, not 29 percent, for 1860–1870, with similar optimistic adjustments for regions. Correspondingly, the increase from 1870 to 1900 should be lower.

Second, the decline across the Civil War was worse than that cited for the whole decade

(Table 2 continued)

	1960	1970	1980	1986
N. England	116.53	205.66	323.14	409.22
M. Atl.	163.00	279.40	475.06	373.41
E.N.C.	217.55	271.05	568.72	260.63
W.N.C.	102.41	128.46	269.86	120.24
S. Atl.	184.71	258.06	450.27	337.47
E.S.C.	127.76	206.25	372.80	269.23
W.S.C.	160.78	223.44	302.04	237.53
Mountain	81.90	102.12	191.56	127.03
Pacific	354.92	362.48	514.20	417.99
NATIONAL	134.84	178.28	336.53	205.87
	4.57	2.83	6.56	−7.86
	6.76	5.66	14.88	−3.37

Sources and Notes to Table 2 and Figure 2: The real values of improved land per acre from 1850 to 1986 are derived from the unadjusted Census and USDA figures through a series of steps described at length in Appendices B–D.

The adjustments begin with state-level values of farm land per acre, already excluding the value of buildings, as in the national averages in Table 1 and Figure 1 above. Regional and national averages are re-adjusted so as to remove the effects of geographic (interstate) drift in America's farms. Before World War I the westward drift had a strong tendency to reduce average land quality. Removing its effects on land value thus raises the rates of growth of the prices of land at fixed sites. See WP, Appendix C for details.

Next, region-specific cost-of-living deflators are used to convert nominal to real values. The deflators refer to urban working-class consumer bundles. The resulting real (deflated) values show the purchasing power of an acre of land when the proceeds from selling it are spent in a city of the same region. WP, Appendix C gives details, and WP, Appendix B notes the similar behavior of farm-product price deflators for individual states.

Finally, the effects of land improvement are removed from land values. The adjustment uses the shares of farm land that are "improved," as measured in the Census between 1850 and 1950, and the value ratios for (improved/unimproved) land as estimated by Tostlebe for several regions. For 1950–1986, movements in each state's land quality are those estimated by Willis Peterson, "Land Quality and Prices," *American Journal of Agricultural Economics* 68(1986)4:812–819. For details, see WP, Appendix D.

1860–1870. A special survey by the USDA gave percentage changes in the nominal value of "farm land" from 1860 to 1867 for each state east of the Rockies, except for the Dakotas and Oklahoma. [See U.S. Department of Agriculture, *Report of the Commissioner of Agriculture . . . 1867* (Washington: GPO, 1868), pp. 102–119.] Weighting states by their 1860 acreage yields these regional changes in nominal land values:

New England	+18%	South Atlantic	−46%
Mid Atlantic	+27%	East South Central	−37%
East North Central	+39%	West South Central	−43%
West North Central	+58%	All of these regions	− 7%

Deflating the all-region change by the change in the urban cost of living yields a drop of 41 percent in the real value of farm land by 1867 (assuming that the land values and the cost of living either shared a greenback basis or shared a gold basis).

The author thanks Paul Rhode for stressing the currency issue and pointing out the USDA data for 1860–1867.

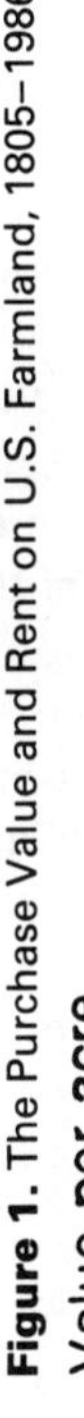

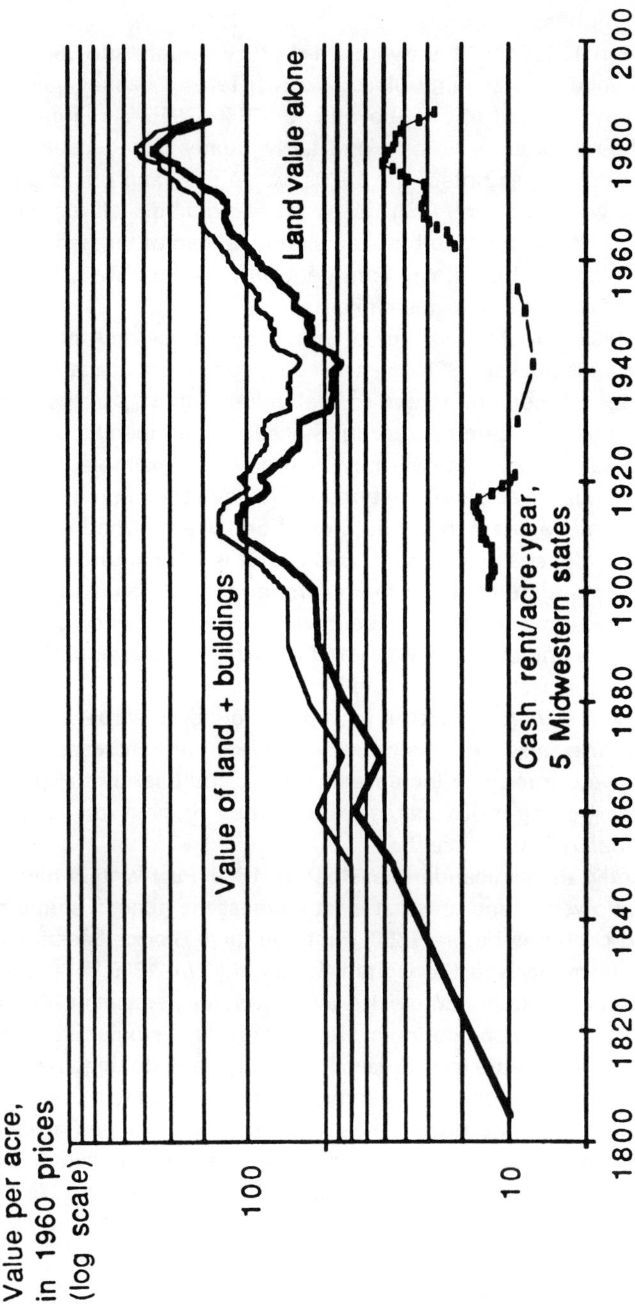

Figure 1. The Purchase Value and Rent on U.S. Farmland, 1805–1986

and 1930, before the New Deal government farm programs could have favored southern landlords.

Between 1940, near the twentieth-century trough, and 1980, the peak, real farmland prices quintupled. The real rate of capital gain was 4.11 percent a year, well above the rate for 1850–1915. As Figures 1 and 2 show, the rise was monotonic and fairly steady. The decade of greatest capital gain was, as noted, the 1970s, when real land prices gained 6.56 percent a year. The forty-year rise was spread across all regions, though New England and the South pulled slightly ahead of the national average in this period. Section III will comment on what little we know about the causes of this impressive sustained rise.

The 1980s brought the sharpest drop in farm prices in American history. From the 1981 peak to 1986, the average acre of U.S. farm land declined 39 percent (or 9.5 percent a year) in real value (Table 1), virtually all of the value loss being a pure price drop. Since it had jumped 83.3 percent over the previous nine years, we may well have an example of the kind of speculative boom and bust that observers have seen in the rise and fall of nominal land values during and after World War I. Whether the rise and fall were really speculative excesses or not is a topic to which we return below. The decline of those five years, at any rate, was more severe in percentage terms than any other five-year decline since the Civil War, including the end of World War I and the Great Depression. One need not dig far into the past to study acute farm crisis.

Land Values before 1850. If land prices quadrupled between 1850 and 1915, had they also been rising earlier? Were there increases in land values that were true price increases? That is, could land of a given quality have been getting more scarce for a century or even longer despite the pushing back of the frontier?

We enter a statistical darkness before 1850. Past writers have lit a few candles, however, and these at least suggest the general shape of things over two centuries before 1850. First, we have Blodget's estimate of the value of farm lands in the United States in 1774–1805.[11] Blodget left us some estimates but not others. His average value per acre in 1805, undocumented but worth considering in view of his relative expertise, implies that the real average value of U.S. farm land may have doubled between 1805 and 1850, again despite the opening of cheaper land to the west. Blodget also offered separate price series for two extreme types of land in the United States of his day. Figure 3 graphs the estimated values per acre of cultivated lands and land "in their natural state" between 1774 and 1805.

11. Samuel Blodget, *Economica: A Statistical Manual for the United States of America* (New York: Augustus M. Kelley reprint of 1806 edition, 1964).

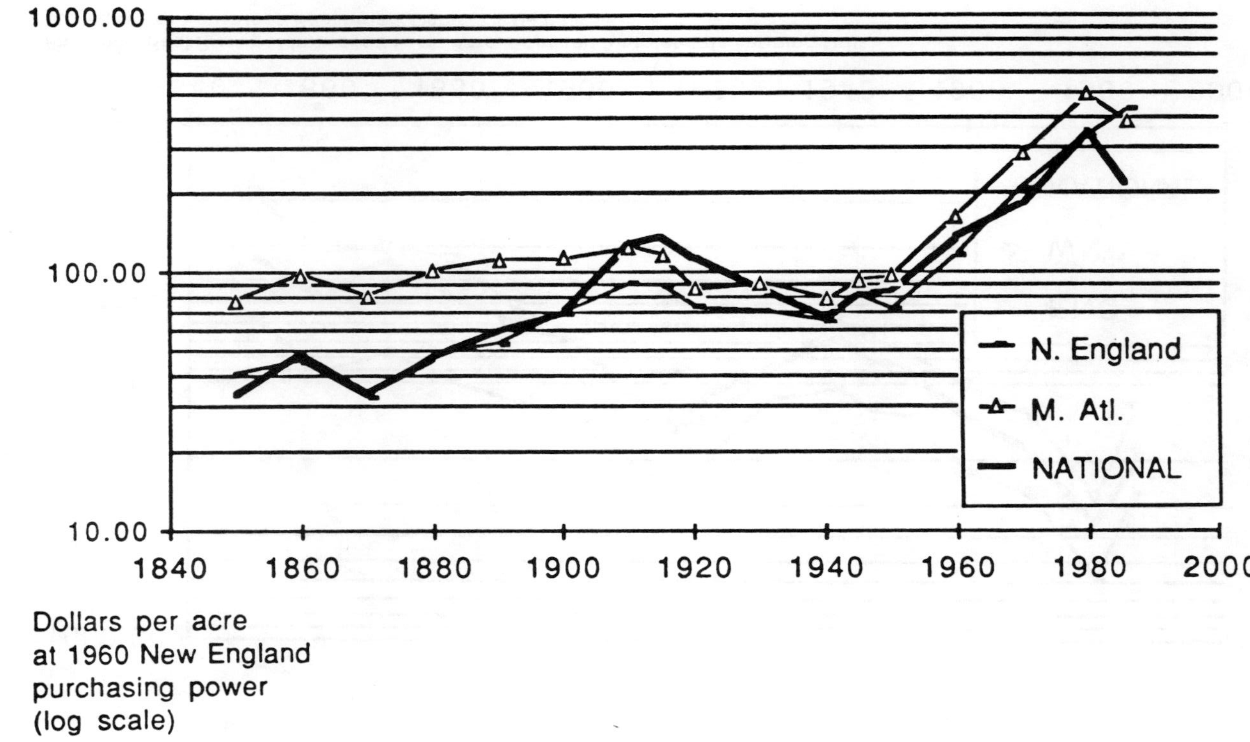
1000.00
100.00
10.00
1840
1860
1880
1900
1920
1940
1960
1980
2000
N. England
M. Atl.
NATIONAL
Dollars per acre
at 1960 New England
purchasing power
(log scale)

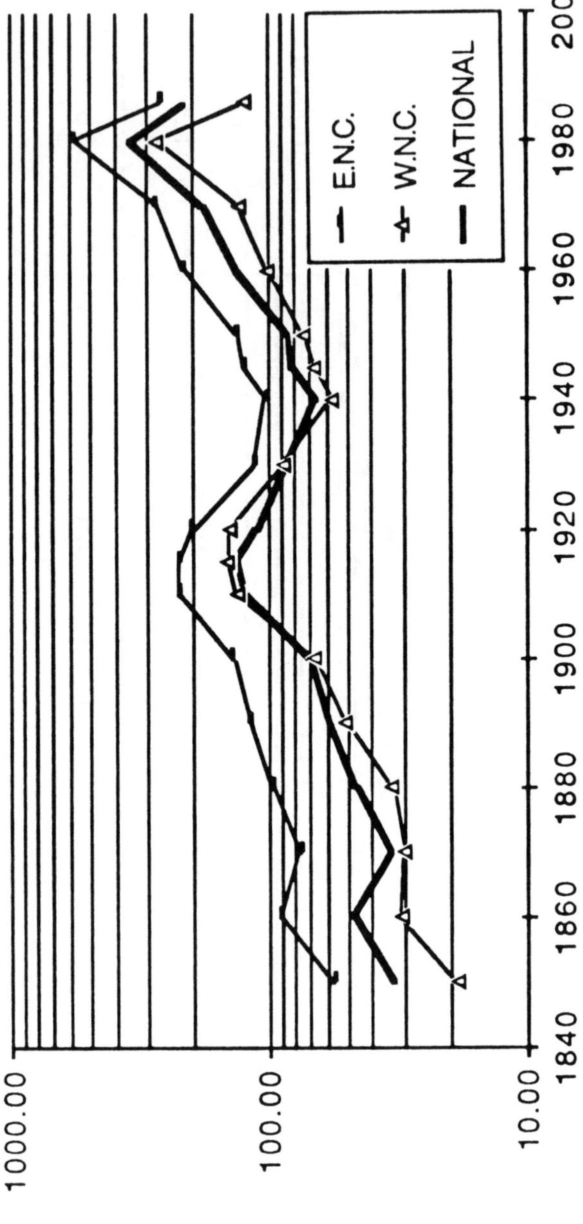

Figure 2. Real Price per Improved Farm Acre, Adjusted for Interstate Farming Drift, by Region for Seventeen Benchmark Dates, 1850–1986

Both series suggest that land values were depressed during the Revolution but soared after it, in the period 1790–1805.

The distinction between land-price changes and land-quality changes must again be given its due. In frontier America, land shifted rapidly from being altogether unworked to being "unimproved, but in farms" (untilled grazing land, forest, etc.), to being fully cultivated. The rate of development of the land slowed down progressively. The share of improved land in total farm land did not peak until 1910, and the total land area in farms peaked only in 1950. Of the land ultimately in farms around 1950, only a quarter was in farms a century earlier, in 1850, and even less was developed before that.[12]

Careful accounting makes it possible to show what Blodget's estimates imply about changes in land price and changes in land quality over the periods 1774–1805 and 1805–1850.[13] The precise magnitudes depend on what one assumes about the unimproved land area in farms, a magnitude on which Blodget offered no guesses. The range of implications is still narrow enough, though, to conclude that there were noteworthy capital gains on farm land in both periods. Between 1774 and 1805, Blodget's figures imply that the real price on fixed-quality land rose at least 121 percent, i.e., more than doubled. Since Figure 3 shows that all the real gain came in the subperiod 1790–1805, this era might have competed with the 1970s as the era of greatest capital gains on U.S. farms. From 1805 to 1850, the real price of fixed-quality land more than doubled again, and possibly tripled, depending on how one reads Blodget's estimates for 1805. The whole chain of land price increases from 1790 to 1915—at least doubling (1790–1805) followed by more than a doubling (1805–1850) and more than a quadrupling (1850–1915)—raises the obvious rhetorical question

12. Defining the 1158.6 million acres in farms in 1950 as the ultimate supply of farmland in the 48 states, historical data give us the following percentages:

(land areas as a % of area in farms in 1950)

	1774	1805	1850	1870	1910	1950	1985
"Cultivated" or "improved"	1.8	3.4	9.8	16.3	41.3	45.9	..
Total land in farms	..	16.3	25.2	35.2	75.9	100.0	87.7
Share "cult." or "improved"	..	21%	39%	46%	54%	46%	..

[*Sources:* Blodget, *Economica,* pp. 60, 196; Ivery D. Clifton and William D. Crowley, Jr., *Farm Real Estate Historical Series: 1850–1970,* USDA, Economic Research Service, Bulletin ERS 520 (Washington, D.C.: U.S. Department of Agriculture, 1973); U.S. Census of 1850; Alvin S. Tostlebe, *Capital in Agriculture: Its Formation and Financing since 1870* (Princeton: Princeton University Press for the National Bureau of Economic Research, 1957); USDA, *Agricultural Statistics, 1985.*]

13. The detailed decompositions of land-value changes for 1774–1805 and 1805–1850 are presented in WP, Appendix D, which also discusses ambiguities and inconsistencies in Blodget's data on improved area and land values.

about the pre-1850 figures: Can we believe the early estimates? Fresh archival data are needed to test Blodget.

If farmland prices rose so rapidly after independence, had they also been rising fast across the colonial era? Our few dim candles in the darkness suggest not. Let us start with an awkward comparison of seventeenth-century Maryland with the 13 colonies/states for 1774–1805. As Figure 3 shows, the average value of all unimproved farm land in the United States (east of the Mississippi River) in 1790 seems to have been no higher than the value of unimproved Maryland farm land in the late seventeenth century. Even the price of improved land did not rise as fast from the 1690s to 1774 as it did after 1790. The awkward comparison hints that land values increased more slowly from the late seventeenth century to the late eighteenth than they were to rise after independence.

The colonial period is complicated, of course, by the paucity of data and the heterogeneity of land. Land varied enormously in value. Land in coastal New England was much more valuable than even coastal land in Maryland or the South, as shown for the seventeenth century by Figure 3 and for the nineteenth by Sturm.[14] Prices were much lower in the interior. One's view of colonial land-price trends will also depend on whether one wants to follow the prices of only those lands actually farmed in colonial times or the prices of all lands in the peak farming area of 1950.

Again, more archival research on land prices is needed. I would hypothesize, however, that the average value of farm land rose much more slowly from the late seventeenth century until Independence than it was to rise later, in the first century of independence.[15]

Rents, Land Values and the Rate of Return. Our overview of the purchase price of land has suggested some conclusions about long-run trends in the scarcity of American farm land. There are some weak hints of only a gentle price rise in the colonial era, followed by rapid price increases after independence, and continued price gains across the late nineteenth century. The main disturbance in this tapered rise was the 29 percent capital loss across the Civil War decade. The twentieth century still stands out as an era afflicted by wide swings in land price: a noticeable rise up to 1914, a 59 percent drop in average value from 1914 to 1942, a tripling from 1942 to 1973, and an unprecedented boom-and-bust swing since 1973. To what extent were these movements caused by forces

14. James L. Sturm, "Investing in the United States, 1793–1893: Upper Wealth-holders in a Market Economy," (unpublished Ph.D. dissertation, Madison: University of Wisconsin, 1969).

15. Here I refrain from conjecture about trends in the market value of *potential* U.S. farm land, most of which was unsettled until the late nineteenth century. There was presumably a decline in another concept of land price not pursued in the text: the cost of settling on farm land and preparing it for use. As the frontier advanced, this presumably dropped.

Figure 3. Some Early-American Series on Land Values per Acre, 1642–1805.

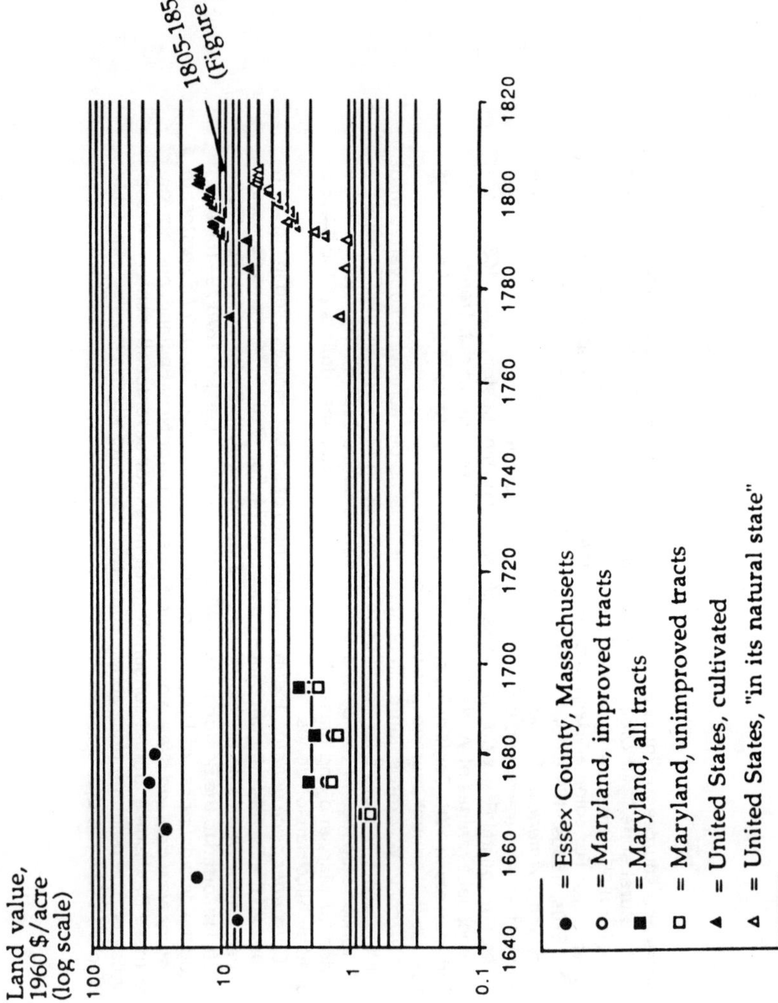

Notes and sources to Figure 3: ▲, △: Values for cultivated lands and for lands "in their natural state," *All U.S., 1774–1805:* Blodget, *Economica,* 60. The extra data point for 1805 is the starting point of the land value series plotted in Figure 1.

■, ○, □: For colonial *Maryland, 1664–1699,* these are decadal averages (1664–69, 1670s, 1680s, 1690s) of V.J. Wyckoff's ("Land Prices in Seventeenth-Century Maryland," *American Economic Review* 28(1938)1:82–88) values per acre for improved tracts, all tracts, and unimproved tracts, respectively.

●: For coastal *Essex County, Massachusetts,* north of Boston, *1642–1682,* these are decadal averages of William I. Davisson's ("Essex County Price Trends: Money and Markets in 17th Century Massachusetts," *Essex Institute Historical Collections* 103(1967) 2) annual-average land values per acre.

Currency: Wyckoff's figures in pounds of tobacco per acre were converted into sterling using Menard's farmgate price of tobacco in Maryland or the Chesapeake (*Historical Statistics of the United States,* vol. II, Series 2583-4). The sterling price series for Maryland and Massachusetts were converted into dollars at the $4.15/£ used by Alice Hanson Jones.

Consumer price deflator: Extending the chain of spliced indices used in Figure 1, I followed the David-Solar price index back to 1774. For 1720–1774, it was necessary to use Cole's wholesale prices (*Historical Statistics,* vol. II, Series 2557). For 1700–1720, I resorted to the Boston price of wheat in units of silver. For 1670/79–1690/99, I used the decadal prices of Terry L. Anderson, *The Economic Growth of Seventeenth Century New England* (New York: Arno Press, 1975) for New England, and hooked to these Davisson's annual consumer price index for seventeenth-century Massachusetts. The resulting price series, converted back into silver, was also applied to Wyckoff's Maryland land values.

changing the true productivity of land in current farm production? To what extent do they reflect "speculative" forces, or changes in the interest rate at which land values are capitalized, or tax changes, or other changes in the nature of farm property rights?

The quickest way to progress toward answers is to compare the purchase price of farm land with its annual rental value. Movements in rental value generally reflect influences on the marginal productivity of a farm acre of given quality, since rents should tend to equal the marginal product of land. The other forces—speculative expectations, interest rates, taxes and property rights—tend more to influence movements in the rent/price ratio (or the reciprocal of the capitalization ratio, called the number of "years' purchase" in English historical terminology). To give the different forces their due, we turn to farm rents, in three stages: (1) a look at documented twentieth-century trends in real rents, (2) judgments about the rent/price ratio in this century, and (3) indirect inferences about trends in real rents before 1900, using a pattern in the rent/value ratio observed for the twentieth century. Knowing rents as well as rates of capital gain allows us some limited comments on trends in the overall rate of return to owning farm land.

Within the twentieth century, data on the rental value of farm land for five Midwestern states[16] extend back to 1900, thanks to the classic study by Clyde Chambers, the censuses of 1930–1954 and USDA annual surveys since 1962.[17] Unfortunately, the data are not a consistent series, but three juxtaposed series that probably differed in coverage.[18] Worse, the three series do not overlap in time, preventing direct tests of their comparability. Figure 1 above shows the time path implied by Chambers' figures, the agricultural census data for 1930–1954 and the USDA. The rental series shows the same broad trends as the purchase-price series for either the United States or for the Midwest, but with some differences. Midwestern land rents rose from 1900 to 1915. Real rents followed real land prices downward until World War II, though the decline in rents was less severe. After World War II, both real rents and real prices rose. Both jumped in the 1970s and fell in the 1980s, but here two differences stand out. First, real rents started falling after a peak in 1977, but real farmland prices did not start falling until after a peak in 1981. Second, farmland prices gyrated much more than rents, so that the rent/value ratio dropped sharply from 1974 to 1980, then jumped.

It is not surprising that rents and prices generally moved together over the century as a whole. They should always be linked by asset markets.

16. Illinois, Iowa, Minnesota, Ohio and Wisconsin.

17. Clyde R. Chambers, *Relation of Land Income to Land Value,* U.S. Department of Agriculture Bulletin no. 1224 (Washington, D.C.: G.P.O., 1924).

18. See WP, Appendix E for details on the three rental series.

Investors can choose whether to hold farm land or to hold such other assets as interest-earning bonds or common stocks. Rents are part of the income from holding farm land, the other part being the net gain on price appreciation and government supports for landowners (minus repairs, depreciation, taxes and a risk premium). If rents rose relative to farmland prices for some outside reason, there would be a tendency to invest more in land, bidding up the price until the ratio of rent to price returned to a level consistent with interest rates, price expectations, and other forces. For similar reasons, a drop in rents should be accompanied by a drop in prices.

Yet the ratio between farmland rents and prices has not been a constant. In each of our three data periods (1900–1920, 1930–1954, and 1961–1986) the rent/value ratio gyrated in a way suggesting a *systematic pattern in forecast errors:* the further the real price of farm land is above (below) its previous longrun trend, the greater the likely overoptimism (overpessimism) about the subsequent price trend.

The fall of the rent/value ratio early in this century, for example, cannot be explained by interest rates, or taxes, or government subsidies, or any good reason why the risk of landholding should have seemed to drop. It must be explained by a rise in farmers' expectations about future farmland prices and rents. Were they right in expecting further increases? They were clearly wrong in the one year 1920, when the fall in the rent/value ratio was immediately followed by a drop in both rents and prices. Other writers have documented their naiveté about price trends around 1920.[19] Their behavior in 1920 certainly *looks* like a speculative bubble, and I shall continue this traditional interpretation for the year 1920, though economists rightly insist that seeing investors make a wrong forecast does not strictly prove irrationality or instability in their behavior.

By 1940, on the other hand, investors seem to have been too pessimistic about farmland values. Interest rates and returns on common stocks were very low, and the onset of war was already ushering in a period of inflation, especially inflation in the relative price of agricultural products. With hindsight, we find it a good year for investing in farm land. Farmers tended to be more pessimistic the lower was the real price relative to its prior trend.

The movement of farm rents since 1962 also suggests that investors' forecast errors were pro-cyclical. The rent/price ratio fell from around 7 percent to around 5 percent at the 1980 trough, and then jumped back up in the crisis of the 1980s, regaining its early-1960s level. Again, as in the two earlier periods covered by the rent data, it appears that the market's price-forecast error was positively related to deviations in land price from

19. See William G. Murray, "Iowa Land Values, 1803–1967," *Palimpsest* 68 (1967) 10: 461–468; and, again, Shideler, *Farm Crisis;* and Johnson, "Postwar Optimism."

prior long-run trend. The trough in the rent/value ratio in 1980 implies that it was then, with price at a cyclical peak but real rents already sliding, that investors most expected acceleration of rents in the future.

Was the fall and rise of the rent/value ratio a speculative land-price bubble? In the wake of the farm-value collapse of the 1980s, economists are drifting back toward an affirmative answer. The econometric study by Moore and Meyers rejects a variant on the rational expectations hypothesis in the market for farm land, even for time series ending at the 1981 peak in land values.[20] Burt finds the farm land market responding to revelations of its mistakes with dampened swings.[21] Featherstone and Baker find overshooting and bubbles in US data on all farm assets for the period 1910–1985.[22]

The belief in speculative bubbles in farm land, while plausible, is not necessary for present purposes. All that is needed is the systematic twentieth-century pattern in the market's revealed forecast errors: the higher the value of land relative to prior long-run trend, the greater the tendency to overestimate the future paths of rents and land values. Knowing that such a pattern has shown up throughout this century, plus a little asset-market theory, generates a working hypothesis about how the rent/value ratio, and farm rents themselves, moved in the period 1850–1900, for which we lack rent data. The small infusion of theory concerns asset-market equilibrium. At any point, investors will link rates of return as follows:

expected rate of return on owning land	=	rent/value ratio *plus* expected rate of land price gain *minus* taxes *minus* cost and disamenity premium on land	=	rate of interest on the alternative asset

Re-expressing the equilibrium relationship so as to focus on the rent/value ratio yields:[23]

rent/value ratio	=	interest rate *minus* actual land-price inflation *plus* tax rate	+	cost and disamenity premium on land *minus* price forecast error
		measurable factors		*unmeasurables (a net residual)*

20. Kevin C. Moore and William H. Meyers, "Predictive Econometric Modeling of the U.S. Farmland Market: An Empirical Test of the Rational Expectations Hypothesis," CARD Report 133 (Ames: Iowa State University, 1986).

21. Oscar Burt, "Econometric Modeling of the Capitalization Formula for Farmland Prices," *American Journal of Agricultural Economics* 68 (1986) 1: 10–26.

22. Allen M. Featherstone and Timothy G. Baker, "An Examination of Farm Sector Real Asset Dynamics, 1910–1985," *American Journal of Agricultural Economics* (August 1987): 532–46.

23. For a more detailed treatment, see WP, Appendix E.

Breaking down the rent/value ratio into two parts is useful because the two parts are documented differently and reflect different kinds of behavior. Movements in the first term can be followed in the data, even for the nineteenth century. Movements in the second are likely to follow the forecast-error pattern just noted for the twentieth century.

Knowing trends in Midwest farm land values, interest rates, and taxes for the second half of the nineteenth century, we can now infer what probably happened to the undocumented rents of that era. Between 1850 and 1860, investors reaped nominal capital gains on Midwest farm land at the rate of 9.44 percent a year. We would have to believe that rents were also rising unless something could have made the rent/value ratio drop faster than 9.44 percent a year. That was surely not the case. On balance, the rent/value ratio probably did not decline between 1850 and 1860. It would have risen, and real rents would have doubled over the decade, if investors in 1860 had foreseen the upcoming capital losses of the Civil War decade, even though interest rates did decline somewhat.[24] The likelihood that investors were overoptimistic on the eve of the Civil War means that they probably bid the value/rent ratio higher—the rent/value ratio lower—than objective and measurable factors would have warranted. The best working hypothesis about the 1850s is that real rents rose, though they did not double, as the changes in measurable factors would have implied.

For the Civil War decade and the farm-protest period 1870–1900, the same reasoning lets us see that real farm rents were probably stable, or rose only slightly, despite wider swings in the purchase value of farm land. In the Civil War decade, the fall in land values was probably due to a shift toward investor pessimism, reinforced by a slight rise in nominal interest rates. The rent/value ratio should therefore have risen enough to cancel the drop in real value, leaving no great change in real farm rents. For 1870–1900, real rents probably again failed to follow farmland values. The rise in values was sufficiently nurtured by a drop in interest rates. With interest rates declining strongly, the rent/value ratio should have declined, and again there should have been no strong upward trend in real farm rents, except perhaps in the 1880s.

In sum, real Midwest farm rents probably moved as follows:[25]

1850s—real rents up, though they probably did not double;
1860s—no clear change in rents, unlike the drop in land values;
1870s—no clear trend in rents; 1880s—rents up, perhaps by 1.6 percent a year;

24. Allan G. Bogue, "Land Credit for Northern Farmers," in Vivian Wiser, ed., *Two Centuries of American Agriculture* (Washington, D.C.: Agricultural History Society, 1976), 76–87.
25. Again, see WP, Appendix E for more details.

1890s—no clear trend in rents;
1900–1915—rents up 0.9 percent a year;
1915–1940—rents down, perhaps 2.0 percent a year;
1940–1977—rents up strongly, perhaps 3.6 percent a year;
1977–1986—rents down 4.8 percent a year.

If rents and capital gains on farm land were both subject to increasingly dramatic swings, and the swings were nearly contemporaneous, it should follow that the same was true of the overall rate of return on owning farm land. That is the case.

The rate of return on investments in farm land can be traced for parts of the twentieth century, and limited inferences can be made about its movement in the second half of the nineteenth. First, however, it is wise to step back and reflect on what a measure of realized rates of return would tell us. Obviously, they reveal something about the real resources that end up in the hands of landowners, to be spent or reinvested. Yet for some debates, one is more concerned with the ex ante expected rate of return, relative to both the expected return on other assets and the subsequent realized return on farm land. Unfortunately, the expected rate of return cannot be observed. Even if each investor were in asset-market equilibrium, the expected rate of return could depart from the expected (or even certain) return on other assets for several reasons that cannot be separately quantified. Consider the role of unobservable concepts in these equations, where only the italicized terms are observable in data: (1) *realized (ex post) return on farm land* equals expected (ex ante) return on farm land plus net forecast error; and (2) expected return on farm land minus *interest rate on an alternative asset* equals a pure disequilibrium gap plus a relative-disamenity premium on owning farm land due to risk aversion, etc., minus the amenity value, the joy and comfort, of being a farm land owner. With so many concepts unobservable, the overall gap between the realized return and the alternative interest rate can tell us little.

Note, however, that all those unobservable differences between the things we can measure are usually thought of as unchanging over the decades. The forecast error should always have a mean value of zero, however wide its year-to-year swings. So should any pure portfolio-disequilibrium gap. The disamenity of risk aversion and the amenity of owning your own farm are values that are supposed to be inherent in the farm sector, changing slowly or not at all. Therefore, if we observe sustained systematic swings in the departure of realized returns on farm land from alternative interest rates, we should ask which of the "unchangeables" actually changed.

Figure 4 offers clues on historic swings in the rate of return and what might have caused them. Note first the solid line representing the rate of current return (excluding any capital gains) on all productive farm assets. Its numerator is the rental stream that the market price of farm assets, dominated by real estate, is supposed to capitalize. Its trend is slightly downward in this century, and it has become more stable. The rate of current return on productive farm assets has remained near the short-term interest rate except during the inflationary surprises of the World Wars, which seem to have slightly buoyed real current returns on farm assets, while cutting the real rate of interest. The stable downward drift in current returns suggests that the net disamenity of owning farm assets drifted downward, perhaps because of the reduction of risk as proxied by movements in the current-return series.

From the turn of the century to World War II, the market for Midwestern farm land followed current rental returns and the real interest rate in the way it should, according to what data we have.[26] The shocks of World War I and the Great Depression had only a limited effect on the capitalized value of farm land. The worst years for real capital loss were 1916–17 = 18.8 percent, 1917–18 = −7.7 percent, 1931–32 = −9.5 percent, and 1932–33 = −13.6 percent. Against these losses must be counted the positive rents, like those positive real returns on all farm assets referred to in the preceding paragraph. None of these losses, nor the gains documented for 1919–1920, is badly out of line with either the interest rate or the subsequent movement of current returns. The same reasonableness of farm land movements seems to have been characteristic of the latter half of the nineteenth century.

Since World War II, on the other hand, the rate of return on Midwestern farm land (and probably all US farm land) has shown more dramatic swings. Capital gains sent the rate of return on farmland ownership well above its current-production return, as Figure 4 shows. What expectations could have made investors raise the capitalized price so much faster than rents (current returns) were rising? As noted, the rents (current returns) that farm asset values are supposed to capitalize were not very volatile and were gradually declining relative to the asset values themselves. The take-off of capital gains on land above current returns on all farm assets, and above real interest rates, was already in evidence in the period 1950–1972. From 1973 to 1979, the real net return on (Midwestern) farm land reached unprecedented heights, with an average rate of 15 percent a year. Between 1980 and 1985, the same real rate of return averaged −13 percent a year. Small wonder that the literature on farm land pricing should now

26. This paragraph and the next are based on a reading not just of Figure 4, but also of the capital gains for additional years covered in Table 1.

be shifting from capital-asset pricing and efficient-market models toward seeing bubbles and rejecting rational-expectations hypotheses.

The striking part of the instability is that it is recent. Earlier swings in farm land values, for all their fame, cannot compare with the gyrations just experienced. The key question for further research, again, is what has changed to make land prices gyrate when current returns do not. That research might benefit from the statement of this *working hypothesis:*

> The dramatic capital gains on U.S. farm land between 1945 and 1972 reflected the progressive realization that the future of government farm supports would be more generous and secure than investors had believed in the gloom of the Great Depression and in World War II. After 1973, farm land prices took on their own dynamic, due to some mixture of an endogenous speculative bubble and a misreading of the post-1973 inflation shocks.

Land Appreciation and Capital Accumulation

The estimates of price appreciation on farm land open up avenues for new research. Particularly ripe for re-examination is the role of capital gains, on land or other assets, in the overall process of modern capital accumulation. This section points out two avenues for further research on American capital accumulation in the nineteenth century.

Land appreciation and the rise and fall of America's savings rate. Scholars have long puzzled over the jump in America's rate of conventional capital formation across the Civil War era—almost as much as they have fretted over the decline of personal and national savings since the early 1970s.[27] There was a marked rise not only in the conventional rate of capital formation (investment/national product, at fixed prices), but also in the conventional rate of national saving (saving/national product, at current prices). The private savings rate remained high and steady for the first

27. Lance E. Davis and Robert E. Gallman, "The Share of Savings and Investment in Gross National Product during the 19th Century in the U.S.A.," in F.C. Lane, ed., *Fourth International Conference in Economic History* (Paris: Mouton, 1973), and "Capital Formation in the United States during the Nineteenth Century," in Peter Mathias and M. M. Postan, eds., *The Cambridge Economic History of Europe,* vol. VII, Part II (Cambridge: Cambridge University Press, 1978), 1–69; Paul A. David, "Invention and Accumulation in America's Economic Growth: A Nineteenth-Century Parable," in Karl Brunner and Allan H. Meltzer, eds., *International Organization, National Policies and Economic Development* (Amsterdam: North-Holland, 1977); Jeffrey G. Williamson and Peter H. Lindert, *American Inequality: A Macroeconomic History* (New York: Academic Press, 1980), Ch. 12; Roger L. Ransom and Richard Sutch, "Domestic Saving as an Active Constraint on Capital Formation in the American Economy, 1839–1928," University of California Project on the History of Saving, Working Papers No. 1 and 2, December 1984.

three quarters of the twentieth century, so that the national saving rate fluctuated only with shifts in government saving, and then dropped off after 1973.[28] Why that happened is still a subject of considerable debate.

Taking capital gains into account will accentuate and reshape the rise and fall of the American savings rate. The savings rate can be defined as the share of resources coming available in the current period that becomes a change in perceived net worth—i.e., is devoted to future, rather than present, consumption. The resources coming available are not just currently earned income but also any property gains, in the form of transfers (e.g., an inheritance) or capital gains on already-held assets. The savings rate can thus be measured as the ratio (conventional savings plus capital gains)/(current income plus capital gains). Confining our view of possible capital gains to those on farm land, we find the following approximate effects on the national saving rate:[29]

Year	Conventional net national rate of savings out of NNP, current prices	Net national savings rate augmented by annual capital gains on farm land
1849/50	12%	15%
1879/80	15	16
1899/1900	19	22
1912	15	16
1929	10	9
1939	4	5
1950	14	15
1973	11	12
1978	9	10
1986(prelim.)	2	2

28. Paul A. David and John L. Scadding, "Private Savings: Ultrarationality, Aggregation and 'Denison's Law,' " *Journal of Political Economy* 82 (1974) 2: 225–49.

29. The savings rates were calculated by making a number of adjustments to estimates available in Raymond W. Goldsmith, *Comparative National Balance Sheets: A Study of Twenty Countries, 1688–1978* (Chicago: University of Chicago Press, 1985). Goldsmith provided estimates of all the magnitudes used here, but it was necessary to improve on his figures for land values and GNP. Farmland values are taken from the present study. For nonfarm land values, Goldsmith assumed a share of land in total real estate value (land plus buildings) that bounced from ⅙ for 1805 to ⅓ for 1850 and back to ⅙ for 1880, ignoring Winnick's estimate of a 40 percent share for residential real estate as of 1890; his ⅙ shares were changed to ⅓ in side-calculations used indirectly here. Finally, Goldsmith's use of Berry's national-product estimates seems inferior to the Gallman estimates for 1850 and 1880, though his choice of Berry over Davis-Gallman ("Capital Formation in the United States," and Paul A. David, "The Growth of Real Product in the United States before 1840: New Evidence, Controlled Conjectures," *Journal of Economic History* 27 (1967)2) made no difference for 1805.

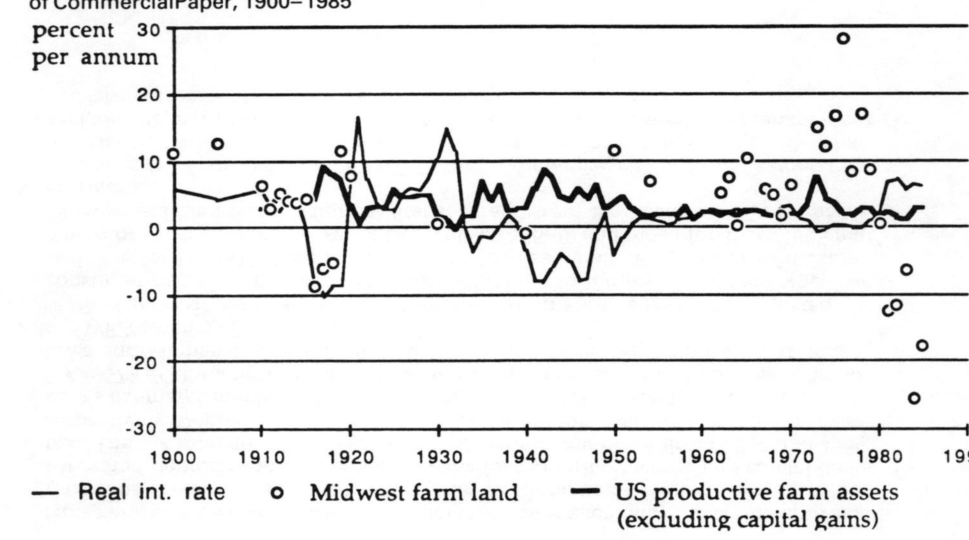

Figure 4. Real Rate of Annual Return on Midwest Farm Land, Compared with Real Annual Current Income Rate on All U.S. Productive Assets and the Real Interest Rate of CommercialPaper, 1900–1985

Notes and sources for Figure 4: *Midwest farm land:* each rate equals the rent/value ratio plus the net value appreciation (relative to the cost-of-living deflator) minus estimated depreciation and property taxes (but ignoring income and capital-gains tax). The tax rates for 1900–1905 and 1985 were extrapolated from 1910 and 1984, respectively. Depreciation rate = 0.40% per annum for 1900–1954, based on Chambers's estimates for circa 1920, and 1.25% for 1962–1985, based on USDA estimates for 1967–1970. All figures refer to five Midwestern states (Ohio, Wisconsin, Illinois, Iowa and Minnesota), averaged by total areas in farms. For numbers and details, see WP, Appendix E.

Real interest rate: equals the 6-month commercial paper rate adjusted for consumer-price inflation from that year's average to the next year's average. The real rates for 1910–1984 were pre-calculated by Featherstone and Baker, "An Examination of Farm Sector Real Asset Dynamics," 545–6, and those for 1900, 1905 and 1985 were added from *Historical Statistics* and from the *Economic Report of the President.*

Real rate of current return on all US farm productive assets: from Featherstone and Baker, "An Examination," 545–6. This measure, based on work by Melichar, excludes capital gains and allegedly excludes all farm income from management and labor (a famously difficult separation to make).

The figures suggest a possible re-dating of the postbellum rise in the savings rate. The "grand traverse" across the Civil War, here from 1850 to 1880, was less impressive in the perceived savings rate than in the real investment share. By cutting the rise in the savings share further, farm capital gains reinforce a revision in the literature already being advanced by recent estimates of Ransom and Sutch.[30] The farmland capital losses of the Civil War era further underline the importance of the war and emancipation on wealth and the incentive to re-accumulate, another point stressed by Ransom and Sutch. The higher savings rates, however, should perhaps be dated at the turn of the century or the first decade of this century, thanks in large part to the farmland value boom. The re-dating of America's peak rate of national savings must, however, await better estimates of overall capital gains, not just those on farm land.

Within this century capital gains on land reinforce the impression of a fall in rates since 1973. The effect of farmland price gyrations by itself is small, because farm land is less than 6 percent of national net worth. But the period 1950–1973 saw capital gains of about 6 percent a year on nonfarm land, raising the augmented savings rate by another 3 percentage points.[31] The capital-gains adjustments suggest a net national savings rate that averaged 15 percent from 1940 to 1973, and a much lower rate since then. For recent years, as for the nineteenth century, the history of American savings is still in flux.

Farmland values and the Habakkuk debate. The fact that capital gains on land can satisfy the savings motive gives another twist to a rich, enjoyable and curious debate in Anglo-American economic history. H. J. Habakkuk's famous book on *American and British Technology in the Nineteenth Century* spawned two decades of debate over why American industry, and the whole American economy, was more mechanized and capital-intensive than Britain's around 1850.[32] After a dozen fine articles with alternative theories of American capital-intensity, it fell to Alex Field to show that American production was *much less* capital-intensive and mechanized than British, and to James and Skinner to note that Habakkuk had not really asserted otherwise![33] Both Field and James-Skinner offered reasons why land abundance should have lowered capital-intensity.

What needs to be added to the resolution of the Habakkuk debate is an

30. Ransom and Sutch, "Domestic Saving."

31. Lindert, "Land Scarcity and American Growth," Appendix Table 1.

32. H. J. Habakkuk, *American and British Technology in the Nineteenth Century* (Cambridge: Cambridge University Press, 1962).

33. Alexander J. Field, "Land Abundance, Interest/Profit Rates, and Nineteenth-Century American and British Technology," *Journal of Economic History* 43 (1983) 2: 419–20; John A. James and Jonathan S. Skinner, "The Resolution of the Labor-Scarcity Paradox," *Journal of Economic History* 45 (1985) 3: 517.

explanation of why Americans had accumulated and installed less capital overall, and not just in certain manufacturing industries, by 1850. The usual explanations about competing for labor against the vast frontier do not explain why Americans and America's foreign creditors accumulated so little tangible reproducible capital in the United States. Nor do the models offered by Field or by James and Skinner.

The other side of abundant farm land, its attraction as a form of savings, can play a supporting role in the revision led by Field and by James and Skinner. Agricultural land was half of America's tangible assets in 1774 and 1805 and still 36 percent of tangible assets in 1850. In Great Britain, by contrast, it dropped from 51 percent in 1760 to 20 percent in 1860, and the decline was even sharper in England and Wales alone, which the debate had meant to contrast with America.[34] The portfolio shares gap suggests that Americans deferred more of their demand for savings into farm land. In this extra way, the farm frontier played a role in reducing America's accumulation of conventional capital, even though it was less important than the simple fact that America was still undercapitalized overall, in the sense of a low national ratio of all tangible assets to GNP.

The Main Determinants of Trends in Farmland Prices

The price of pig is something big
   because its corn, you'll understand
   is high-priced too,
   because it grew
   upon the high-priced farming land.

If you want to know why
   that land is high, consider this:
   its price is raised
   because it pays
   thereon to raise
   the costly corn, the high-priced pig.

—H. J. Davenport

To explain the diverse movements in land scarcity is obviously a complex task. Everybody who has lived through, or written about, changes on the land knows of several forces that probably played some kind of role. To decide which of these have been most important, one needs a framework for weighing different possible explanations against each other, giving each its quantitative due.

The trained economic historian must choose between two familiar paths in deciding how to pursue a quantitative accounting. One path, the one not chosen here, is formal quantitative modeling. It has the virtue of explicitness: its assumptions and its final judgments are transparently clear to the trained scholar. The kind of formal model most appropriate to

34. Goldsmith, *Comparative National Balance Sheets,* 122–23.

explaining long-run movements in a market price of farm land would be a computable general equilibrium (CGE) model dividing the economy into sectors that use land differently. Such models have been applied to similar tasks elsewhere.[35] CGE models, however, have only limited persuasive powers. They are not easy to read. Furthermore, most readers retain doubts about their simplifying assumptions, doubts that are hard to dispel because the CGE models are seldom easy to test. Even when tested, they seem to perform only respectably at best, never brilliantly. For all their power of suggestion, they tend to persuade only those readers who use such models themselves.

The less formal path chosen here begins with a review of the forces that seem likely to have had the most influence on farmland prices. As a rough preliminary test of their explanatory power, we shall then examine how well they seem to correlate with historical movements in land prices over several periods of U.S. history since 1800. The objective here is modest. I shall seek only patterns in a few leading forces that seem to explain the direction of contrast between any two historical periods. No attempt is made to develop a concrete percentage-point accounting of the observed price trends.

Even a crude view of rough historical correlations, however, yields a useful and simple perspective: with minor exceptions, *the kinds of forces that have given farm landowners higher land prices are those that slow down the advance of real wage rates and GNP per capita.* While more harmonious events are conceivable in theory, in practice there tends to be a conflict between the farm landowning interest and general progress.

The first major influence on farmland prices and rents is the *terms of trade* faced by farmers, or the ratio of the prices they receive to the prices they pay. Whenever some outside force raises the ratio, shifting the terms of trade in farmers' favor, the real value of owning or using farm land rises. Indeed, it should rise by an even greater percentage than the percentage shift in the terms of trade, according to the "magnification effect" of modern trade theory. If a sector of the economy is favored by, say, a 10 percent increase in terms of trade, the inputs used in that sector will benefit unevenly. Those used in similar proportions across all sectors of the economy will not benefit much. At the other extreme, those inputs that are highly concentrated in this sector will gain large increases in price or profit. They will necessarily gain more than 10 percent because the weighted average of the large and small effects on different input suppliers must average out to a 10 percent gain.

---

35. The working paper version of Lindert, "Land Scarcity and American Growth;" Williamson and Lindert, *American Inequality.*

The terms of trade ratio for farmers must be viewed as a force that is itself affected by other things. It is affected, firstly, by transportation improvements that, in effect, move farms economically closer to their urban and foreign markets. It is also affected by price conditions in those distant markets, and by government policies designed to change the terms of trade (tariffs, taxes, subsidies, etc.). It is not fully exogenous to the model sketched here, though it will be discussed as an additional force for expositional convenience.

A second leading force is *government subsidization of agriculture,* including the provision of public goods net of taxes. It is a difficult force to measure because it takes so many different forms. To the extent that it affects the prices farmers receive and pay, its influence is already built into the terms of trade. But the terms of trade ratio fails to reflect some kinds of government aid, such as payments to compensate for low prices, (e.g., "deficiency payments") and payments for leaving land uncultivated or at least not cultivated with certain crops (e.g., "acreage retirement" payments). To be sure that these extra dimensions are given their due, we shall look at a crude measure of government subsidization, namely the ratio of government farm payments to net farming income. Such payments should raise the price of farm land though only payments going to tenants would raise rents.

Population growth is likely to raise the price of farm land through two channels. First, to the extent that it raises the size of the labor force, it supplies more labor, especially more unskilled labor. The resulting downward pressure on unskilled wage rates would enhance the return to owning farm land. Second, faster population growth tends to raise the share of non-working dependents in the population. The extra dependents tend to affect product demand in the same way that extra poverty would affect it: they shift demand back toward food and other staples. This demand shift would raise the demand for, and the price of, agricultural land. Both the labor-supply and the product-demand effects could raise the price of farm land even apart from any influence transmitted through the terms of trade, by cheapening the supply of labor and other inputs to the farm sector.

Productivity growth in agriculture is likely to raise the price of land, as long as one sets aside its effect on the relative price of farm products (the terms of trade). For any given terms of trade, farm productivity improvements raise the marginal product of land, bidding up both rents and purchase prices. The opposite result would hold in the case of "immiserizing progress," in which farm productivity improvements lowered the price of farm products so much as to worsen the returns to owning farm land, but this perverse possibility seems less likely in an economy open to foreign trade, and is not pursued here. Productivity growth gives a still stronger stimulus to farm values if it has a *land-using bias,* an effect estimated below.

A greater *supply of farm land* should clearly lower its price. On this point there should be little controversy. The magnitude of this effect is uncertain, however. The larger the direct effect, the more it might be magnified through an influence on savings. If a 10 percent expansion of the supply of farm land directly lowered its price by, say, 14 percent, the real value of farmland wealth would be reduced by about 4 percent. The reduction in wealth would cause individuals having lifetime savings targets (for old age, bequests, etc.) to save more in the form of manmade capital to compensate for the lower value of land. The extra manmade capital should draw extra labor and other inputs away from agriculture, further depressing the price of farm land. Conversely, if a 10 percent expansion of the land supply had less than a 10 percent direct effect on the land price, the extra land wealth would work to dampen the effect on land price. Here we need only argue that the direction, but not the magnitude, of the effect of extra land on the land price is clear.

Finally, economic progress in general, represented by a *rise in income per capita* resulting from capital accumulation and nonfarm technological progress, should have a negative effect on the price of farm land, though the result depends on the degree of technological bias and the effect of such progress on the farm product prices.[36] If product prices (the terms of trade) and interest rates were fixed by international markets or some other outside force, economic progress would bid a nation's labor and other productive inputs away from agriculture, lowering the land price there. It would do so through Engel effects, shifting product demand and therefore input supplies away from agriculture. The opposite result could be imagined, of course. For example, if technological progress were greatly labor-saving, it could release so much labor to the agricultural sector that farm landowners would benefit. I judge such cases to be historically rare, however, and presume that the rise in material living standards comes in a way that draws resources away from agriculture.

American historical experience suggests that the simple list of forces just presented helps explain the historical contrasts in price trends for farm land. To see how, let us follow trends over long periods of time, periods starting and ending with the U.S. economy near full employment, so that we can abstract from the distortions that arise from shorter-run depressions. Table 3 divides the experience since 1800 into seven periods chosen with an eye to data availability and distinctive farmland price trends. A causal history can be sketched, with due caution, period by period.

36. Lindert, "Land Scarcity and American Growth," 870–73.

The Long Rise in Land Prices before World War I. The ten- fold increase in the real price of farm land between 1800 and 1913 was the logical result of trends in the terms of trade and rising population density, which together outweighed the price-depressing effect of the general economic progress. So Table 3 suggests, with its view of conditions common to four prewar periods (c1800–1839, 1839–1859, 1869–1899, and 1899–1913).[37] Over the whole era of more than a century, the terms of trade shifted in favor of agriculture. In fact, the true price trends favored agriculture more than the rates in Table 3 can show. Unmeasured product-quality improvement, which was presumably greater for nonfarm products, means that nonfarm products were cheapened, and farm products made relatively more expensive, at a faster rate than the available price indices can show. In addition, transportation improvements gave farmers better price trends than the usual series can show. The relevant price trends are those at the farm gate, but the available series tend to be more urban. Missing from our view is the 4.68 percent per annum productivity improvement in domestic American transportation between 1815 and 1859, much of which got passed onto farmers.[38] Missing also is the 1.89 percent productivity growth in American railroads, 1870–1910 and the drop in ocean freight rates on American grain shipped to Europe in the late nineteenth century.[39] Even these unknown adjustments, raising the long rise in farmers' terms of trade toward 1 percent per annum (or more?), would leave us with a likely understatement of the effect of better product prices on the value of farm land. We must remember the "magnification effect," through which a 1 percent rise in the terms of trade would eventually raise by much more the price of land highly concentrated in the farm sector. The terms of trade improvement alone could account for perhaps half of the long historic rise in farm values before World War I.

The other main explanation for the long rise would seem to be the fact that population (and the labor force) grew faster than the supply of land, despite some efforts of the U.S. government to make land relatively cheap for would-be yeomen. Both in the early federal period and in the first decade of this century, the frontier did not expand fast enough to match the demographic contributions to labor force and food demand. As long as

37. The Civil War decade has been omitted here, to save on the space required to discuss its special events (slave emancipation, wartime destruction in the South, inflation and deflation, and population slowdown). Its drop in farmland prices can be explained, however, in terms consistent with the present framework.

38. Williamson and Lindert, *American Inequality,* 171, citing Mak and Walton and Fishlow.

39. Albert Fishlow, "Productivity and Technological Change in the Railroad Sector, 1840–1910," in Dorothy S. Brady, ed., *Output, Employment and Productivity in the United States after 1800, Studies in Income and Wealth,* vol. 30 (New York: National Bureau of Economic Research), 626.

Table 3. Movements in Farm Land Prices and Influences on them, Seven Periods, 1800–1984

	early federal c 1800–1839	antebellum 1839–1859	late 19th century 1869–1899	start of 20th c. 1899–1913	long price drop 1913–1945	postwar 1945–1973	post-OPEC 1973–1984
	(Rates of change in percent per year)						
Land Scarcity:							
Price of farm land per acre	1.99	2.86	2.40	4.89	−1.62	3.31	3.32
Rent on Midwest farm land	rising	rising	no clear trend[e]	0.69	−1.37	≤ 3.35	1.73
Factors that should raise farm land prices:							
Terms of trade for farmers[a]=P_f/P all	0.36[b]	0.59	0.00	1.11	0.24	−0.68	−1.61
[Not fully exogenous]							
Government payments as a % of income	0	0	0	0	0→5.8%	5.8%→7.8%	7.8%→24.3%
U.S. population growth	2.99	3.09	2.19	1.89	1.14	1.49	1.01
Productivity growth in agriculture	0.00	0.00	0.79	−0.24[f]	1.02[g]	1.65	1.83
Land-using bias in productivity growth[h]			0.04	−0.13	0.07	0.46	−0.63
Factors that should lower farm land prices:							
Supply of farm land	0.98[c]	3.32	2.44	0.54	0.74	−0.36	−0.80
Gross national product per capita[d]	1.11	1.59	2.51	2.17	1.99	1.04	1.19

Sources: For farm land prices and rents, see the sources for Figures 2 and 3 above. Figures for the terms of trade, government payments as a percentage of farming income, population growth, the supply of farm land from 1839 on, and the growth of GNP per capita since 1869 are all from US Census Bureau publications (*Historical Statistics of the United States, Statistical Abstract of the United States*) and *The Economic Report of the President.* Estimates of the growth rate for total factor productivity in agriculture are discussed in Williamson and Lindert, *American Inequality,* chapter 7, 10, 11, extended to 1984 with the help of USDA, *Agricultural Statistics, 1985.* The land-using bias of productivity change in U.S. agriculture is from Yujiro Hayami and Vernon W. Ruttan, *Agricultural Development: An International Perspective* (Baltimore: Johns Hopkins University Press, 1985), 191. The supply of farm land for c1800–1839, actually 1805–1850, is from Blodget, *Economica,* and USDA (1973). The growth rates for product per capita, c1800–1839 and 1839–1859, are from David, "The Growth of Real Product in the United States."

Notes:

[a] The terms of trade for farmers are the ratio of an index of the prices farm operators receive (P_f) to an index of the prices they pay for productive inputs and family living (P_{all}). For the years up to 1910, this is the ratio of wholesale prices of farm products to the overall wholesale price index, usually in New York (but see footnote b in this table, and the discussion in the text). For 1910 on, this is the official "parity" ratio: prices received by farmers divided by the prices they pay, including wages, taxes, etc.

[b] Cincinnati, Berry's ratio of the price of products of Northern agriculture to the prices of all products. The corresponding ratio in New York had no trend at all from 1800 to 1839.

[c] 1805–1850

[d] As noted in the text, the growth rate in GNP per capita is used as a summary measure standing for nonfarm technological progress and all capita accumulation per person, with the side-effect of shifting demand away from food products (according to Engel's Law).

[e] Between 1869 and 1899, interest rates dropped considerably. This should have lowered the ratio of rent to purchase price, perhaps enough to imply no trend in rents despite a rise in purchase prices.

[f] 1899–1909.

[g] 1909–1948.

[h] The numbers in this row refer to the following periods: 1880–1900, 1900–1915, 1915–1945, 1945–1975, and 1975–1980.

the unit effect of each percent growth in labor supply matches that for land supply, as experiments with computable models suggest, the net effect should have been to raise the price of farm land further. To be sure, the Engel effects and other labor-absorbing effects of the rise in per capita income probably acted to retard the advance of farmland prices, but this effect, formal models suggest, should have been weaker for each percent of income growth than the effects just discussed. We have at least the core of an explanation for the long rise.

The long rise was not steady, of course. The start of the twentieth century stood out, with the real price of farm land rising faster than five percent a year. What was different about this period? Above all, the product-price "scissors" cut in favor of farmers, in America as in other countries. Indeed, 1910–1914 still stands out as the all-time peak in farmers' relative prices, the era still fondly cited as an index base for discussing "parity" in discussions of farm price policy. The improvement in prices was apparently due to an acceleration of population growth worldwide without an acceleration in the growth of income per capita. The other factor pushing up real farm prices in this era was the accelerated rise of the population/land ratio, as immigration reached flood tide and the land frontier was closed.

The Long Price Drop. The long drop in real land prices across the two world wars and the intervening decades can be explained, more or less, though some side-questions must remain unanswered. Our explanatory variables behaved differently in the 1913–1945 era of price decline, as compared with earlier experience.[40] For most of the period, the terms of trade moved against the farm sector, though they jumped back up in the late 1930s and World War II. Population growth slowed down, primarily because America imposed tight restrictions on immigration after 1924, while the supply of land actually grew a bit faster than before. Both changes contributed to the softening of the market for farm land.

Starting in 1933, the government made a determined attempt to counteract the price decline, with programs to cut agricultural supply. Yet by World War II, the impact of the new programs was still quite limited. Table 3 might mislead in this respect by showing that government payments to

40. It should be noted that the present analysis is somewhat sensitive to the choice of 1913 and, especially, of 1945, as period boundaries. The basic conclusions suggested here would seem to hold up if 1940 were used in place of 1945, but 1940 was a year of high unemployment, complicating the present interpretation, which focuses on long-term market pressures that should show up at full employment. It seemed wiser to span well across the Great Depression of the 1930s, as argued in an earlier article, Lindert, "Land Scarcity and American Growth," 877.

If 1948 or 1950 were chosen instead of 1945, the present interpretation of the role of income/capita growth would be somewhat undermined. More research should be devoted to fitting trends, to avoid such sensitivity to the choice of period boundaries.

farmers were as much as 5.8 percent of net farming income. In fact, that share of support was a temporary phenomenon around 1945, with most of the rise in government transfer payments to farmers yet to come.

A somewhat different question about the long price decline is harder to answer without further study. Instead of comparing 1913–1945 with earlier experience, let us compare it to a "no-change" economy in which none of the explanatory variables in Table 3 changed at all. Relative to such a null economy, why should 1913–1945 have been one of farmland price decline? The question poses a difficult challenge. The terms of trade did not shift against agriculture for the period as a whole. Population growth slightly exceeded the expansion of farm land. Newfound productivity advances in agriculture should have raised the productivity of land further. So why the price decline? The best tentative answer is that in 1913–1945, relative to a no-change economy, income per capita continued to advance, albeit very unevenly, at an overall rate of about two percent a year. For the moment the best answer seems to be that this drew enough labor and other resources away from agriculture to undermine farmland prices. But this is hardly a firm conclusion.

The Postwar Rise—and Decline? The impressive postwar rise in farmland prices requires a different kind of explanation, because more things were changing. Let us begin with the period 1945–1973, which is given separate treatment in Table 3. The terms of trade moved seriously against farmers, making the rise in land prices look paradoxical. Other forces did move in the direction necessary to help explain an impressive price rise: the population/land ratio re-accelerated, agricultural productivity continued to accelerate and now shifted toward a land-using bias, income per capita grew more slowly and the government continued the march toward subsidization. The fact that the terms of trade declined in the face of such price-raising forces is best explained by the wordwide shift from wartime agricultural scarcity (World War II and Korean War) to increasing productivity as yield-raising advances spread from country to country. We can take only temporary comfort, though, in knowing more factors raising price than lowering it in this period. Until a better way is found for weighing the different effects quantitatively, we can only conjecture that the forces just listed were the reasons for the postwar price rise.

The post-OPEC period 1973–1984 is especially hard to analyze in terms of long-term forces. Eleven years is a short period, and it was an unstable 11 years at that. Table 1's display of net changes for this era can only suggest what might be the trend underlying the recent boom and bust in farmland prices. Basically, 1973–1984 had thus far been an accentuated (and unstable) version of 1945–1973. Again the terms of trade have dropped, again because the acceleration of agricultural productivity is worldwide (and possibly because the growth of world population has begun to slow down, cancelling the likely Engel effects of the income deceleration caused by the oil shocks). Yet again the population/farmland ratio has risen, along with U.S. agricultural productivity and government payments to farmers. There is the start of an explanation here, but only the start.

Conclusions

The long history of farmland prices in the United States shows some striking changes in trend. After what may have been a trendless colonial era, the value of farm land rose impressively across the nineteenth century, despite the expansion of the land supply as a continent was settled. Soon after its fastest rise, in the early twentieth century, the real price of farm land went into a thirty-year slump. It then resumed its impressive upward trend after 1945. The most unstable period by far has been the boom and bust since 1973. Neither the Civil War decade, nor the famous bust in farmland speculation in 1920, nor the Great Depression, brought as sharp a change in trend as that since 1980.

The historic price movements were "real" in more than one sense. They were movements in relative-price ratios and not just nominal-price inflations and deflations. They were not just, or mainly, the result of shifts in the quality of farm land—instead, they were genuine price movements. And they were shared by all regions of the United States, despite some regional differences noted above.

The twentieth-century behavior of real rents and the rate of return on midwestern farm land paralleled those in its purchase value. Again, there was a rise before World War I, a long decline to World War II, a rise into the 1970s, and a serious decline thereafter. Each swing of rents, however, was more restrained than the nearly contemporaneous swing in purchase prices. The overall rate of return also moved with both rents and prices, though with still wider fluctuations. The market for farm land has shown what look like endogenous bubbles, especially in the boom and bust of 1973–1986.

Charting the long-run movements in farmland prices, rents and returns reopens two issues relating to capital accumulation in American history. First, incorporating capital gains and losses on farm land into measures of savings accentuates and slightly re-dates the famous rise and fall of American savings. Second, the attraction of investment in farm land may have played a role in explaining the lower capital-intensity of the American economy relative to the British in the middle of the nineteenth century. Both conjectures about accumulation, however, require further research.

A tentative explanation of the peculiar movements in American farm-land prices focuses on six factors: the terms of trade (farm prices divided by the prices farmers must pay), government farm supports, population growth, productivity growth in agriculture, the supply of farm land, and general economic "progress," proxied here by the growth of GNP per capita. There is a pattern to the roles played by these separate forces: those changes that historically acted to raise the price of farm land are changes that should harm the advance of wages and average income. The farmer landowner's best friends in the play reviewed here were such forces as rapid population growth, the loss of the frontier, government subsidy payments, and the advance in farm productivity. All but one of these land-price-raisers tended to reduce the growth rate for real wages and average income. The one exception, which probably raised wage rates and national product while also raising farmland values, was the advance of agricultural productivity growth, aided by government research and extension subsidies. This factor aside, the interests of farm landowners have been in conflict with general economic progress. So says a tentative history of three centuries of American farmland values.

An Overview of California Agricultural Mechanization, 1870–1930

ALAN L. OLMSTEAD

PAUL RHODE

Today California is by far the nation's leading agricultural state, accounting for roughly 10 percent of the total value of the agricultural output produced in the United States. No other region can boast its superb combination of climate, soils, institutions, and highly educated farmers. In the nineteenth century this advantageous mixture of agricultural inputs transformed California from a frontier society in 1860 into a virtual cornucopia by 1900. A hallmark of California agriculture has been its highly mechanized farms. Nineteenth-century observers watched in awe as cumbersome steam tractors and giant combines worked their way across vast fields. In the twentieth century California farmers led the nation in the adoption of gasoline tractors, mechanical cotton pickers, tomato harvesters, and dozens of less well-known machines.

This paper will analyze the evolution of mechanization on California farms emphasizing developments in power sources and in the mechanization of small grain production. We argue that California's distinctive role as an innovator and early user of new technologies represented a rational response by the state's farmers and mechanics to the area's peculiar economic and geographical conditions. In addition, we argue that there was a dynamic process whereby early innovations paved the way for subsequent developments. First, we will show that many innovations were complementary in that the extra horsepower acquired to power one device was avail-

ALAN L. OLMSTEAD is Director of the Institute of Governmental Affairs and Professor of Economics, University of California, Davis. PAUL RHODE is a research associate at the Institute of Governmental Affairs, University of California, Davis, and a graduate student in the Department of Economics at Stanford University. Research on this paper was supported by the California Agricultural Experiment Station Project #CA-D-AHC-45733 and by the National Endowment for the Humanities Independent Study and Research Fellowship.

able at low marginal cost to power yet another new machine. Second, we argue that learning-by-doing by both farmers and manufacturers, as well as the regular interaction of these two groups were essential ingredients in the explanation of the actual path of mechanization and economic growth. Herein lie important lessons for economic development.

The story of agricultural mechanization in California illustrates the cumulative and reinforcing character of the invention and diffusion processes. Mechanization of one activity set in motion strong economic and cultural forces that encouraged further mechanization of other, sometimes quite different, activities. On-farm mechanization was closely tied to inventive efforts of local merchants. Specialized crops and growing conditions created demands for new types of equipment. Protected by high transportation costs from competition with large firms located in the Midwest, a local farm implement industry flourished by providing Pacific Coast farmers with equipment especially suited to their requirements. In many instances the inventors designed and perfected prototypes that later captured national and international markets. Grain combines, tracklaying tractors, giant land planes, and sugar beet harvesters, to name but a few, emerged from California's shops revolutionizing production processes around the world.

Early successes bred other innovations as skilled mechanics and other specialized inputs flowed into the new area of opportunity. The maturation of distribution networks and repair facilities, along with increasing farmer familiarity with one type of machinery, lowered the barriers to adopting the next generation of equipment. The result of this dynamic process was to accelerate mechanization and strengthen the impact of agricultural development in California as backward linkages stimulated urban employment and manufacturing. Like a magnet, agglomeration economies encouraged producers to locate near each other, reducing costs and enhancing the exchange of ideas. In particular, Stockton emerged as an important manufacturing center.

To provide a context for understanding mechanization, it would be useful to have an overview of the major trends in California agriculture. Figure 1 shows the acres harvested by major crop classifications. The information contained here summarizes several important phases of California agricultural evolution. By the time of the American Civil War in 1861, California was already an important grain-producing state, and by 1890 it ranked second to Minnesota in wheat production. The growth of grain output in California fostered the rise of the state's equipment industry which turned out a number of labor-saving machines. These inventive efforts continued even after small grain acreage rapidly declined at the turn of the century. The effect on the hours of labor required to produce an acre of wheat was revolutionary. As a rough estimate, the hand methods

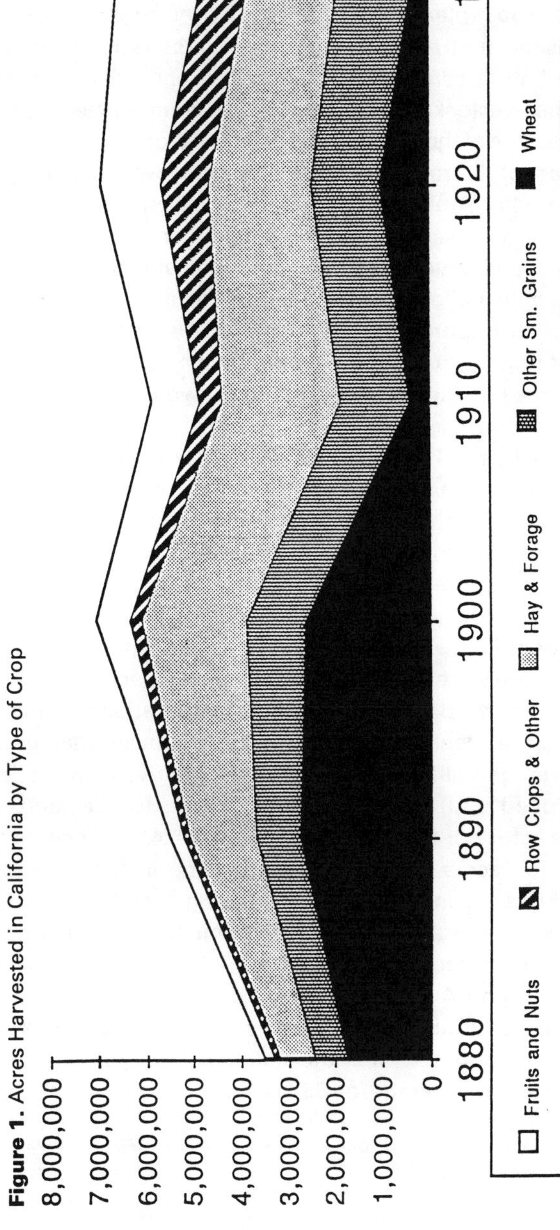

Figure 1. Acres Harvested in California by Type of Crop

prevailing at the middle of the nineteenth century took at least 60 hours per acre. The machine methods of 1900, including the use of combined harvesters, cut the time to less than 3 hours in California.[1] Other trends are of equal significance. The replacement of horses by motor vehicles in urban areas and, to a lesser extent, the early adoption of tractors in the twentieth century led to a decline in acreage devoted to hay and other forage crops. The growth of the livestock and dairy industries counteracted these forces and hay remained one of the state's most valuable crops.

The impact of the rapid adoption of gang plows, combined harvesters, tractors, and other equipment on reducing the demand for labor was more than offset by a massive shift into labor intensive, irrigated crops. The growth in acreage devoted to row crops indicates the spread of cotton in the San Joaquin Valley. World War I led to a boom in cotton planting. Cotton grew in importance during the 1920s and, by 1950, it was the state's most important crop (measured by the value of output). Since that date, California has ranked among the top two or three cotton-producing states in the nation.

Between 1870 and 1930 the state's area in tree and vine crops increased from about 100,000 acres to about 2 million acres. The relatively small amount of acreage in orchards and vineyards depicted in Figure 1 hides the enormous value of the output emanating from these lands and the large number of seasonal workers employed in the harvest. As the acreage in orchards and vineyards expanded, California agriculture became increasingly identified with these specialty crops. But, as was the case with truck crops, there were few advances in harvest mechanization.

The change in crop mix had important implications for the size of California farms. Fruit ranches were typically far smaller than grain farms. In 1880, the average California operation had nearly 460 acres (of improved and unimproved land) compared with 134 acres for the nation as a whole. By 1930, the acres per farm in the state fell to 224, far closer to the national average of 157 acres per farm.[2] This decline in farm size in California should not be taken as a sign of deteriorating conditions. Throughout this period, the state's farms remained, by almost any standard, prosperous and highly mechanized.

Figures 2, 3 and 4 help illustrate this point. As a rule of thumb, the average value of output and the average value of implements on California farms was two-to-three times the national averages. In 1930 California farms produced an average of $3,108 in output and had $676 worth of

1. U.S. Department of Labor, *Thirteenth Annual Report: Hand and Machine Labor,* 2 vols. (Washington, D.C.: GPO, 1898), Vol. II, 470–73.

2. U.S. Bureau of the Census, *Historical Statistics of the United States: Colonial Times to 1970,* 2 pts. (Washington, D.C.: Government Printing Office, 1975), Pt. 1, 461.

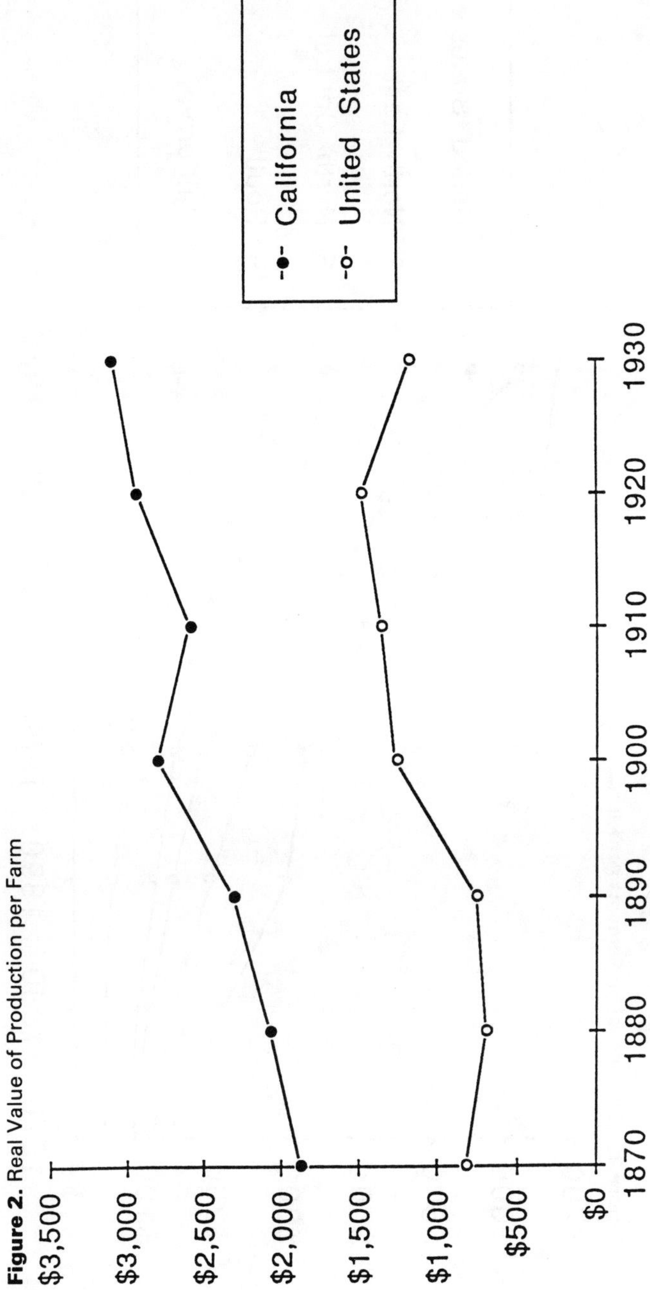

Figure 2. Real Value of Production per Farm

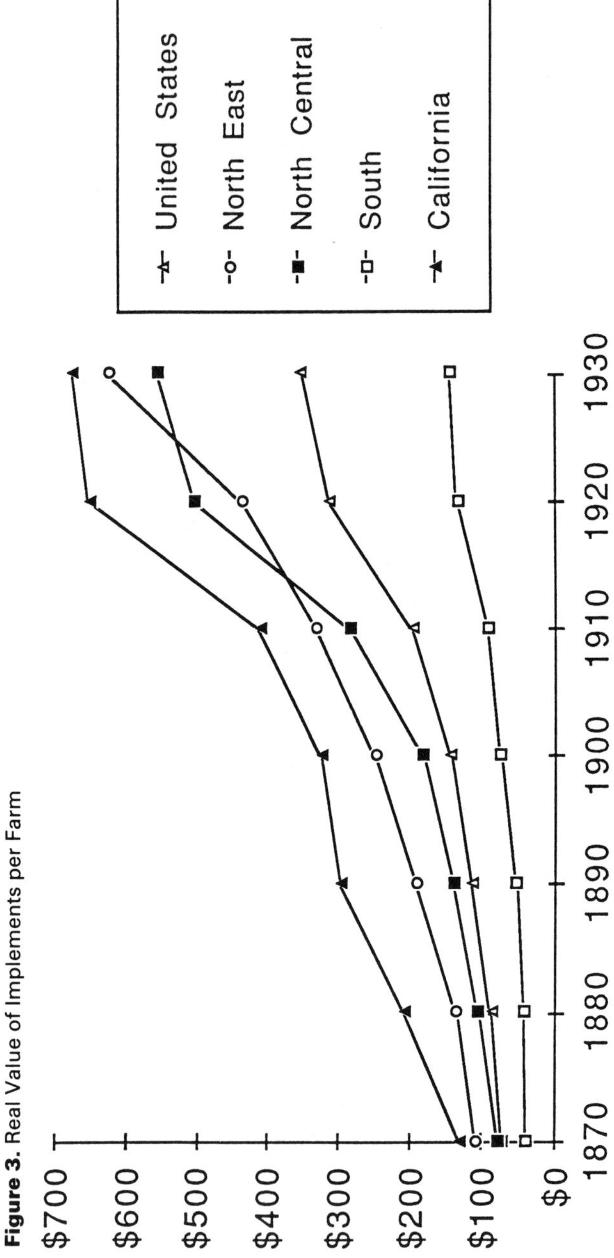

Figure 3. Real Value of Implements per Farm

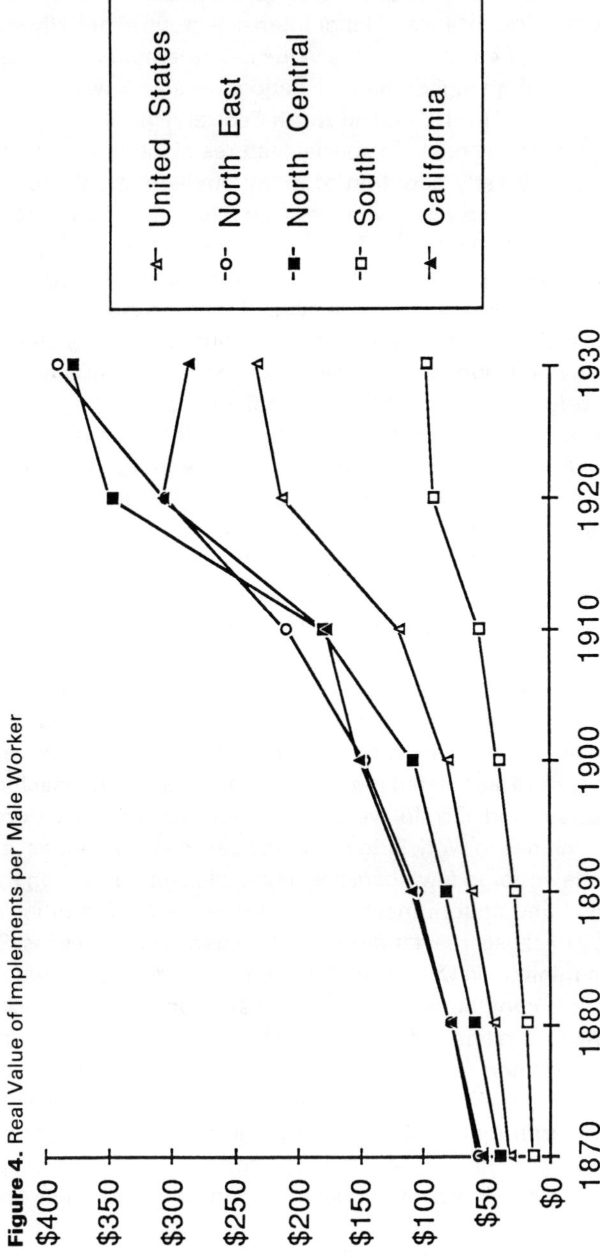

Figure 4. Real Value of Implements per Male Worker

Implements. The national averages were \$1,184 and \$354 respectively. Because of the importance of labor intensive crops especially suited to the state's Mediterranean climate, the value of implements per male worker in California, although higher than the national average, was not significantly different from the North East and North Central regions.

We should briefly note the special features of California agriculture that help explain the early adoption of many implements. California farmers were highly educated and inventive. As we have just indicated, they were more prosperous than farmers in many areas of the country. This advantage gave them the financial wherewithal to support their penchant for tinkering. Nowhere was this more evident than on the bonanza ranches of the state which often served as proving grounds for harvester prototypes that later gained popularity elsewhere. There were numerous cases of growers designing and building new implements.

The large scale of many California farms offered a special incentive for mechanization because it allowed farmers to spread the fixed cost of expensive equipment. The scarcity of labor in California meant relatively high wage rates and periods of uncertain labor supply. During the 1870s farm wage rates in California averaged 70 percent above those prevailing in the Midwest.[3] From the 1860s California farmers developed a distinctive set of labor relations. They specialized and expanded their acreage beyond what family labor could manage, thereby becoming increasingly reliant on hired workers. Whereas the family farm in the Midwest supplied most of its own labor, this was not the norm in California. Here "factories in the field" depended on a migrant proletariat to gather the harvest.[4] These economic factors stimulated the adoption of labor-saving machinery.

The climate and terrain were also favorable. Extensive dry seasons allowed machines to work long hours in near-ideal conditions, and the flat Central Valley offered few obstacles to wheeled equipment. In the cases of small grains and cotton, mechanization was delayed in other regions of the country because freestanding moisture damaged the crops. Such problems were minimal in California. All things considered, the state's climatic and economic conditions were exceptionally conducive to mechanization.

The new generation of farm equipment of the nineteenth century relied increasingly on horses and mules for power. In this era the substitution of horse power for human labor and other forms of animal power was a key sign of modernization. Horses on any one farm were essentially a fixed asset. A stock of horses accumulated for a given task was potentially available at a relatively low variable cost to perform other tasks. Thus,

3. George K. Holmes, "Wages of Farm Labor," *USDA Bureau of Statistics Bulletin 99,* Nov. 1912, 29–43.

4. For more information, see Paul S. Taylor and Tom Vasey, "Historical Background of California Farm Labor," *Rural Sociology* 1 (September 1936).

once a farmer increased his pool of horses, he was more likely to adopt new power-intensive equipment. For these reasons an examination of horses on California farms will yield important insights into the course of mechanization.

Figure 5 offers a first indication of the dramatic differences between California and other regions within the United States. Most modern machinery was designed to be pulled at a rapid pace by horses. Oxen were either too slow, as in the case of plowing, or could not do the job, as in the case with harvesting grain. Thus the ratio of horses and mules to oxen is a rough proxy for the transition in agricultural practices in the nineteenth century. As early as 1850 this ratio was almost twice as high in California as in the nation as a whole; by 1870 it was 6 times as great; and by 1890 it was roughly 30 times higher. Clearly California farmers adopted the horse-based technologies more rapidly than their counterparts elsewhere.

Figures 6 and 7 further illustrate the differences in the use of horsepower on California farms. In 1870 the average number of horses and mules on a California farm was almost three times the national average, and the number of horses and mules per male worker was more than twice the national average. Comparing California with the prosperous North Central states does not dramatically change the picture—California farmers were using an enormous amount of horsepower in the nineteenth century.

California's convergence toward the national norm was caused in large part by changes in the state's cropping patterns as grain farms gave way to specialized crops.[5] The new orchard and vineyard operations used twice as many horses per acre as grain farms, but they were far smaller and used an enormous amount of harvest labor. This accounts for the marked declines in draft animals per farm and per worker. After 1910 the decline in work animals per farm in California accelerated because of the rapid adoption of gasoline tractors. Whereas the high ratio of work animals per farm in the nineteenth century was a sign of modernization, by the 1920s it was beginning to be a sign of backwardness.

California was a leader in early adoption of tractors. In 1917, when the first official data became available, there were over 1,350 tractors in the state.[6] By 1920 the number had jumped to 13,850; over 10 percent of California farms had tractors compared with 3.6 percent for the nation as a whole. In 1925, nearly one-fifth of California farms reported tractors, proportionally more than in Illinois or Iowa, and just behind the nation-leading Dakotas. These figures understate the power available in California because the tractors adopted in the West were, on average, Substantially larger than

5. In 1900 the average hay and grain farm in California had 367 improved acres; the average fruit ranch had 56 improved acres.

6. USDA, *Monthly Crop Report,* April 1917, 33.

Figure 5. Horses and Mules per Oxen

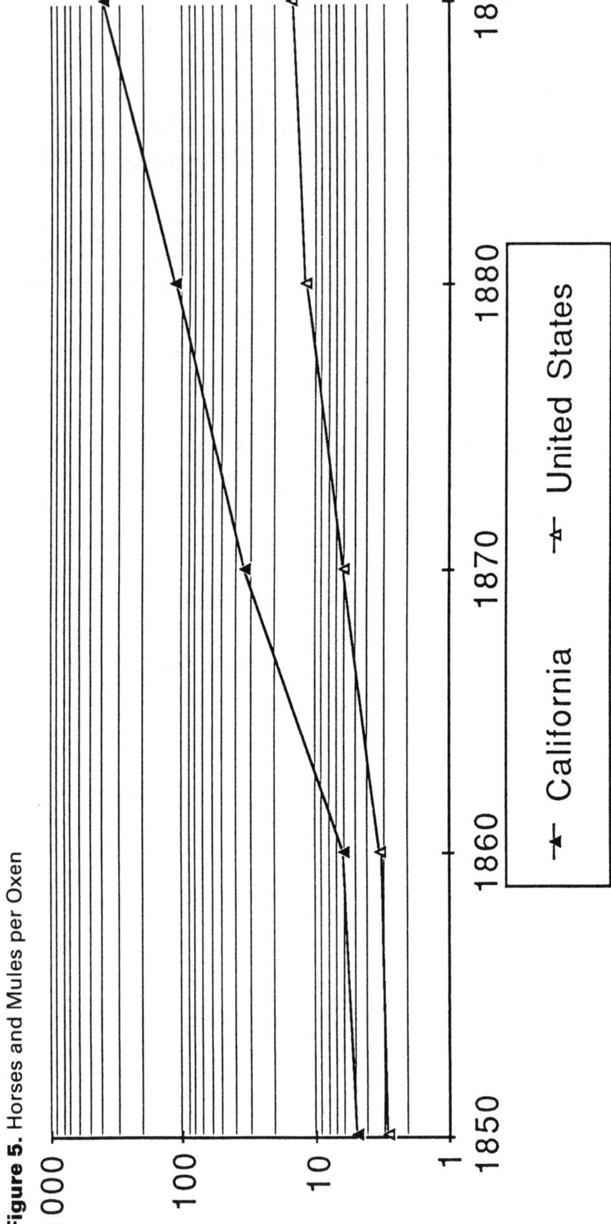

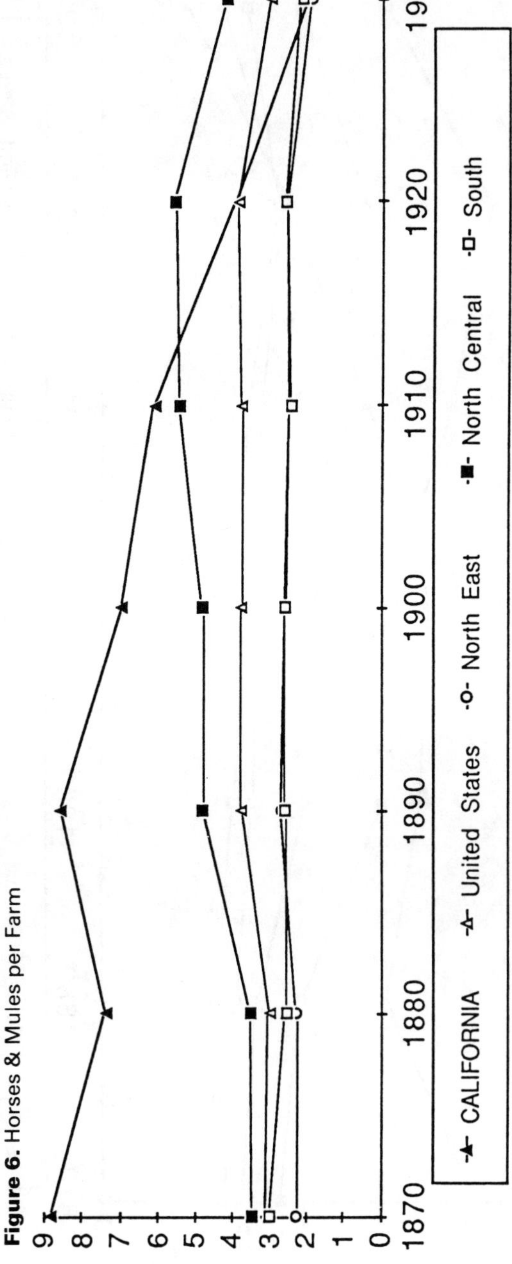

Figure 6. Horses & Mules per Farm

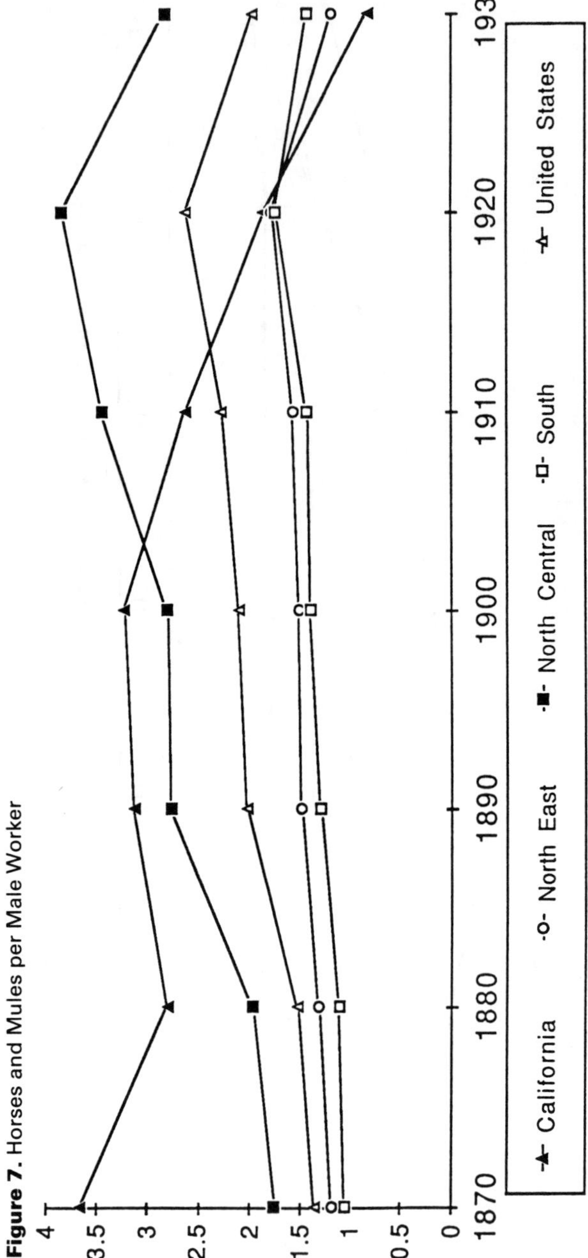

Figure 7. Horses and Mules per Male Worker

those found elsewhere.[7] In particular, Western farmers were the predominant users of enormous tracklaying tractors, which were invented in California. Surveys in the late teens and early twenties suggest that about one-half of the tractors sold in California were crawlers.[8] In 1930 tracklaying tractors accounted for 30 percent of tractors in the state compared with less than 5 percent for the nation as a whole.[9]

Finally, California farmers were the nation's pioneers in the utilization of electric power on farms. The world's first purported use of electricity for irrigation pumping took place in the Central Valley just before the turn of the century. Consistent data on rural electricity use are not available until 1929. At that time California led the entire nation. Over one-half of California farms purchased electric power compared with about one-tenth for the United States as a whole.[10]

The best proxies for the long-term trend in electrification in the state are the expansion of irrigated acreage, and the number of agricultural pumps. The upward march in irrigated acreage shown in Figure 8 was made possible by an enormous increase in the number of pumps (see Figure 9). Whereas California accounted for roughly one-fourth of the irrigated acreage in the United States, it had roughly 70 percent of all agricultural pumps. Some of these pumps were gasoline or diesel driven, but a growing majority were powered by electricity.

Table 1 presents data on the relative importance of these different power sources in 1924. California was clearly on the cutting edge. It was the only state with more tractor power than animal power and ranked first in terms of absolute horse power of trucks and stationary motors and engines.[11] Overall, the state had two to three times the national average of horsepower capacity per worker and per farm in 1924. Only the Great Plains states ranked higher. And because California had far higher rates of capacity utilization, it was the virtually unchallenged leader in terms of horsepower hours used per farm and worker.

7. U.S. Bureau of the Census, *Sixteenth Census of the United States: 1940,* Vol. 1, *Agriculture,* Part 6 (Washington, D.C.: Government Printing Office, 1942), 676–77.

8. It is worth noting that during this period roughly one-half of all tractors (crawlers and wheeled) enumerated in California were products of the state's factories. B.D. Moses, "Tractors on California Farms—1922," F. Hal Higgins Collection, Department of Special Collections, Shields Library, University of California, Davis.

9. "Crawler Tractors for Farm Use, Sept. 15, 1932," F. Hal Higgins Collection.

10. *Electrical Times,* January 2, 1948, p. 5; and U.S. Bureau of the Census, *Fifteenth Census of the United States: 1930, Agriculture Vol. II, Pt. 3-the Western States* (Washington, D.C.: GPO, 1932), 23, 53.

11. The latter category included stationary gas engines, electric motors and windmills. Data from 1930 indicate that electric power overwhelmingly dominated this category in California whereas gas engines were most important elsewhere. See W. M. Hurst and L. M. Church, "Power and Machinery in Agriculture," *USDA Misc. Publ. No. 157,* April 1933, 16–17.

Figure 8. Irrigation in California

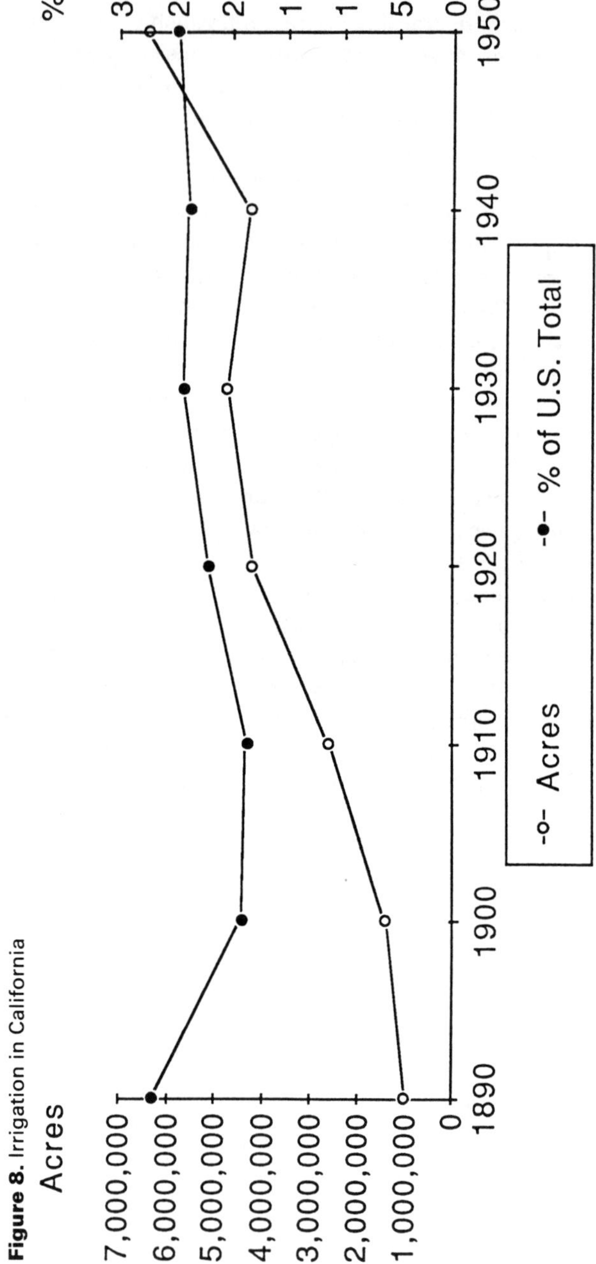

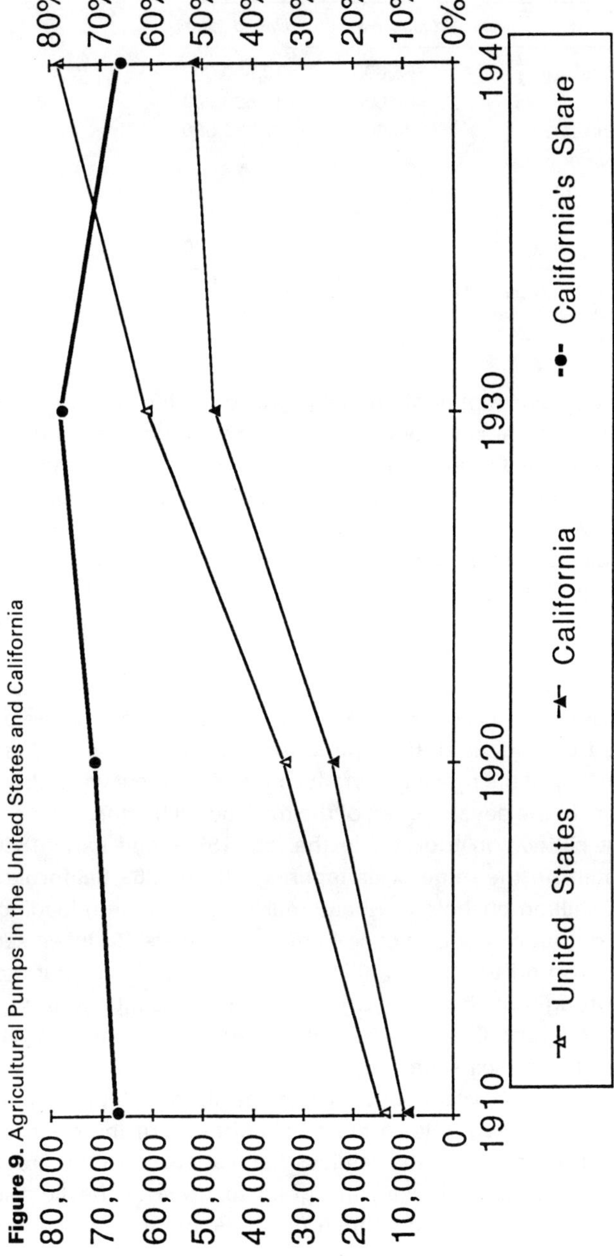

Figure 9. Agricultural Pumps in the United States and California

Table 1.   Primary Horsepower on Farm, 1924

	California	United States
Work Animals	356,000	19,800,000
Tractors	602,000	10,500,000
Motor Trucks	500,000	7,120,000
Stationary Engines, Windmills, and Electricity	870,000	10,000,000
Total Horsepower	2,328,000	47,420,000

Source: C.D. Kinsman, "An Appraisal of Power Used on Farms in the United States," *USDA Department Bulletin No. 1348,* July 1925, pp. 53–55.

The extensive supply of animal power on California farms encouraged local manufacturers to produce new types of equipment, and the development of new and larger implements in turn often created the need for new sources of power. This process of responding to the opportunities and bottlenecks created by previous technological changes provided a continuing stimulation to innovation. Tracing the changes in wheat farming technology will illustrate how the cumulative technological changes led to a distinctively different path of mechanical development in the West as compared to that which occurred in the Midwest.[12]

For example, the evolution of tractor inventions in California was a logical consequence of the mechanization of the grain harvest, arising from the tremendous draft requirements of the early combines. For this reason, many of the most important combine builders were also important innovators in the development of the tracklaying tractor.

Wheat cultivation took off in the late 1850s and early 1860s just as employment in the mines was tapering off. In 1869 California produced about 16 million bushels of wheat, making it the eighth leading state. By 1889, as output climbed to over 40 million bushels, it ranked second. But, with the drop-off to only 6 million bushels in 1909, the state fell to 23rd place. Throughout the second half of the nineteenth century and early twentieth century, California wheat farmers have consistently led in the adoption of large-scale machinery.

As noted earlier, a number of factors specific to California created conditions extremely favorable to the mechanization of the grain harvest. Of special importance was the rainless harvest season that stretched from June to November, allowing the grain to ripen in the field longer. By contrast, in the Midwest the threat of winds and hail, which would lodge

12. For further development of these general themes see Nathan Rosenberg, *Inside the Black Box: Technology and Economics* (Cambridge: Cambridge University Press, 1982).

and shatter the grain, put a premium on early harvesting. There, sweating the cut grain was also necessary before it was dry enough to be threshed.

Almost immediately after wheat cultivation began in the state, its farmers developed a distinctive set of cultural practices. Plowing the fertile California soil was nothing like working the rocky soils in the East or the dense sod of the Midwest. In California ranchers used two, four, and even eight-bottomed gang plows, cutting just a few inches deep. Leo Rogin, author of the classic study of nineteenth-century grain machinery, provides an excellent summary of the use of plows in the state during that era:

> In California the nature of the soil and the superficial character of the seedbed preparation made for the utilization of gang plows almost from the beginning of the extensive cultivation of wheat. The *Pacific Rural Press* for 1871 stated that a large number of gang plows were being built and used in the state. Except on the largest bonanza farms, two-bottom gang plows drawn by four horses predominated during the seventies. Brewer's survey for 1879 disclosed that eight-horse units drawing, presumably, plows equipped with two gangs of three to four bottoms each were employed in portions of that state. But, except on the bonanza farms, the two-bottom implement was so predominant that the British Royal Commission on Agriculture reported that, in California, two-bottom riding plows drawn by four to six horses or mules were "universally" employed. With the speeding up of agricultural operations in the eighties as a result of the introduction of the harvester-thresher, the eight-horse unit drawing two gangs with four bottoms each came to be employed to an increasing extent, and by the early nineties, may be considered as typical of the large scale, not necessarily bonanza, farms of the San Joaquin and Sacramento valleys.[13]

A study of cereal growers reported in the *Tenth Census* shows that the average acreage plowed by one man increased dramatically as one moved from the East Coast to the Midwest and Prairie states and then on to California. East of the Ohio River the norm in 1880 was to plow one-and-one-half acres per day or less. In most of the prairie regions, two-and-one-half acres was a good day's work. In California, it was common for one man with a gang plow and a team of eight horses to complete six to ten acres per day. The tendency of California's farmers to use larger plows continued into the twentieth century. In 1913 the state's farmers on average used 4.2 horses per plow (for all crops) while the national average was 2.7. After tractors came on line, the state's farmers were also noted for using larger models

13. Leo Rogin, *The Introduction of Farm Machinery in its Relation to the Productivity of Labor in the Agriculture of the United States During the Nineteenth Century*, University of California Publications in Economics, Vol. 9 (Berkeley, CA.: University of California Press, 1931), 42–44.

and larger equipment. Once again, this pattern influenced subsequent manufacturing and cultural decisions.[14]

The preference for large plows in California stimulated local investors and manufacturers who vied to capture the specialized market. The state's firms could compete with superior Eastern plows because they concentrated on the plow frames rather than on the bottoms which, in fact, often were supplied by Eastern companies. As evidence of the different focus of their innovative activity, note the statement of the Agricultural Commissioner of the United States: "patents granted on wheel plows in 1869 to residents of California and Oregon largely exceed in number those granted for inventions of a like character from all the other states of the Union."[15] Between 1859 and 1873 California accounted for roughly one-quarter of the nation's patenting activity for gang or multi-bottom plows. By way of contrast, the state's contribution to the development of small single-bottom plows was insignificant.[16]

An example of one of the important tillage implements to originate in California was the Stockton Gang Plow. These plows typically featured four, though sometimes three to six, bottoms with small moldboards of little curvature, mounted on a heavy triangular wooden three-wheeled frame. Collectively, they were called Stockton Gang Plows because their major producers operated in the Stockton area, the leading center of the California farm equipment industry. H.C. Shaw and Co., a major Stockton firm which began building plows in 1849 and continued to sell implements into the 1950s, advertised in 1886 that over 20,000 of its gang plows were in use in the state.[17] Other important developments took place in the perfection of listers, harrows, levelers, and earth-moving equipment.

The adoption of distinctive labor-saving techniques carried over to grain sowing and harvest activities. By the early 1870s, the machinery and methods that California grain farmers employed to seed and harvest their crops differed dramatically from those used in other important wheat producing states. In the 1870s progressive wheat farmers in the Midwest were rapidly adopting grain drills which saved considerable seed, provided better yields, and did more uniform work than broadcasting. But drills also required more labor and horsepower. An 1875 survey conducted by the United States Department of Agriculture showed that more than 50 percent of the farmers in the Midwest used grain drills, but

14. U.S. Bureau of the Census, *Tenth Census,* Vol. 3, 434; USDA, *Monthly Crop Report,* February 1918.

15. USDA, *Agricultural Report 1869,* 326.

16. U.S. Patent Office, *Subject-matter Index of Patents for Inventions Issued by the United States Patent Office from 1790 to 1873, inclusive. . .* (Washington, D.C.: GPO 1874).

17. *Pacific Rural Press,* October 30, 1886, 373; also see Reynold M. Wik, "Some Interpretations of the Mechanization of Agriculture in the Far West," *Agricultural History* 49 (1975): 78.

that virtually all California grain farmers sowed their grain.[18] California farmers were sometimes accused of being careless and slovenly for using sowing, a technique which was also common to more backward regions such as the American South. However, California's farmers use of broadcast sowers reflected a rational response to their own factor price environment.

The methods employed in California in fact bore little resemblance with the hand-sowing techniques practiced in the South. Among the broadcasting equipment used in California were advanced high-capacity endgate seeders of local design. By the 1880s improved models were capable of seeding up to 60 acres in one day. By contrast, an 8- to 10-foot drill could seed about 15 acres a day and a man broadcasting by hand could seed roughly 7 or 8 acres a day.[19] The use of labor-saving techniques was most evident on the states bonanza wheat ranches where some farmers attached a broadcast sower to the back of a gang plow and then attached a harrow behind the sower thereby accomplishing the plowing, sowing, and harrowing with a single operation.[20]

California wheat growers also followed a different technological path in their harvest operations by relying primarily on headers instead of reapers. This practice would have enormous implications for the subsequent development of combines in California. The header was one of the many harvesting machines that emerged in the Midwest during the explosion of inventive activity before self-raking reapers became the dominant design east of the Rockies. The header cut only the top of the straw. The cut grain was then transported on a continuous apron to an accompanying wagon or header box. Headers typically had larger cutting bars and, hence, greater capacity than reapers, but the most significant advantage was that headers eliminated the need for binding. This feature saved the labor of about five binders. The initial cost of the header was about 50 to 100 percent more than the reaper, but its real drawback was in humid areas where the grain was not dry enough to harvest unless it was dead ripe. This involved enormous crop risks in the climate of the Midwest; risks that were virtually nonexistent in the dry California summers. For these reasons California became the only substantial market for the header technology. Although headers were the most important machine for harvesting grain in California by the end of the 1860s, the national production was not

18. USDA, *Agricultural Report,* 1875, 42.

19. Rogin, *Introduction of Farm Machinery,* 206–12; R. L. Adams, *Farm Management Notes for California* (Berkeley, CA.: U.C. Associated Students' Store, 1921), 121–22, 136.

20. For example, Reynold Wik, *The Mechanization of Agriculture and the Grain Trade in the Great Central Valley of California,* Pioneer Museum Project Grant Proposal (1974), p. 9. A copy is held in the F. Hal Higgins Library of Agricultural Technology at U.C. Davis; Leo Rogin, *The Introduction of Farm Machinery,* 204.

too significant. In 1869, 2,500 headers were produced compared with about 120,000 reapers.[21]

In the early 1850s, the header did not differ greatly in design from the reaper. The machines possessed similar reels and cutting bars, though the header made a shorter cut of grain and was generally pushed rather than pulled through the fields. The paths of subsequent development diverged, leading to significantly different machines and little opportunity for borrowing between the two technologies. In order to gain perspective it will be useful to review briefly the evolution of the reaper technology in the East. Commercially successful reapers date back to the patents of Obed Hussey (1833) and Cyrus McCormick (1834). The essential elements of the reaper as it emerged in the 1850s included a cutting bar, ranging from four to six feet in length, with reciprocating knives. The grain fell onto a platform from which it was manually raked to the ground. The machine, which did the work of about seven men with cradles, required a driver and raker. By the 1850s both men could ride on the reaper. In addition a crew of about six workers was needed to bind and shock the cut grain. The tendency in the Midwest was to develop machines using fewer and fewer horses. Many earlier reapers required four horses, but by the 1860s design and construction improvements reduced the machine's draft. Two horses became the rule, and it was not necessary to change teams as often.[22]

A major innovation, beginning in the mid-1850s, was the appearance of "self-raking" machines. These were reapers with rotating mechanical arms that raked the grain off the machine's platform at appropriate intervals. By the end of the Civil War, self-raker sales represented a sizeable portion of the market.[23] Other important advances occurred after 1860. In 1864 the Marsh Brothers of Illinois began marketing "harvesters" with a continuous apron. The apron delivered the cut grain to a work station on the machine where two binders tied the sheaves by hand. The next step was to automate the binding. Wire binders appeared in the 1870s and twine binders (of the Appleby patent) in the mid-1880s. Self-binders dispensed with about five workers compared with a self-raking reaper. There were few significant design changes after the mid-1880s. This entire evolutionary path, focusing on easing and eventually eliminating the tasks of rakers and binders, was largely irrelevant to the header. The heading machine had done away with raking, binding, and shocking from the start.

The header technology evolved in a different direction, leading directly to the development in California of a commercial combined harvester.

21. Rogin, *The Introduction of Farm Machinery,* 106.
22. Alan L. Olmstead, "The Mechanization of Reaping and Mowing in American Agriculture," *The Journal of Economic History* 35 (1975) 2: 344–51.
23. William T. Hutchinson, *Cyrus Hall McCormick: Harvest, 1856–1888* (New York: D. Appleton-Century Company, 1935), 96, 393.

From the starting point of the header, it was quite simple and natural to add a thresher pulled along its side. There had been numerous attempts in the East and Midwest to perfect a machine that reaped and threshed in one operation. Among those which came closest to succeeding was Hiram Moore's combine built in 1835. Moore operated his machines on his own and neighboring farms in Kalamazoo, Michigan for about five seasons, yet it did not gain lasting acceptance in the Midwest. Combining suffered from the same problems with moisture that plagued heading in humid areas. In 1853 Moore's invention was given new life when a model was sent to California to work on the ranch of John Horner near San Jose. Although it was destroyed by fire after harvesting only 600 acres, Moore's machine served as a prototype over the next three decades of combine development on the Pacific Coast.[24]

The subsequent development may be usefully divided into two periods, breaking in the mid-1880s. The first phase was one of extensive and nearly continuous experimentation. During the early period more than twenty firms engaged in some form of small-scale combine production and there were numerous additional examples of farmers and local blacksmiths taking the relatively simple steps to attach a heading machine to a thresher. By the mid-1880s workable designs were available and the period of large-scale production and adoption began. Most of the innovating firms, including the two leading enterprises—the Stockton Combined Harvester and Agricultural Works and the Holt Company—were located in Stockton.[25]

Table 2 shows the rate of diffusion of the combined harvester in California. It presents the best available estimates on the number of combines and the percentage of the state's small grain crop harvested by this method at various dates. During the harvest of 1880 "comparatively few" machines operated in California, and agricultural authorities, such as Brewer and Hilgard, clearly suggest that even those machines should be considered as experimental. In 1881 about 20 combines were being built in Stockton.[26] By 1888, between 500 and 600 were in use. The first truly popular model was the Houser, built by the Stockton Combined Harvester and Agricultural Works. In 1889, its advertisements claimed that there were 500 Houser machines in use, and that they outnumbered all of the competitors put together.[27] Soon thereafter, the Houser was overtaken by

24. F. Hal Higgins, "John M. Horner and the Development of the Combine Harvester," *Agricultural History* 32 (1958): 14–24; *Farm Implement News,* 1888, 18.

25. Alan L. Olmstead and Paul Rhode, "A Survey of the Development of California Agricultural Machinery, 1860–1960," Agricultural History Center Working Paper No. 11, 1983, University of California, Davis, 65–69.

26. U.S. Bureau of the Census, *Tenth Census of the United States: 1880, Agriculture Vol. 3,* 458.

27. Rogin, *Introduction of Farm Machinery,* 121–23; William H. Brewer, "Cereal Report," U.S. Census of 1880, Vol. 3, *Agriculture* (Washington, D.C., 1883), 458.

Table 2.   Grain Combines in California

	Number	Percent of All Small Grains Harvested
1881	> 20	< 1
1888	500–600	15
1900	n.a.	67
1914	> 3000	> 90
1928	2600	n.a.
1936	3000	n.a.

Sources: 1881: U.S. Census 1880, Vol. 3, p. 458.
1888: "Development of the Combine" *Farm Implement News,* Vol. II, No. 49, Dec. 6, 1928. Percent generated based on an assumption of 1000 acres harvested per season.
1900: USDA, *Statistical Bull.* No. 20.
1914: *Economist,* Nov. 28, 1914.
1928: *Combine Yearbook,* 1919, 33.
1936: *Pacific Rural Press,* Feb. 20, 1937, 266.

machines in the Holt line. The innovative products of the Holt company, which included in 1893 the first successful hillside combine, became dominant on the West Coast. By 1915 Holt's publicity boasted that over 90 percent of California's wheat crop was harvested by the 3,000 Holt combines in the state.[28] It is important to recognize that adoption of combine-harvesters east of the Rockies was only in its infancy at this date.

The combines used in California were much larger than any equipment common in the Midwest during the nineteenth century. The standard combine of the 1880s had a cutting bar of 14 to 16 feet in length, was pulled by 20 to 24 horses or mules, and had a capacity of 20 to 30 acres a day. By contrast, the most common self-binding reapers had 6-foot cutting bars and a capacity of 10 to 12 acres per day using 3 horses. The standard headers used in California had a 12-foot cutting bar and could harvest between 15 to 25 acres daily. It is interesting to note that the horsepower requirements of the combine mode were not greater than that of the header mode once the power needs of the threshing operation are considered. Indeed, they were probably smaller. The largest combines adopted in the Midwest in the 1920s and 1930s were slightly smaller than the standard California machine of the 1880s. Yet Western combines continued to grow in size. By the turn of the century the standard California combine had a cutting bar of 16 to 22 feet, a daily capacity of 22 to 30 acres, and was pulled by a team of 25 to 35 horses.[29]

These enormous teams of horses were one of the most impressive if

28. *Economist,* Nov. 28, 1914.
29. Rogin, *Introduction of Farm Machinery,* 147–49.

problematic features of the combined harvesters. The awe this practice inspired in the East is reflected in the numerous descriptions and photographs of combines appearing in national farm publications and in travel books about the Pacific states. One postcard of 5 Holt side-hill combines with a total of 165 horses and mules working on a wheat ranch in the Pacific Northwest was reportedly among the all-time best sellers.[30] Farmers outside the West seldom teamed more than eight horses and did not use large-scale equipment in the field until after the introduction of the gasoline tractor.

The primary reasons for the differences were undoubtedly cost and scale considerations, but the prejudice in the East that large teams were unworkable and the lack of practice probably also played important roles. In California the opposite attitudes were said to prevail. The *Pacific Rural Press* boasted "(i)f one man could drive all the mules in the State it would be the acme from one point of view."[31] California farmers had gradually developed their ability to manage large teams as a result of their experience with gang plows and headers.

Working out how to run and how to stop the machines with large teams was among the critical initial problems of combine development. Runaway teams were a major threat. The first combine in California, built for John Horner by James Patterson in 1853, was destroyed on its inaugural run when its team of 22 mules became frightened and bolted.[32] This led early combine builders to change the method of draft and improve the braking and gearing systems. Holt's key initial innovations involved replacing the clumsy, noisy, tight-gearing methods then in use with a more flexible belting system. This substantially reduced both normal wear and the enormous damage caused by runaways.[33]

The difficulties and costs of large teams of horses also induced efforts to use steam power. As early as 1871, B. F. Cook built a horse-drawn combine with an auxiliary steam engine to power the thresher. In the mid-1880s G. S. Berry of Lindsay, California, constructed and operated a self-propelled straw-burning steam combine.[34] Throughout the 1890s and 1900s Daniel Best and Benjamin Holt continued to perfect huge steam tractors to pull their even larger harvesters. Steam tractors did not substantially lower labor costs and replaced the threat of runaways with the greater danger of sparks and fire. While steam-driven combines never

30. Caterpillar Tractor Company, *Fifty Years on Tracks* (Peoria, Illinois: Caterpillar Tractor Co., 1954), 6–7.

31. Graeme Quick and Wesley Buchele, *The Grain Harvesters* (St. Joseph: American Society of Agricultural Engineers, 1978), 119.

32. F. Hal Higgins, "John M. Horner and the Combine Harvester," 14–24.

33. F. Hal Higgins, "A Century of Combining," June 24, 1931 (Caterpillar Tractor Company), Higgins Collection.

34. Olmstead and Rhode, "*Survey of Development,*" 69.

came into vogue, these innovative efforts did have one highly important by-product—the tracklaying tractor. Once again, we see the cumulative nature of the inventive process; an attempt to solve a problem within the context of a specific technology led to the emergence of an entirely new product.

Wheat cultivation in California at the turn of the century expanded only in new areas—hilly regions and recently drained lands—where declining yields had not set in. Steam tractors were completely unsuited to the undulating terrain opened to profitable cereal growing by the hillside combine. Yet with modifications they proved valuable on reclaimed soils which were too soft or too broken to use horses. For example, on the enormous ranches of the newly drained Tulare Lake Basin, steam combines, many driven by tractors mounted on wheels eight feet in diameter, were employed exclusively.[35] Steam tractors were also tried on the extraordinarily productive peat soils reclaimed from the Sacramento-San Joaquin delta.

The first practical tracklaying farm tractors were initially developed to operate on these soft soils. Culminating several years of effort, Benjamin Holt successfully tested in 1904 a 40-horse-power crawler on a delta island owned by his family near Stockton.[36] The principle of the tracklayer was to enlarge the area of contact, spreading the weight of the heavy engine and increasing traction. Earlier machines, using enormous wheels to avoid sinking into the peat, lacked maneuverability and power. Although the crawlers were first designed to solve essentially a local problem, this innovation was of global significance. The Caterpillar Tractor Company (formed by the merger of the Holt and Best enterprises) would build larger more powerful equipment that rapidly spread throughout the world.

The reoccurring pattern of one invention creating new needs and opportunities that lead to yet another invention offers important lessons for understanding the development or lack of development in other times and places. The key to explaining the progression of innovations in California was the close link between manufacturers and farmers that facilitated constant feedback between the two groups and the keen competition among producers that spurred inventive activity. Entrepreneurs seeking their fortunes were in close tune with their potential customers' needs and vied with one another to perfect equipment that would satisfy those needs.

Protected by high transport costs, some small manufacturers developed specialized niches in the local market and then perfected their prod-

34. Olmstead and Rhode, *"Survey of Development,"* 69.
35. Rogin, *Introduction of Farm Machinery,* 149–51.
36. Caterpillar Tractor Company, *Fifty Years,* 12.

ucts and captured sufficient economies of scale to allow them to enter national and international markets. Often this meant physically moving production facilities to new locations or being absorbed by existing national firms and in turn transferring the locus of production to be closer to the large Midwestern market and to sources of raw materials. But as the nature of California agriculture changed creating new localized demands, the process of small firms responding to potential market opportunities repeated itself.

The California experience mirrored the process that took place in many northeastern and midwestern states where local manufacturers were consistently responding to regional needs and to changing cropping patterns. The familiar story of John Deere perfecting a polished steel moldboard so that his plows could scour the heavy, sticky prairie sod is the prototypical case.

The experiences of California and many other northern states stands in stark contrast with the histories of other regions. The emergence of a dynamic manufacturing sector based on satisfying the demands of agriculture did not occur in the American South, nor has it taken place in most less-developed countries. In these areas the burdens of history severed the potential backward linkages so important for economic development. In the South the absence of a local manufacturing base aware of regional needs delayed the development of machinery especially suited to the area's geographical conditions and factor prices. In Gavin Wright's perceptive analysis, the South was a "colonial economy" dependent on northern firms for its technologies.[37] As a result, both its agriculture and industrial sectors languished. The same can be said for most less-developed countries where immense class barriers separate manufacturers from peasants limiting the exchange of ideas and depressing demand. In California, the cumulative force of history worked in the opposite direction.

Appendix: Sources and Notes for All Figures

Figure 1: "Acres Harvested in California by Type of Crop"
  Fruits and Nuts
    1870: Transactions of the California State Agricultural Society.
    1872: Number of Trees, by state and county; does not distinguish between bearing and non-bearing.
    1880: Data not available, 1880 figure is a straight average of 1870 and 1890.

---

37. Gavin Wright, *Old South, New South: Revolutions in the Southern Economy Since the Civil War* (New York: Basic Books, 1986), 156–64.

1890: 11th Census. of the U.S., 1890, *Report on the Statistics of Agriculture,* pp. 498–99, 583–91.

1900 & 1910: Abstract of the 13th Census, 1910, pp. 635, 650.

1920: California State Commission of Horticulture, *Monthly Bulletin,* April 1919. 14th Census of the U.S., 1920, Vol. 6, Pt.3, pp. 343.

1930: 15th Census of the U.S., 1930, Vol. 2, Pt.3, pp. 556–561.

Notes: Trees to Acres Ratio taken from Adams, *Farm Management,* 1921, for almonds, lemons, olives, and oranges. Other sub-tropical trees, orchard trees, and vines per acre from 12th Census of the U.S., 1900. Walnut trees per Acre from California State Board of Horticulture, 1901. All other ratios taken from U.S. Census of 1930. For 1900 a weighted average (by 1930 acres) was used for number of peaches and nectarines.

Crop Production excluding Fruits and Nuts

Compendium of the 10th Census, 1880, pp. 746–47; *11th Census. of the U.S.,* 1890, *Agriculture,* Vol. 5, *Report on Statistics of Agriculture,* pp. 355, 421; *Abstract of the 13th Census,* 1910, p. 631–34. Data 1900–1910; *15th Census of the U.S.,* 1930, *Agriculture,* Vol. 2, Pt.3, pp. 538–43, and Vol. 4, pp. 728–802; Census of Agriculture, 1950, Vol. 2, p.57. Data for 1880–1950.

Notes: 1880 potato acres not available, estimate used based on 1880 Output and 1890 Yields. Other small grains= oats, barley and rye; buckwheat included 1870–90. Hay and Forage= All tame or cultivated grasses, or other crops grown for forage. Row crops= vegtables, potatos, corn, cotton, and sugar beets. Other= eatable legumes, beans, seeds, rice, kafir, milo, peanuts, and small fruits; buckwheat after 1890.

Figure 2: "Real Value of Production per Farm"

1930: 15th Census, 1930, *Agriculture,* Vol. 4, p. 712.

1920: 15th Census, 1930, *Agriculture,* Vol. 4, p. 712.

1910: 13th Census, 1910, Vol. 5, p. 545.

1900: 13th Census, 1910, Vol. 5, p. 545.

1890: 11th Census, 1890, Vol. 5, p. 84.

1880: 10th Census, 1880, *Statistics of Agriculture,* p. 102.

1870: Compendium of the 9th Census, 1870, p. 692.

Note: Deflated using Wholesale Price Index (1910–14=100) of Farm Products, *Historical Statistics,* Pt. 1, pp. 199–201.

Figure 3: "Real Value of Implements per Farm"

1930: 16th Census, 1940, *Agriculture,* Vol. 3, pp. 48–49.

1920: 14th Census, 1920, Vol. 5, p. 50.

1910: Abstract of 13th Census, 1910, pp. 276–77.

1900: Abstract of 13th Census, 1910, pp. 276–77.

1890: 11th Census, 1890, Vol. 5, p. 84.

1880: 10th Census, 1880, *Statistics of Agriculture,* p. 102.

1870: Compendium of the 9th Census, 1870, p. 690.

Note: The value of implements is deflated on a base of 1910–14 using Alvin S. Tostlebe, *Capital in Agriculture: Its Formation and Financing since 1870* (Princeton: Princeton Univ. Press, 1957).

Figure 4: "Real Value of Implements per Male Worker"

1930: 15th Census, 1930, *Population,* Vol. 4, pp. 19–20.

1920: 14th Census, 1920, *Occupations,* p. 56.

1910: 13th Census, 1910, *Population,* p. 90.

1900: 12th Census, Population, 1900, Vol. 2, Pt. 2, p. 510.

1890: 11th Census, 1890, *Population,* Pt. 2, p. 306.

1880: 10th Census, 1880, *Population,* p. 776.

1870: 9th Census, 1870, Vol. 1, p. 674.

Figure 5: "Horses and Mules per Oxen"

Stanley Lebergott, *The Americans,* (New York: W.W. Norton & Co., 1984), Table 24.1, p. 299.

Figure 6: "Horses and Mules per Farm"

Number of Farms: Historical Statistics, Pt. 1, p. 459.

Horses and Mules:

1930: 16th Census, 1940, *Agriculture,* Vol. 3, pp. 40–41.

1920: 14th Census, 1920, Vol. 5, p. 530.

1900 & 1910: Abstract of 13th Census, 1910, p. 320.

1870 through *1890: 11th Census,* 1890, Vol. 5, p. 84.

Figure 7: "Horses and Mules per Male Worker"

Number of Farms: Historical Statistics, Pt. 1, p. 459.

Agricultural Male Workers:

1930: 15th Census, 1930, *Population,* Vol. 4, pp. 19–20.

1920: 14th Census, 1920, *Occupations,* p. 56.

1910: 13th Census, 1910, *Population,* p. 90.

1900: 12th Census, Population, 1900, Vol. 2, Pt. 2, p. 510.

1890: 11th Census, 1890, *Population,* Pt. 2, p. 306.

1880: 10th Census, 1880, *Population,* p. 776.

1870: 9th Census, 1870, Vol. 1, p. 674.

Figure 8: "Irrigation in California" and Figure 9: "Agricultural Pumps in the United States and California"

1920 and 1930: 15th Census, 1930, *Irrigation of Agricultural Lands,* pp. 13–15, 85–87.

1910: Samuel Fortier, *Use of Water in Irrigation,* (New York: McGraw-Hill, Inc.). p. 2.

1900: Bureau of the Census, Bulletin 16, Irrigation in the United States: 1902. Washington: G.P.O., 1904).

1890: 11th Census, Agriculture, 1890.

Quantification in American Agricultural History, 1850–1910: A Re-examination

WILLIAM N. PARKER

> All, all of a piece throughout:
> Thy chase had a beast in view;
> Thy wars brought nothing about;
> Thy lovers were all untrue.
> 'Tis well an old age is out,
> And time to begin a new.
> —*J. Dryden, Chorus from
> the Secular Masque, ca. 1688*

Between 1958 and 1962, an effort was begun at the University of North Carolina to employ quantification in the historical study of the principal branches of the American agricultural industry: (1) the plantation and small cotton farms of the South; (2) the mixed grain and livestock farms in the Northeast and North Central (Middle West) states; (3) the specialized wheat farms and cattle ranches of the Great Plains; (4) the specialized dairy and hay farming in the market areas of the coastal cities of the Northeast and around the Great Lakes. The decades 1840–1860 and 1900–1910 were chosen as end points to catch these regional groups of specialized producers at significant points in their modern history. The cotton regions were to be shown first, at the height of the slave labor system and then after the full evolution of the post-bellum organization of free farms as share tenancies and share and wage labor plantations. For the grain and meat producers the dates captured the organization before and after the great expansion from the eastern edge of the Plains to the Rockies, and the evolution of the complex market system of open range with seasonal drives and shipment through the feeding and

WILLIAM PARKER is Professor of Economics at Yale University. The author is indebted to David Weiman and other members of his dissertation workshop at Yale University for corrections and reassurance on this paper, and to Heather Salome of the Council on West European Studies at Yale for preparation of the diagram.

fattening areas to central slaughtering points, to be combined with animals from the older mixed grain and livestock farms. The study of dairying was designed to cover the period of market growth through the rail deliveries in urban milksheds as the heavy industrial development in the North Atlantic seaboard and upper Middle West took place.

The data for this work were found in the three main sorts of sources mined by the "fathers" of American agrarian history, Malin, Shannon, Phillips, Gray, Gates, Webb and Bidwell, and Falconer: (1) the manuscript returns of the federal censuses, available for 1850–1880; (2) the published compilations of the Census, the earlier statistics and reports made to the agricultural office in the U.S. Patent Bureau and after 1863 to the U.S. Commissioner (later Secretary) of Agriculture; (3) data yielded as part of "literary" evidence: plantation records, travellers' accounts and farm diaries, published proceedings of agricultural societies, reports of state Boards and Commissioners of Agriculture, and most especially, the collections of that peculiarly American journalistic enterprise—the farm journals and periodicals which appeared from the presses in a number of major cities adjoining the farming areas, and whose combined circulation in the 1880s must have totalled, on conservative estimate, at least several hundred thousand. Among the news of new developments in farming, endless letters from readers, moralistic sermons, poems about the farm home, accounts of the world's natural and man-made wonders and other similarly entertaining, instructive and self-revealing information, are contained occasional bits of concrete fact: acreages, yields, animal weights and diet, uses of labor time, effect of various practices and improvements.

The manuscript census materials for 1860 drawn in the UNC (Parker-Gallman) sample have since been greatly extended by the later researches of Swan, Foust, Fogel and Engerman, Wright, Sutch, Ransom, and others. Most recently, a sample has been drawn for the northern areas in 1860 by Bateman and Atack.[1] A large body of data now exists to survey and compare the state of U.S. agriculture, North and South, as of 1860.

The other sources described above were utilized in a research effort designed to measure the changes in labor productivity in the various operations of crop production, livestock raising and farm capital formation in the major producing areas. This work made a wide survey of farm journals and government publications to derive sets of contemporaneous observations on man-hours in each specific operation sufficient to mark central tendencies in the different regions in the two terminal periods. Per acre yields and, for animals, average weights, milk yields, and slaughter rates were compiled. Each commodity study, when completed, was designed

---

1. Jeremy Atack and Fred Bateman, *To Their Own Soil: Agriculture in the Antebellum North* (Ames: Iowa State University Press, 1987).

thus to measure the impact of three factors in the productivity growth: (1) changes in the distribution of the output over regions of differing original natural character, i.e., westward movement; (2) changes in mechanical techniques in operations on the soil or in livestock handling over the period; and (3) changes in bio-chemical technology and knowledge affecting fertility of the soil, genetic features of seed and breeds, insect control, and the like. Data on production and yields were drawn largely from published Census and USDA sources, by county or state.

Studies on the following products then were planned: (1) Farm-formed capital: land clearing, fencing, construction; (2) Dairy products: raw milk, cream, butter, cheese, buttermilk; (3) Field crops: cotton, corn, wheat, oats, hay; (4) Power: numbers and utilization of draft animals—horses, mules and oxen, both on the farms and in cities; (5) Meat: pork and beef, with the mix changing toward the latter following changing tastes and relative costs. Production data were gathered on sheep and wool, tobacco, potatoes, poultry and other minor crops, but the effort stopped far short of measuring labor inputs in this heterogeneous body of output.

Of these studies, those on capital formation, dairying (in terms of raw milk), and the grains were completed by the UNC group so far as the data allowed. A partial study on cotton was issued much later. A list of these publications is appended below. Much data was collected for studies of livestock but the work was held up by two seemingly insuperable obstacles: (1) the measurement of supplies of wild forage in animal diets in the earlier period (the so-called "grass and acorn problem") and (2) the complexity of the systems of range and feed-lot prevailing in the Middle West in the later period. Left unpublished also was a crowning effort to combine all the studies to show the complete time budget for farm labor in the United States as it changed between the earlier and later periods, and to trace the influence of each of the three fundamental sources of productivity change in producing the composite result, i.e., the four-to-nine-fold expansions of output of the various products under conditions of constant or rising yields per acre and per animal accompanying a growth of the farm labor force from 3.8 million in 1850 to 9.1 million in 1900—a rise of about 150 percent above its 1850 level.

The present paper reviews what appear in retrospect to have been the most significant findings of this uncompleted effort with respect to the influence of the three major historical factors affecting labor productivity growth in these agricultural operations. I regret having to defer—probably forever—the completion of the livestock and cotton studies and any effort to add the results on the individual crops up into national totals. The aging of research work sometimes gives new perspective on what was done, but it cannot make new data grow where none grew before, or convert what would be sheer guesswork into defensible estimates. An effort is made, however, to understand and take into account the criticism of—and improvement on—the study of the major grains contained in a paper of F. M. Fisher and P. Temin, and to use that discussion as an opening to a wider investigation in the realms of a highly speculative counterfactual history of American development.[2]

It would be idle to re-examine and criticize in detail the individual studies—either those accomplished and those stillborn—that constitute the literary remains of this ambitious project. In all the cases the method of research and estimation was roughly identical. Contemporary sources from the middle and the end of the century were sifted to find specific statements about the man-hours used in specific farm tasks, given per acre or animal, or per unit of final output, (L/A) or (L/O), and to complement these with statements about product yields per acre or per animal, (O/A).[3] Labor productivity in the crop was then expressed from these three sets of estimates in some variant of the following equation:

$$\frac{O}{L} = \frac{1}{(a)\dfrac{L/A}{O/A} + (1-a)\dfrac{L}{O}}$$

Where a is the proportion of total cost devoted to pre-harvest, or in dairying, pre-milking operations.

In the three completed sets of studies—land clearing, milk and grain, the producing areas of the nation as of mid-century were divided into regions. For land clearing, the division was simple—between the forested and the nonforested areas. For dairying, the mid-century data were available only on a state basis, but here the distinction appeared between the Northeast where commercial milk markets already existed and rural areas where small local sales prevailed, or where the cow was simply a part of the family subsistence economy. For the grains, regions were defined on the basis of yields, following the usual sectional divisions. For each region, then, at mid-century and again at the end of century (up to 1910) the data required to estimate the equation were arrayed. Averages, or more usually medians, were taken and their significance assessed by inspection of the range and clustering. In only a few cases were the data plentiful enough to allow or

2. Franklin M. Fisher and Peter Temin, "Regional Specialization and the Supply of Wheat in the United States, 1867–1914," *The Review of Economics and Statistics* 52 (May 1970): 134–49.

3. Labor input is expressed in the sources per acre or per animal (L_1/A) for most tasks, requiring then a separate yield figure (O/A) to show productivity (O/L_1). For operations directly on the crops (cotton picking, corn husking, and shelling, grain threshing, sometimes milking) the figure is given relative to the volume of output (L_2/O).

suggest the computation of standard statistical measures of dispersion. Yet in the three completed groups of studies, the eyeballing techniques, done in an honest and orderly fashion, with some help from elementary statistics, produced plausible results. Since no substantive hypotheses were being tested, the temptation to skew the results one way or another was not present.

Combining the regional estimates of productivity in each period into a national average where the regions—Northeast, South, and West (R_1, R_2, R_3)—were weighted by relative shares in acreage or animal herd gave the basis for a national index of productivity change over the period. Extracted from the three completed studies, this measure was as shown in Table 1. The basic formula:

$$\frac{O_{US}}{L_{US}} = \sum_{R_1}^{R_3} V \; \frac{1}{\dfrac{L_1}{L} \times \dfrac{L_1/A}{O/A} + \dfrac{L_2}{L} \times \dfrac{L_2}{O}}$$

(where V is a region's share of national output) also made apparent that this composite change was the arithmetic result of changes, individually and in interactive combination, among the three variables—the regional weights or shares in total output (V), regional yields (O/A), and regional labor inputs per acre (L_1) or in some operations directly per output unit (L_2). Having gone this far, we found the temptation irresistible to try to give a quantitative assessment of the relative importance of these three factors in producing the composite result. This game seemed the more important since—as stated above—each of the three factors could be related to a significant element in America's agrarian history:

> regional weights—westward movement and regional specialization;
> yields per acre and animal—improvements in farm practices, genetic changes, fertilizers, pest control, etc.;
> labor per acre or directly per output unit—improvements in tools, machinery, and power.

The solution was suggested by a statistician, Leo Katz of Michigan State University, on a model used in fertilizer experiments in Britain.[4] In fullest

---

4. See William N. Parker and Judith Klein, "Productivity Growth in Grain Production in the United States: 1840–60 and 1900–10," in *Output, Employment and Productivity in the United States after 1800* (NBER, Studies in Income and Wealth, 1966), vol. 30. The reference is to F. Yates, *The Design and Analysis of Factorial Experiments,* Imperial Institute of Soil Science, Technical Communication 35, Harpenden, England, 1937.

Table 1.   Indices of Labor Productivity Change in Three Groups of Farm Products

		1900's
Land clearing:	acres per man-day (1850's = 100)	340
Milk:	pounds per man-hour (1850's = 100)	83
Grains:		
Corn	bushels per man-year (1840/60 = 100)	365
Wheat	bushels per man-year (1840/60 = 100)	417
Oats	bushels per man-year (1840/60 = 100)	363

extension the model in the study on grains yielded a set of eight indices, combining the values of each of the three variables alternately at their Period 1 (1840/60) and Period 2 (1900/1910) levels.[5]

As Mancur Olson—the political theorist and economic historian—is said to have said and has himself clearly exemplified, "To a boy with a new hammer, the whole world looks like a nail." I should have appreciated the risks of assumptions implicit when a method—a rather primitive one at that—of statistical testing is transferred from controlled physical experiments and tests to the data of an historical record where not only seeds, weather, and life processes were involved, but also the infinite variation derived from the behavior of the society of two-footed human animals. Most unreal was the assumption that the 1910 outputs could have been produced under the 1850 regional distributions among regions without sharply rising marginal costs in the 1850 regions. This problem and its implications for statistical studies of causation and influence in history are taken up in the next section. First, however, I wish to examine what the data of three completed sets of studies did seem to show even where not put to the extreme torture of counterfactual quantification.

We have then studies of three products and in each three factors responsible in combination for the result. Disregarding statistical refinements, a look simply at the average values computed from the data suggests a very curious observation. In each case there is a distinction between an active factor where the difference between 1850 values and those of 1910 is pronounced, and two relatively passive factors whose values either remain stable and simply prevent serious declines in productivity despite the growing quantities of output and other changes that did take place. In each case, this active factor is a different one of the three. Thus:

5. See Parker and Klein, "Productivity Growth," 529, and for an example, 532–33.

| | | Factor | |
| | *1* | *2* | *3* |
Products	*Regional shifts*	*Farm Practices*	*Machinery*
1. Land clearing	Active	Passive	Passive
2. Milk	Passive	Active	Passive
3. Grains	Passive	Passive	Active

Land clearing provides the simplest, starkest case since the 'crop' involved is simply cleared acreage and Factor 2—yields, genetic improvement and subtle changes in practices—is not involved. Here, as in Table 2 drawn from Primack's study shows, the difference between labor costs for clearing forested areas are of the order of 20 to 30 times the costs of the first plowing of the grasslands.[6]

In the mid-century, 80 percent of the nearly 5 million acres of land cleared annually lay in the forested areas whereas in the 1900s, of a newly cleared acreage nearly 50 percent greater, 70 percent was nonforested. The whole operation was estimated to have used up 11.6 percent of the farm labor force in the 1850s, and thanks largely to the movement West, only 2.3 percent in the 1910s. Set against this, the modest changes in techniques in either of the two types of cover have a distinctly minor influence. However one weights the regions or computes the indices, labor saving in this operation comes principally from the movement into the open West.[7]

6. "Before 1850, American farms were cut from the forest, and the work of forming a farm took time. Five acres of forest clearing in a year in addition to current crops was about the limit for a farm family. Even farmers who specialized in clearing land for sale might count on two hundred acres or so of forest clearing as the labor of a lifetime. Like nearly everything else, they existed in Yankee and southern variants whose use, however, was not confined to either region. Clearing was done by controlled burning in most cases; the differences lay in the method of killing the trees and the time allowed after drying the fallen timber. With the so-called Yankee method, probably adopted from Swedish settlers along the Delaware, the trees were felled by the ax into poles for burning and allowed to dry on the ground over a period of months or years. A first firing would leave large limbs and trunks to be cut into manageable lengths, repiled and burned again. The repiling for the second burning was often the occasion of a frontier cooperative activity, the "log-rolling." An alternative method, adopted from the Indians and favored in the South, replaced the labor of the arm and ax by the action of wind and time. Here the larger trees were killed by girdling—the removal of a wide circle of bark from completely around the trunk. If this was done early enough in the season, the land could be ready for cultivation beneath the dead trees in the spring. As limbs and trees rotted or were blown down, they could be collected a few at a time and burned. Girdling had a low immediate labor cost and permitted quick cultivation over a large acreage, but livestock and crops were damaged by falling branches in the phantom forests, and ultimately the wood had to be collected and burned." (Martin L. Primack, "Land Clearing Under Ninteenth-Century Techniques: Some Preliminary Calculations," *Journal of Economic History* 22 (1982): 485–97.

7. This is not to say that in the absence of any technical change the move on to the grasslands would have gone ahead at the same rate. Grassland farming required not only the

Table 2. LABOR COSTS IN LAND CLEARING, 1850'S AND 1900'S

	Median	Range	Range Divided by Median
		(man-days/acre)	
1850's			
Forest	33*	+ 17, − 7	+.50, −.21
Nonforested Area	1.5	±.5	±.33
1900's			
Forest	25*	+ 8, − 10	+.32, −.40
Nonforested Area	.75	+.25, −.50	+.33, −.67

*Including stump-pulling at thirteen man-days in 1850's, and five man-days with stump-pullers in 1910's.

Fred Bateman's studies of dairying also present some individual and rather surprising features. Over the whole period the index of productivity per labor hour declines. The two major influences allowing the expansion of output both relate to the greater value of milk due to the growth of the commercial demand from cities. Part of the growth came through a lengthening of the average milk year for a cow from 237 to 300 days—an adjustment which does not economize on milking labor or on feed input and is only possible—indeed went on half-unperceived by farmers—because of the use of family and off-peak labor in the barn. A second way—upgrading of the herds to give yields closer to the best yields of mid-century created a 20 percent rise in daily yields per cow, a saving of fixed labor costs in care, feeding, and—to a large degree—in milking as well. Improved stock came through diffusion of standard breeds, but there was little or no advance in the best levels of yield. The American herd in 1910 was still far from a collection of highly bred specialized dairy cattle. Against such improvement as occurred, there came stricter sanitary requirements in feeding, barn care, and milking which actually raised labor costs of the milk as it left the farm. Dairying responded then neither to improved technology nor to westward movement as such, but rather to the widespread growth of markets which added value to the product, thus inducing more work, care and attention in how dairying was conducted.

Studies of beef and pork, could they be completed, would almost certainly show a similar absence of striking technical change. Spatial re-

---

improved steel breaking plow but all the numerous adaptations, especially windmills and barbed wire, described in W.P. Webb's classic study, *The Great Plains.* Except in the very laborious operation of stump removal, clearing techniques on forest cover remained almost unchanged until well into the twentieth century. Blasting powder to remove or break-up stumps, introduced in the 1880s was virtually the only improvement.

arrangement following the railroad and centralized stockyards was important and improvements of breed and feed would leave a mark. But such
improvements would have to be set against much higher costs which the
shrinking of the woodland and open range necessitated. Among the field
crops, too, the tentative study of cotton shows the predominant influence
of geographical shifts over technical change. The only major improvement
was the substitution of the horse (or mule)-drawn cultivator for the hoe—a
very great improvement indeed but a very simple change made perhaps in
part as a response to the disappearance of slavery.

Only in the grains, and more in the cereal grasses than in corn, was
mechanical technology dominant in saving labor. To show what was done
here, it is simplest to expand the list of variables as follows:

$$
\begin{aligned}
L_1 &= \text{Preharvest labor (man-hours)} \\
L_2 &= \text{Harvest labor (man-hours)} \\
L_3 &= \text{Postharvest labor (man-hours)} \\
R_1 &= \text{Northeast} \\
R_2 &= \text{South (wheat and oats)} \\
R_{2a} &= \text{Middle east (corn)} \\
R_{2b} &= \text{South (corn)} \\
R_3 &= \text{West (of the Ohio River and the Mississippi)} \\
A &= \text{Area (acres planted)} \\
O &= \text{Output (bushels of threshed grain)}
\end{aligned}
$$

$$
\begin{aligned}
a &= L_1/A & y &= O/A \\
b &= L_2/A & v &= O/\Sigma O \\
c &= L_3/O & w &= A/\Sigma A \\
1 &= \text{period 1 (1840–60)} \\
2 &= \text{period 2 (1900–10)}
\end{aligned}
$$

$$
\sideset{_{R_1}^{R_3}}{}\Sigma \left(\frac{a+b}{y} + c \right) v = \text{national average labor input per bushel}
$$

Using these variables, Table 3 shows the averaged data for the four main
cost variables arrayed by region with the regional output weights. It is
evident by inspection that in wheat and oats the sharp fall occurred in
harvesting and threshing costs in all regions to about a quarter of their
earlier level with the introduction of the mechanical reaper, thresher, and
combines. The strong shift toward regional specialization (from a roughly
equal distribution of the acreage among the three regions to an 87 percent
concentration in the West in the second period) affected second period
costs in the "passive" sense of accomodating the great increases in output, but unlike the situation in land clearing, costs were not so different
among the regions as to produce a strong, direct effect on the observed
indices. Labor-saving in corn was largely confined to preharvest opera-

Table 3. Labor Inputs, Land Yields, and Weights, Periods 1 and 2

Region	a 1	2	b 1	2	V 1	2	c 1	2	v 1	2	w 1	2
					WHEAT							
R_1	19.1	11.6	15.0	3.0	14.5	17.5	.73	.19	.334	.046	.259	.037
R_2	11.3	10.7	12.5	3.0	8.4	12.3	.73	.29	.342	.075	.459	.085
R_3	12.4	4.7	15.0	2.3	13.0	14.0	.73	.19	.324	.879	.282	.878
U.S.	13.6	5.5	13.9	2.4	11.3	14.0	.73	.20				
					OATS							
R_1	14.3	9.3	12.8	3.4	28.5	29.7	.40	.23	.422	.087	.316	.077
R_2	8.8	9.5	11.0	4.5	13.9	17.0	.40	.24	.332	.044	.506	.068
R_3	8.8	3.9	12.8	2.6	29.3	26.5	.40	.10	.246	.869	.178	.855
U.S.	10.5	4.7	11.9	2.8	21.3	26.1	.40	.12				
					CORN							
R_1	98.3	46.4	13.0	13.0	33.5	36.8			.097	.029	.057	.020
R_{2a}	52.0	26.7	10.1	10.1	21.8	24.4			.344	.099	.310	.106
R_{2b}	67.3	21.3	4.3	4.3	11.8	16.1			.279	.175	.465	.285
R_{3a}	46.2	14.2	13.0	7.6	32.7	31.0			.280	.697	.168	.589
U.S.	60.8	18.2	8.1	7.0	19.6	26.2						

tions with the substitution of the cultivator for the hoe, as in cotton. The development of direct feeding in the field or from the shock reduced harvesting costs in the West, (R_3), but it is not clear where the growth of silage in dairy districts shows up in the cases on which the statistics are based. Overall in corn, as in dairying, technical change appears relatively unimportant compared to many small changes in practice made in response to the greater value of the fodder as feed for commercial meat production. It is noteworthy, too, that in all the grains, per acre yields do not rise significantly although adaptation to new soils and climates in the corn and wheat belts is presumably of decisive importance in keeping yields steady under great expansions of acreage and output.[8]

The three groups of studies, fragmentary and incomplete as they are, thus point to the quite diverse influence of the three productivity elements in one line of production relative to another. But the three studies, taken together, have a deeper meaning for American agricultural history as a whole. By lucky coincidence, each illustrates a different fundamental aspect of agricultural development, and indeed of modern economic history

8. See the discussion surrounding Table 4, below.

in the large. The pattern is simple and its logic is compelling.[9] In the Western world as a whole, population growth even in the twelfth to thirteenth centuries, then again in the late fourteenth and fifteenth, did not meet at once the diminishing returns that led in the models of Ricardo and Mill to a stationary state or the periodic recurrence of Malthusian crises. Beginning in 1750, presence of rich lands on the frontier created growing surpluses which even in the late eighteenth century became available for trade. The American frontier in particular was full of bonanzas—minerals, deep soils, the treeless grasslands, and these presented not an obstacle but a huge productive opportunity accelerating the advance into the continent. Overlapping the influence of population growth and richer resources, the expansion of the market continued the movements. The dairy study shows the self-reinforcing opportunities for growth as the market put a value on the time of farm labor, diffused known practices and increased the possibilities of regional specialization. The productivity gains discovered by Adam Smith in specialized industrial production appeared in specialized agricultural regions as well. And finally, as Smith also predicted, intense specialization in the grains under the strains induced by growing market opportunity drew on technical innovation, specifically mechanical invention, a spin-off from mechanical industry, to continue its thrust forward. When this eco-technic process got underway, agriculture was carried with industry into the high tech of the twentieth century—the age of Schumpeter. After the period of our studies closed, the recurrent twentieth century technological revolutions—in farm power, in mechanical improvements, in insect control, in soil chemistry, plant and animal bio-chemistry and genetics have enabled the American agricultural labor force to produce itself almost to extinction.

The conclusions just drawn are those which, twenty years later I would take from the studies. But at the time they were published, much attention was focused on the statistical operations imposed on the regional averages for each variable in the effort to give a precise measure of its effect, alone and in combination with the others.

In this effort, the basic measure, combining the variables for each region, was calculated with alternative values of the variables on the Period 1 base. This was done most elaborately in the grains study. Here the productivity indices shown in Table 1 above were produced from the basic equation, written with Period 2 values, divided by the same index with

9. I have developed these ideas in *Europe, America, and The Wider World*, Vol. 1, *Europe and the World Economy* (Cambridge: Cambridge University Press, 1984), Chapter 11. It should be noted that in the 1850s, ignorance of how to handle prairie soils and grasses did cause settlers to go around them to seek out the familiar conditions of forested areas.

Period 1 values. Then six intermediate indices were constructed on the base of the period 1 index = 100 by varying each variable alone and in conjunction with each of the others, as shown in Table 4.

Following the method of Yates in the agricultural experiments, the study then computed the percentage contribution of each variable, alone and in interaction with each of the others to the total productivity growth. Adding the indices in which each variable and pair of variables appears in Period 2 values and subtracting those where it appears in Period 1 values,[10] the separate influence of each variable and of each pair of variables was calculated. Reducing the whole array in percentage terms yields the following measures of relative influence (Parker and Klein, 1966):

	Wheat	*Oats*	*Corn*
v	*.170*	*.287*	*.207*
abc	*.598*	*.506*	*.562*
y	*.082*	*.005*	*.084*
v(abc)	*.158*	*.234*	*.120*
vy	−.049	−.026	−.007
(abc)y	*.046*	*.004*	*.032*
v(abc)y	−.005	−.010	−.002

This measure shows specifically the principal conclusion derived by inspection from the data, i.e., that the variable (abc)—technology in the field operations—accounts statistically for between 50 and 60 percent of the whole productivity growth, and that this effect is especially powerful when taken in conjunction with the shift of all these crops, into the Midwest [v(abc) or index (i_6)]. Further torture of the data separated the independent effect of the three mechanization variables: (a) improved plows, seed drills, and in corn the cultivator, (b) mechanization in reaping, (c) mechanization in threshing, to show that (b) and (c) accounted for nearly all the change in wheat and oats, and (a) alone in corn. These offer simply a more precise statement of the conclusions arrived at earlier.

In retrospect, this work appears neat, tidy and precise—publishable research. But when one reflects upon its meaning and implications, the ground beneath it begins to tremble. At several points it appeared even at the time particularly vulnerable. First, the measurements had been at fixed points along the regional cost curves without reference to the volumes of output as they changed between the periods. Second, I had ignored factor substitution, assuming in effect two alternative technologies, each with

---

10. This is a rough approximation to Yates' exact method. See Parker and Klein, "Productivity Growth," Appendix C, and reference to Yates, there cited.

Table 4.   Labor Requirements (U.S. Average and Indexes) As Affected by Inter-regional Shifts, Regional Yields, and Regional Labor Inputs per Acre

| | | Period for Values of | | Labor Requirement (L/O)* | | | Productivity (O/L) Index ($i_1/i_8 \times 100$) | | |
| | | | | Wheat | Oats | Corn | Wheat | Oats | Corn |
Index	v	y	abc	(1)	(2)	(3)	(4)	(5)	(6)
i_1	1	1	2	3.17	1.45	3.50	100	100	100
i_2	1	2	1	2.68	1.37	2.94	118.3	105.8	119.0
i_3	1	1	2	1.29	0.78	1.54	245.7	185.9	227.3
i_4	2	1	1	2.90	1.18	2.70	109.3	122.9	129.6
i_5	1	2	2	1.05	0.72	1.32	302.1	201.2	265.2
i_6	2	1	2	0.84	0.39	1.06	377.3	371.7	330.2
i_7	2	2	1	2.69	1.23	2.45	117.8	117.9	142.9
i_8	2	2	2	0.76	0.40	0.96	416.7	362.6	364.6

Source: Parker and Klein [1966], Table 2.

$$*L\left(\frac{a \pm b}{y} + c\right) v.$$

fixed coefficients with only the possibility of a shift from one to the other. This violated two sacred allied assumptions of economics: factor substitutability and diminishing returns. Moreover, the mere statistical measure of an interaction effect between technical change and Westward movement did not include a "feedback" effect from one variable to the other. The fact that the combine was introduced at a time when wheat was moving into lands particularly suited to its use gave the two shifts a particularly powerful effect in conjunction with each other. But it did not in a sense credit the combine with encouraging the movement into just those lands, nor take into account that the presence of such lands may have incited just such a technological development.

On both these points it may be argued that the indices which tried to measure the effect of Westward movement of the crops in the absence of technological change (i_4 and i_7) were less seriously flawed than those which did the reverse. To give an imaginary historical meaning to the indices, sheer common sense plausibility had to be taken into account. With or without new agricultural technology the westward moment of agriculture would have taken place. Mankind had been making similar moves for 10,000 years. That it occurred at the speed it did, in the presence of commercialization and with the release of labor that allowed industrialization to proceed in mid-America was due then to the technical environment—notably the railroad and farm machinery, improved plows, and some improved knowledge about the crops. And these technological advances, being part of a movement across the whole Atlantic economy were essentially independent of the terrains and soils of the regions into which they moved.

The high levels of the indices including technical change (i_3 and i_5) and absent western movement compared to the low levels of those on the reverse assumption (i_4 and i_7) seemed a valid reason for stressing the importance of technical change in the growth.

Fortunately for the cause of truth, these conclusions did not go unchallenged. Without Westward movement and regional specialization, the whole commercialized growth in output, even in the presence of technical change could not have occurred under conditions of constant costs. Lands poorer in quality and harder to farm would have been drawn on in the older regions. Moreover, the counterfactual here seemed less plausible than the contrary one. Cost curves in the western region might be imagined to be essentially flat in the prairies (especially if the railroad is taken as available) even under the older technology. But it is much harder to imagine that, had this huge output growth been confined to its mid-century regional distribution, cost curves would not have violently risen, even under the new technology.

This point was made in painfully exacting detail and completeness in an article by F. M. Fisher and P. Temin in the 1970 *Review of Economics and Statistics.* The problem was recast by them in the form of a classic problem in farmers' response to price. In doing this, they were able to make use of the USDA annual estimates of acreage (planted or harvested), output and farm price by states for the 47 years, 1867–1914. No doubt the recent availability of Nerlove's technique of estimation, using lagged output to capture farmers' adjustment over time, was a powerful incentive to the work. This permitted them to calculate something which could be called long-run supply curves for these grains by state, assuming rational responses in these decisions about crop choice. The conclusions, made on the basis of the share of acreage planted to wheat relative to the price of wheat relative to other grains formed a kind of lower limit to the rise in supply costs since the need to increase the supplies of all products simultaneously in the East would have pushed all prices up much further. They found, even within the low levels of acreage in which production actually occurred east of the Mississippi, that the long-run elasticity of supply was less than 2.0, meaning a tripling of price to extract a six-fold increase in output. This, they called, "a gross underestimate of what costs would have been."

So ingenious an econometrical study clearly constituted a further move down the track of better specified counterfactual history. To one who has not used the technique himself, lingering, atavistic doubts remain, partly from one's dim recollection of these matters from Marshall's text in graduate school days. The project appears to assume farmers' response to price, but is that assumption the most relevant one in understanding the history?

What about the decision to shift technologies, or the decision to move West, or to leave farming altogether? Did these all depend on the lagged prices of wheat? Perhaps a suspicious economist, who is only semi-neoclassical, can be forgiven for suspecting that when those using these techniques come on the scene, they are in the end only using the "rationality" assumption in the same old short-run context for the nth time. If the farmers again pass that test, what is left to help us explain all the environmental changes to which they are making so rational a response? These, it is said, are all compounded in a "shift" variable t—for time, which seems a shorthand for everything in which a historian is interested. *Timeo Danaos et dona ferentes.* Still one must be grateful both for the correction and for the spirit in which it was given. My study itself will have served a valuable social purpose if it should lead two such careful scholars and exact thinkers a little way down the slippery slope of freer imaginative exercises.

For, what in this case is the real counterfactual? What would American history have been like without the West? A number of lines of adjustment—four in demand, one in supply—suggest themselves. Those affecting demand include:

(1) Reduced population growth, both through lower birth rates and curtailed immigration;

(2) An altered diet, feeding to humans the vast quantities of corn fed to hogs;

(3) Conversion of part of the eastern cotton lands to feed crops, as has occurred since 1940. Given the strong position of U.S. cotton supplies on world markets, this might have caused no loss in foreign exchange.

(4) Countless readjustments in foreign trade, allowing industrial exports to make up for the loss of the exported farm "surpluses" no longer available to weigh down farm prices.

Altogether these adjustments, especially that in diets, might not have eaten seriously into the country's real income growth. Diets without the West would have had less beef, pork, and wheat, more cornmeal and potatoes. Are we prepared to say that was a "bad thing"? And unless an animal protein diet is essential to a development of heavy industry, industrialization would not need to have been much affected. Chicago would have been much smaller (also possibly not a bad thing?), but supplies of iron and steel could have been just as great, and the industrial and urban development to its east below the lakes, to Pittsburgh, and through central New York to the western shore of the Atlantic, could have been unchanged. Without western railroads, the iron and steel industry might have found demand reduced, but a few less railroads and a few less steel plants for the "robber barons" to profit from might have made life (and the economic historian's job) much simpler and cleaner. Perhaps the main effect of a truncated geography would not have been in agriculture at all, but in the loss of the Texas-Oklahoma oil fields after the automobile was invented. To contemplate the United States without Texas *or* California may indeed give one pause.

A final line of argument might be that without the Prairies and Plains, the direction of agricultural technology would somehow have altered. This possibility was contemplated in the paper on grains. As Fisher and Temin write, "Parker and Klein were aware of this problem (of supply elasticity) but they did not pursue it. They said instead that the pressure of increased demand might have led to technical change in the East in the absence of the ability to expand in the West. But having mentioned this possibility, they immediately discounted it."[11]

Yet how, pray, is history of technological change—hypothetical or real—to be written simply under the guidance of the economist's great "tool": elasticity and factor substitution in response to relative price? On the reef of hard facts from the supply side in technological history the frail barks of growth models splinter apart. Historians credulous enough to climb aboard, have had to shed most of their baggage to keep afloat. Noah's ark sheltered the trusting animals that boarded it, and kept them alive during a hard time. But surely neither they nor Noah rejoiced when on emerging, they found themselves on Mt. Arafat, in an air so rarified, and at a height so lofty that they still could not make out the ground below. We are still waiting for the theorists to provide us, not an Ark to float us above the floods of facts, but a railroad—or even a canal—to push us through them.[12]

Appendix A: Total Factor Productivity

John W. Kendrick's work on "total factor productivity" was published just as the research described in this paper was coming to an end.[13] It represents the most thorough and definitive effort to measure productivity

11. Franklin M. Fisher and Peter Temin, "Regional Specialization," 138.

12. Theoretical work at the INDUSTRIENS UTREDNINGSINSTITUT (Stockholm), under Gunnar Eliasson and others has something evidently new, though complex and ultimately somewhat mysterious, to say on modelling innovation, following the path Schumpeter, Erik Dohmen, and Nelson and Winter: See R.H. Day and G. Eliasson, eds., *The Dynamics of Market Economies* (Amsterdam: North Holland, 1986), and various IUI working papers by Eliasson and others. It is still difficult to see how such models can be used by an historian to explain in concrete terms the emergence of, say, an electrical technology in the late nineteenth century, without a specification of the path of intellectual development for the underlying science and an understanding of the limits placed on engineering by the state of scientific development.

13. John W. Kendrick, *Productivity Trends in the United States,* (NBER, Princeton: Princeton University Press, 1961.

growth as output, in "real" price-deflated money terms relative to a weighted index of inputs of labor and capital. It is based on two interrelated concepts of modern economics as it stood in the 1950s to 1970s: (1) the concept of an aggregate supply curve, or production function $O = f(L,K)$, shifting through time, (2) a measurement of the changing physical volume of capital by deflating a series of estimated net stock (annual purchases less depreciation allowances) by a price index.

In a study of nineteenth century agriculture, I felt that such measures concealed and confused more than they revealed. Had I had any clear and operational understanding of the concept of a production function outside of the engineering limits of a single plant or plot of ground in which test conditions could be established and experiments could be performed, and of measuring "capital" as a cause and not a result of historical change, I could no doubt have strained credulity and conscience even further than I did, grubbed up the data somewhere, performed the necessary statistical operations and, throwing a bone to Kendrick's lion in the path, I could have tip-toed by.

My uneasiness about value measures, derived from the well-known paradoxes of long-run aggregation and measurement, was compounded in these studies by the many peculiarities of agriculture as an industry. Land—the farmer's principal asset—not only possesses an original and indestructible productive power, but also serves as a store of value, hence it serves as a retirement fund, a promise of future gains and an assurance of social position and good character. Its value is affected by all these considerations. Moreover, as an input, only the annual depletion of land and the expense of restoring its fertility should be measured. The same is true of land improvements, many of which—ditching, fencing, barn-building, road construction—are very long-lived indeed. Furthermore, this whole body of farm capital was created and maintained largely by farmers' own labor. Use of a single value measure of capital added to the usual problems of appropriate depreciation rates and price weights the problem of valuation of farm labor, much of it done in off-hours and at idle seasons in the farm year. Would it not be enough for my study of the nineteenth century then to discover how the productivity of that labor, applied to the tasks of capital formation and maintenance had changed over time?

I would then be working with a mental picture of the industry as shown roughly in Figure 1, below.

In such a treatment, if I could have further developed it, 'capital' would be both an output and an input in current production—a structure or medium, like skills or social organization, through which labor works, and so an intermediate element in the process of long-run output growth.

I was led into looking at labor productivity also through the related

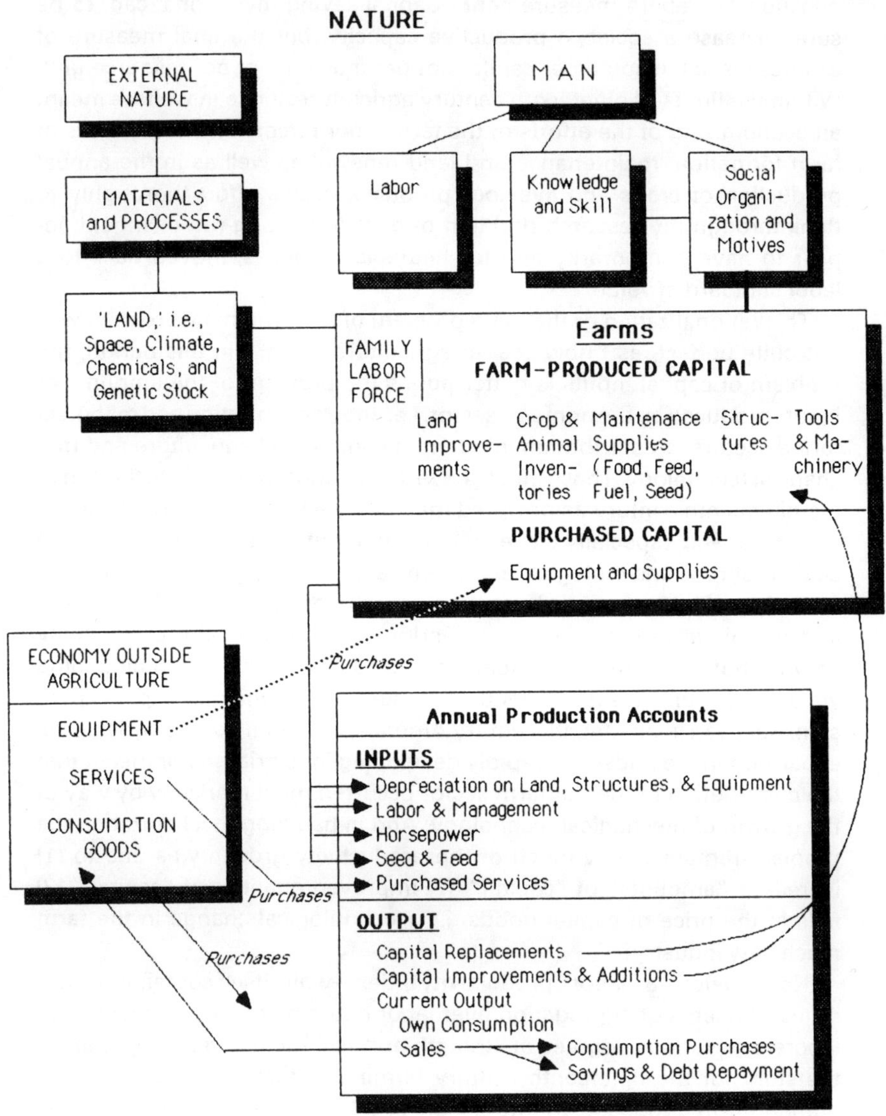

Diagram 1. Input & Output Flows in the 19th-Century Agricultural Sector

question of welfare measurement. 'Capital-saving inventions' can, to be sure, increase a society's productive capacity, but the final measure of progress is still income per capita, not per machine, or per resource unit. Within the limits of nineteenth century agriculture, a rise in welfare meant an economizing of the efforts of the farm labor force, in its heavy tasks of farm formation, maintenance and land renewal as well as in the annual production of crops and livestock products. In this effort thoroughly to think through my research problem of explaining long-run change, I appear to have, temporarily and for heuristic purposes, moved back to a labor standard of value.

This rationalization of the main problem of long-run measurement was not quite perfect, as I now see. In agricultural history in this period, the problem of capital inputs is better put as the problem of the amounts of inputs produced by the nonfarm sector, i.e., the transportation and manufacturing sectors, a contribution inherent in commercial agriculture and mechanical technology. These inputs have grown enormously since the end of the nineteenth century as powered machinery and tractors have replaced the horse, and especially since 1940 when inputs of land and labor have actually shrunk and biological and chemical technology have come in force to the farm. In the nineteenth century, a growth of knowledge of agronomy and of animal breeding and husbandry contributed to the productivity growth, both in permitting expansion on to new lands, and maintaining yields steady on all lands under the pressure of tremendous output expansion. But even in nineteenth century America, the setting of the agricultural expansion in the midst of a rapidly developing industrial sector meant that the contribution of the nonfarm sector occurred most markedly by way of the growth of mechanical technology. And in handling mechanization the problem remains: how much of the productivity growth was due to (1) increased "amounts" of "capital," (2) improved qualities of "capital," (3) falls in the price of capital goods, i.e., technological change in the farm machinery industry?

Nevertheless, a "labor" productivity index—while itself containing enormous problems of aggregation over labor hours of varying intensity, and laborers of various degrees of strength and skill—tells a very large part of the story for the nineteenth century farming. It measures the saving in human effort produced by the action of all the other "inputs" used by labor, both the purchased equipment and materials, and the invisible inputs of labor quality, technical knowledge, specialization and social organization. It is then in a sense a "total factor productivity" measure even though it does not permit a separate assessment of the "contribution" of its several tangible and intangible components.

APPENDIX B: PUBLISHED STUDIES

Atack, Jeremy and Fred Bateman, "Mid-Nineteenth Century Crop Yields and Labor Productivity Growth in American Agriculture: A New Look at Parker and Klein," *Technique, Spirit and Form in the Making of the Modern Economies;* Essays in Honor of William N. Parker. Research in Economic History, Supplement 3, Greenwich, Conn. and London, JAI Press, 1984, 215–242.

Atack, Jeremy and Fred Bateman, *To Their Own Soil: Agriculture in the Antelbellum North,* Iowa State University Press, Ames, 1987.

Bateman, Fred, "Improvement in American Dairy Farming, 1850–1910: A Quantitative Analysis," *Journal of Economic History,* Vol. XXVIII, June 1968, 255–273.

Bateman, Fred, "Labor Inputs and Productivity in American Dairy Agriculture, 1850–1910," *Journal of Economic History,* Vol. XXIX, June 1969, 206–229.

Fisher, Franklin M. and Peter Temin, "Regional Specialization and the Supply of Wheat in the United States, 1867–1914," *The Review of Economics and Statistics,* 52, May 1970, 134–149.

Parker, William N., "Productivity Growth in Grain Production in the United States: 1840–60 and 1900–10," with Judity L. V. Klein in *Output, Employment and Productivity in the United States after 1800,* National Bureau of Economic Research: Studies in Income and Wealth, XXX, 1966.

Parker, William N. "Productivity Growth in American Grain Farming: An Analysis of its 19th Century Sources," in Robert Fogel and Stanley Engerman, eds., *The Reinterpretation of American Economic History,* Harper & Row, 1971, 175–186.

Parker, William N., "A Note on Regional Culture in the Corn Harvest," *Agricultural History,* 46:1, Jan. 1972, 188–189.

Parker, William N., "Labor Productivityin Cotton Farming: The History of a Research," *Agricultural History,* 53:1, Jan. 1979.

Parker, William N. and S. DeCanio, "Two Hidden Sources of Productivity Growth in American Agriculture, 1850–1930," *Agricultural History,* 56:4, Oct. 1982, 648–662.

Primack, Martin L., "Land Clearing Under Nineteenth-Century Techniques: Some Preliminary Calculations," *Journal of Economic History,* Vol. XXII, December 1962, 484–497.

Primack, Martin L., "Farm Construction as a Use of Farm Labor in the United States, 1850–1910," *Journal of Economic History,* Vol. XXV, March 1965, 114–125.

Primack, Martin L., "Farm Fencing in the Nineteenth Century," *Journal of Economic History,* Vol. XXIX, June 1969, 287–291.

Whartenby, Franklee, *Labor productivity in Ante-bellum Agriculture in South Caroline,* Ph.D. dissertation, University of North Carolina, Dept. of Economics, 1961.

Capitalists Without Capital: The Burden of Slavery and the Impact of Emancipation

ROGER RANSOM

RICHARD SUTCH

> Where the capitalist outlook prevails, as on American plantations, this entire surplus value [of slave labor] is regarded as profit. . . . The price paid for a slave is nothing but the anticipated and capitalised surplus-value or profit to be wrung out of the slave.
>
> *Karl Marx, Capital*[1]

Karl Marx recognized the capitalist nature of American slavery long before American historians. The historians, of course, had always known that the essential economic feature of American slavery was that human labor had been capitalized. Black men and women were owned. They could be bought and sold, moved anywhere within the South at their owner's will, and put to any work their owner commanded.[2] But, what Marx also understood was that the slave holding existed to make a profit for the owner. The entire labor product of the slave family, above whatever provision for food and other necessities the owner cared to make, was expropriated. That residual was the owner's profit and the expectation of a continued flow of such returns made slave property an earning asset. The price paid for a slave reflected the consensus of the buyer and seller concerning the potential value of the continuous stream

ROGER RANSOM is Professor of History, University of California, Riverside. RICHARD SUTCH is Professor of Economics, University of California, Berkeley.

1. Karl Marx, *Capital: A Critique of Political Economy,* Vol. 3, "The Process of Capitalist Production as a Whole," Frederick Engles, ed. (London: 1894), 804.

2. For an extended discussion of these points see Richard Sutch, "The Treatment Received by American Slaves: A Critical Review of the Evidence Presented in *Time on the Cross,*" *Explorations in Economic History* 12 (October 1975): 335–438.

of profits that could be extracted from the slave and, in the case of a female, from her descendants as well.

Today, Marx would find no disagreement with American historians on these points.[3] However, this is because there has been a change of view. Until about thirty years ago, American historians denied that there was profit in slave ownership and rejected the idea that a capitalist outlook could have characterized plantation owners. American slavery was instead viewed as a vestige or throwback to feudalism and slaveholders were depicted as proud, even noble, paternalistic landlords tragically out of place in capitalist America.[4] What changed this perception was a successful demonstration based on a sizable collection of quantitative data that slavery was profitable.

Almost three quarters of a century after the publication of volume three of *Capital,* Alfred Conrad and John Meyer explained that the slave system was able to flourish in the midst of a capitalist market economy because it was itself a form of capitalism.[5] The southern plantation system, though, was an unusual species of capitalism in which the capital assets were land and slaves rather than physical or financial capital. Conrad and Meyer argued that slaves were viewed as an alternative to railroad bonds, agricultural land, textile factories, or other manufactories as an outlet for the investment funds of American capitalists. Since slaves were capital, the

3. See the thoughtful discussion of capitalism and slavery by Gavin Wright "Capitalism and Slavery on the Islands: A Lesson from the Mainland," *Journal of Interdisciplinary History* 17 (Spring 1987): 851–70. Even the attackers of the new orthodoxy, Robert Fogel and Stanley Engerman, agree that the slave plantation was a capitalist business enterprise organized for profit [*Time on the Cross,* Two Volumes, (Boston: Little Brown, 1974)]. Fogel and Engerman disagree with the rest of the profession about the significance of this fact for the treatment of slaves and the technical efficiency of slavery. For a full discussion of these issues see Paul David, Herbert Gutman, Richard Sutch, Peter Temin, and Gavin Wright, *Reckoning with Slavery: A Critical Study in the Quantitative History of American Negro Slavery* (New York: Oxford University Press, 1976).

4. See the writings of Ulrich B. Phillips, particularly his book, *American Negro Slavery: A Survey of the Supply, Employment and Control of Negro Labor as Determined by the Plantation Regime* (D. Appleton and Company, 1918). The American historians who have been most influenced by Marx's class analysis have generally stood away from the emerging consensus about the capitalist nature of slavery and tended to support Phillips' view. In this regard they differ with both Marx himself and most economic historians outside the Marxist tradition as well. See Eugene Genovese, *The Political Economic of Slavery: Studies in the Economy and Society of the Slave South* (Pantheon Books, 1965).

5. Alfred Conrad and John Meyer, "The Economics of Slavery in the Ante Bellum South," *Journal of Political Economy* 66 (April 1958). Every subsequent empirical study of American slavery has concluded that slave labor produced a significant surplus for the slave owner. In addition to Conrad and Meyer see Richard Sutch, "The Profitability of Slavery—Revisited," *Southern Economic Journal* 31 (April 1965): 365–77; Roger L. Ransom and Richard Sutch, *One Kind of Freedom: The Economic Consequences of Emancipation* (New York and London: Cambridge University Press, 1977), Appendix A; and Gavin Wright, *The Political Economy of the Cotton South: Households, Markets, and Wealth in the Nineteenth Century* (W.W. Norton, 1978), Chapter 3.

market price of slaves adjusted to bring the expected rate of return from an investment in slaves into equality with the expected rate of return from the alternative investments. Conrad and Meyer assembled quantitative evidence on prices, productivity, and costs establishing that the rate of return to slaveholders indeed equalled what could be earned from alternatives, about 6 to 8 percent per year.

In the 30 years since Conrad and Meyer's article appeared, numerous writers have introduced refinements, extensions, and new data to the analysis. All have employed some variant of the capital-asset pricing model introduced by Conrad and Meyer and all have invariably concluded that slave owners made at least a normal rate of return on their human property.[6]

The conceptualization of slavery as a form of capitalism and slaves as a form of "human capital" is easy to grasp as a logical proposition, but it gained its persuasive force through the accumulation of quantitative evidence. Figure 1 presents a time series of slave prices that spans the period from 1805 to 1860. These price data are developed in detail in the appendix to this paper and refer to the average price of all slaves: men, women, and children together. At the beginning of the period this value is estimated to have been in the neighborhood of $300. By 1860 it had risen as high as $800. There can hardly be a more dramatic demonstration of the profitability of slavery. Eight hundred dollars per slave is ten times as large as the gross value of annual crop output per capita in the South around 1857–1860.[7]

The discovery, or rediscovery, of the capitalist nature of slavery has proven to be a significant development. This breakthrough not only has prompted a complete and far-ranging reexamination of racial slavery in the Americas, it also has stimulated a renaissance in the field of American economic history more generally. While these developments can only be described as positive, the discovery of the capitalist nature of slavery has also had at least one unfortunate consequence. Perhaps because of the general association of capitalism with economic growth, high standards of living, and economic development, there has been a tendency to overlook

6. Major quantitative studies which bolstered and extended Conrad and Meyer's finding of profitability include: Robert Evans, Jr., "The Economics of American Negro Slavery," in Universities-National Bureau Committee for Economic Research, *Aspects of Labor Economics* (Princeton, N.J.: Princeton University Press, 1962): 185–243; Sutch, "The Profitability of Slavery"; and James Foust and Dale Swan, "Productivity and Profitability of Antebellum Slave Labor: A Micro Approach," *Agricultural History* 44 (January 1970): 39–62. One exception to the unanimity is the paper by Edward Saraydar, "A Note on the Profitability of Slavery," *Southern Economic Journal* 30 (April 1964): 325–32. However, Saraydar's calculations were shown to be in error in Sutch, a point ultimately conceded by Saraydar. These essays and several more are collected in Hugh G. J. Aitken, editor, *Did Slavery Pay? Readings in the Economics of Black Slavery in the United States* (Houghton Mifflin, 1971).

7. Ransom and Sutch, "Growth and Welfare in the American South in the Nineteenth Century," *Explorations in Economic History* 16 (April 1979): 213.

Figure 1. Price of an Average Slave

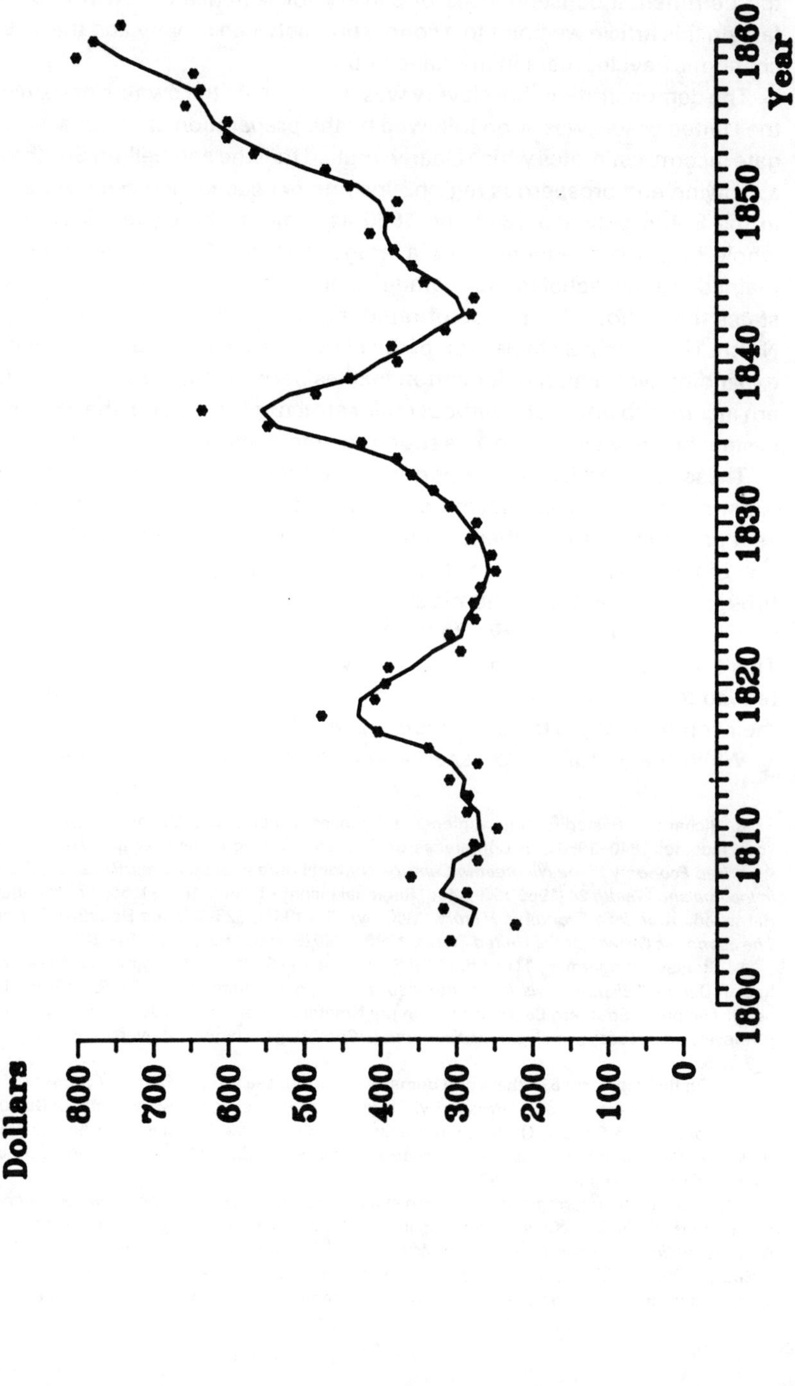

the detrimental consequences of slavery for American growth and welfare. In this article we point to a connection between slavery and the lack of economic development in the slave South.

The demonstration that slavery was profitable to the owners of slaves in the United States was soon followed by the preparation of regional aggregate income estimates which clearly implied that the antebellum South was a growing and prosperous region.[8] Income per capita increased as rapidly in the South between 1840 and 1860 as it did in the United States as a whole.[9] As a consequence, the average income of free southerners remained roughly equal to the average income of those living in the northern states throughout this period of rapid manufacturing development in the North. The principal staple crop of the plantation system was cotton and an expanding world market for cotton textiles permitted the growth of southern incomes to take place without major structural changes in the economic system or the development of a southern manufacturing industry.[10]

These discoveries are important to understanding the dynamics of American history. Had income per capita of free whites in the South not kept pace with their northern counterparts, it might have induced a migration of white labor to the North, have caused slave prices to fall, and have brought considerable economic stress to the slave economy. As it was, the two regions, free and slave, followed parallel but separate growth paths. There was insignificant migration between the two regions, slave prices tended to rise (as illustrated in Figure 1), and the slave owners intensified their commitment to the plantation regime.[11]

While the picture of prosperity and growing average income for the

8. Richard A. Easterlin, "Interregional Differences in Per Capita Income, Population, and Total Income; 1840–1950," in Conference on Research in Income and Wealth, *Trends in the American Economy in the Nineteenth Century,* National Bureau of Economic Research, *Studies in Income and Wealth* 24 (1960): 73–140; "Regional Income Trends, 1840–1950," in Seymour E. Harris, ed., *American Economic History* (McGraw Hill, 1961): 525–47; and Douglass C. North, *The Economic Growth of the United States, 1790–1860* (Prentice Hall, 1961), 91–94.

9. Stanley L. Engerman, "The Effects of Slavery on the Southern Economy: A Review of the Recent Debate," *Explorations in Entrepreneurial History* 4 (Winter 1967): 71–97; "Some Economic Factors in Southern Backwardness in the Nineteenth Century," in John F. Kain and John R. Meyer, eds., *Essays in Regional Economics* (Cambridge: Harvard University Press, 1971): 279–306.

10. On the importance of the world demand for cotton see Wright, *Political Economy of the Cotton South,* Chapter 4, and "Prosperity, Progress, and American Slavery," in Paul David, et al., *Reckoning with Slavery.* On southern manufacturing see Fred Bateman and Thomas Weiss, *A Deplorable Scarcity: The Failure of Industrialization in the Slave Economy* (Chapel Hill: University of North Carolina Press, 1981).

11. The data on interregional migration establishes the absence of labor flows between the two regions; Peter D. McClelland and Richard J. Zeckhauser, *Demographic Dimensions of the New Republic: American Interregional Migration, Vital Statistics, and Manumissions, 1800–1860* (New York and London: Cambridge University Press, 1982). As far as we can judge, there is no reason to believe that there were significant capital flows between the North and South.

white population of the South is statistically accurate, it would be wrong to interpret these facts as evidence of economic development in the South. The slaves, of course, did not share in this prosperity nor did they benefit from the growth. Moreover, the increases in southern incomes per capita— even when attention is focused exclusively on the welfare of the whites— was produced almost entirely by the continuous westward movement of population to new and more fertile land.[12] As a consequence, the growth within any of the subregions of the South was less than for the slave South as a whole. During this period, cotton production was not mechanized, agricultural methods did not change, and productivity on a given piece of land remained nearly constant.[13]

While westward movement required an implicit investment, inasmuch as slaveowners who chose to move west sacrificed a substantial portion of several years' output, such investment was fundamentally different from the type that was propelling northern economic growth. In the North, the stock of physical capital was continually expanding as investors constructed factories and installed machinery. As a consequence, labor productivity in manufacturing was considerably above that in northern agriculture and the manufacturing sector grew rapidly.[14] The productivity gains in the South were achieved not by moving workers from agriculture to an industrial sector made productive with machinery and new technology, but by moving workers to more productive soil. The difference is significant, since a process of physical capital formation could, in principle, continue indefinitely, whereas there is a natural limit to gains that can be achieved from geographical relocation. The South, therefore, grew but did not develop.

Slaveholders were capitalists without physical capital. Their wealth was in the form of slaves and land. Slave capital represented 44 percent of all

12. Richard Sutch, "The Breeding of Slaves for Sale and the Westward Expansion of Slavery, 1850–1860," in Stanley L. Engerman and Eugene D. Genovese, eds., *Race and Slavery in the Western Hemisphere: Quantitative Studies* (Princeton, N.J.: Princeton University Press, 1975), 173–210.

13. Engerman's figures show a growth rate in the South Atlantic Region of 1.21 percent per year and a 1.28 percent growth rate for the East South Central region, but a growth rate for the south as a whole of 1.67 percent ("Some Economic Factors in Southern Backwardness," Table 2, p. 287). We have estimated the rate of growth of crop output per capita to have been no more than 0.64 percent per year between 1839 and 1857; Ransom and Sutch, "Growth and Welfare," 212–17.

14. Paul A. David, "The Growth of Real Product in the United States Before 1840: New Evidence, Controlled Conjectures," *Journal of Economic History* 27 (June 1967): 151–97. Kenneth L. Sokoloff, "Productivity Growth in Manufacturing during Early Industrialization: Evidence from the American Northeast, 1820–1860," in Stanley L. Engerman and Robert E. Gallman, eds. *Long-Term Factors in American Economic Growth,* National Bureau of Economic Research, *Studies in Income and Wealth* 51 (Chicago: University of Chicago Press, 1986): 679–729. Also see the "Comment" in the same volume by Jeffrey G. Williamson, 729–36.

wealth in the major cotton-growing states of the South in 1859, real estate (land and buildings) was more than 25 percent, while physical capital amounted to less than 10 percent of the total. Manufacturing capital amounted to only 1 percent of the total wealth accumulated.[15] The lack of physical capital formation in the South reconciles the quantitative information on overall southern growth rates with the contemporary view that the South was economically backward compared to the Northeast.[16]

A more important point is that the relative shortage of physical capital in the South can be explained by the presence of slavery. In a capitalist society physical capital is owned by private entrepreneurs who are induced to invest in and hold capital by the flow of returns they hope to receive. In the American South slaves were an alternative to physical capital that could satiate the demand for holding wealth. In short, slaves as assets crowded physical capital out of the portfolios of southern capitalists.

There is a related argument about the relationship between slavery and manufacturing development that dates back to the days of slavery itself. Some contemporary observers argued that purchases of slaves, as they put it, "absorbed capital." Simply put, the proposition was that slave-owners invested their savings in slaves and therefore did not have the money available for other forms of investment.[17] Though at first glance this earlier argument appears quite similar to ours, it is actually rather different. Moreover it is wrong; it confuses physical capital with financial capital. The purchase of a slave does not destroy the money paid, it transfers the money to the slave's previous owner who would then be free to invest in physical capital if he so desired.

Conrad and Meyer attacked this old idea in their 1958 article by pointing out that "[i]t is difficult to see how the capitalization of an income stream,

---

15. The estimates are for the states of South Carolina, Georgia, Alabama, Mississippi, and Louisiana. They reflect a slight adjustment of our earlier calculations (Ransom and Sutch, *One Kind of Freedom,* 53) to incorporate new estimates of the value of the slave stock presented in the appendix to this paper.

16. The economic backwardness of the slave South has been stressed by Eugene D. Genovese, *The Political Economy of Slavery* and Douglas Dowd, "A Comparative Analysis of Economic Development in the American West and South," *Journal of Economic History* 16 (December 1956): 558–74. These writers were echoing earlier treatments by J. E. Cairnes, *The Slave Power: Its Character, Career, and Probable Designs: Being an Attempt to Explain the Real Issues Involved in the American Contest* (New York: Harper and Row, 1969), first published in 1862; and Robert R. Russell, "The General Effects of Slavery upon Southern Economic Progress," *Journal of Southern History* 4 (February 1938): 34–54.

17. For contemporary statements of this view see "A Carolinian" [Daniel Reaves Goodloe], *Inquiry into the Causes which have Retarded the Accumulation of Wealth and Increase of Population in the Southern States: In which the Question of Slavery is Considered in a Politico-Economical Point of View* (W. Blanchard, 1846); Frederick Law Olmsted, *The Cotton Kingdom,* edited with an introduction by Arthur Schlesinger (New York: Alfred A. Knopf, 1953), originally published in 1861, and Cairnes, *The Slave Power.*

excellent by contemporary standards, can be said to count as a loss of wealth." They concluded that "capitalization of the labor force did not of itself operate against southern development."[18] Ever since the absorption argument has received little support.[19] We are proposing a more sophisticated notion of absorption of capital.

In the version we advance, it is not the capitalization of the labor force itself, but the increase in assets caused by a growth of the slave population that displaces physical capital and operates against southern development. Our argument is that the growth of the slave population would depress the rate of saving (as conventionally defined). When slaves are wealth and the number of slaves is growing, an increase in the owner's wealth is automatic and the need to save from current income to augment wealth would be attenuated. Thus we argue that the growth in the value of the South's slave population would reduce the rate of growth of the physical capital stock within the region.[20] Consequently economic development would be retarded.

This proposition was first advanced by John Moes in an article which has since received very little attention.[21] Fogel and Engerman attempted to counter Moes's argument with one of their own. They suggested that there should be no reason why the existence of slavery would reduce the number of other investment opportunities or affect the expected rate of return on nonslave investments. To the extent that saving positively responds to the rate of return on investment, wealth creation in the form of increments to the slave population would simply be an additional investment opportunity that, in Fogel and Engerman's view, would have the effect of increasing the propensity to save.[22] Note, however, that this argument implicitly assumes a perfectly elastic supply of saving with respect to the rate of return. Since all empirical studies suggest that saving is inelastic—or at least nearly so—we fail to see the empirical relevance of their criticism.

18. Conrad and Meyer, "The Economics of Slavery," 81–82.

19. For additional criticism of this view from economic historians see Alfred H. Conrad, et al., "Slavery as an Obstacle to Economic Growth in the United States: A Panel Discussion," *Journal of Economic History* 27 (December 1967): 518–60; Robert W. Fogel and Stanley L. Engerman, "The Economics of Slavery," in *The Reinterpretation of American Economic History* (New York: Harper and Row, 1971): 336; George D. Green, *Finance and Economic Development in the Old South: Louisiana Banking, 1804–1861* (Stanford, CA: Stanford University Press, 1972), 60–61; and Bateman and Weiss, *A Deplorable Scarcity,* 74–76.

20. There is another reason why saving would be depressed by slavery. Slaves could not save for themselves and the slaveowner had no reason to save on their behalf. We do not explore the implications of this mechanism in this article, but the effect it would have, if significant, would reinforce the conclusions we draw.

21. John E. Moes, "The Absorption of Capital in Slave Labor in the Ante-Bellum South and Economic Growth," *American Journal of Economics and Sociology* 20 (October 1961): 535–41.

22. Fogel and Engerman, "The Economics of Slavery," 336.

Our contention that a growing slave population and advancing slave prices would reduce conventional saving, slow the rate of growth of the capital stock, and retard development is analogous to the argument that spending by government which is funded through the creation of national debt rather than by taxes will reduce conventional saving.[23] If, as is usually supposed, the saving-income ratio is independent of the rate of interest, then the displacement of physical capital will be dollar-for-dollar. Put somewhat differently, conventional economic theory suggests that there is a natural relationship between the flow of income and the stock of wealth usually expressed as an optimal wealth-income ratio. Because slaves were regarded as wealth they would displace other assets— particularly physical capital—from the portfolios of slaveowners. The growth of the slave population, therefore, would tend to "crowd out" new investment and slow the growth of the stock of physical capital.

Importation of slaves from abroad was prohibited beginning in 1808 but the slave population of the United States continued to grow at rapid rates due to natural increase.[24] Figure 2 displays the trend in the slave population. The data are described in greater detail in the appendix. The overall rate of population growth was over 2.4 percent per year. If, as the new historical consensus assumes, slaves were an investment included in the asset portfolio of the planter/entrepreneur, they helped satisfy the owner's demand for wealth. But unlike most other forms of capital, which depreciate with time, the stock of slaves appreciated. Thus, the growth of the slave population continuously increased the stock of wealth.

To obtain an idea of the order of magnitude of the expansion of slave wealth, we have estimated the total value of the slave population of the United States for each year for the period 1805 to 1860 and present the

23. The national debt argument is made by Franco Modigliani and is based on his well-known "life cycle hypothesis of saving": "Long-run Implications of Alternative Fiscal Policies and the Burden of the National Debt," *Economic Journal* 71 (December 1961): 730–55. Robert Hall first called attention to the equivalence of Modigliani's burden of the national debt and Moes' burden of slavery in a paper written in 1967. Although Hall's paper remains unpublished, it was discussed briefly at the Economic History Association's 1967 Philadelphia meetings; Conrad, et al., "Slavery as an Obstacle to Growth," particularly in the contribution by Sutch. The question is pursued more fully in Ransom and Sutch, "The Long-Run Implications of Capital Absorption in Slave Labor," National Economic Association, New York, December 1982. The Modigliani argument about the national debt is not without its critics, but note that the Moes-Hall argument is immune to the "Ricardian Equivalence" counterargument made by Robert Barro, "Are Government Bonds Net Wealth?" *Journal of Political Economy* 82 (November/December 1974): 1095–117.

24. Perhaps the increase was not so "natural." There has been a debate among economic historians concerning the practice of slave "breeding." The quantitative evidence on this point is presented in Sutch, "The Breeding of Slaves"; for a recent review of the debate on slave breeding see Richard Sutch, "Slave Breeding," in Randall M. Miller and John David Smith, eds. *Dictionary of Afro-American Slavery* (Greenwood Press, 1988).

Figure 2. Slave Population of the United States

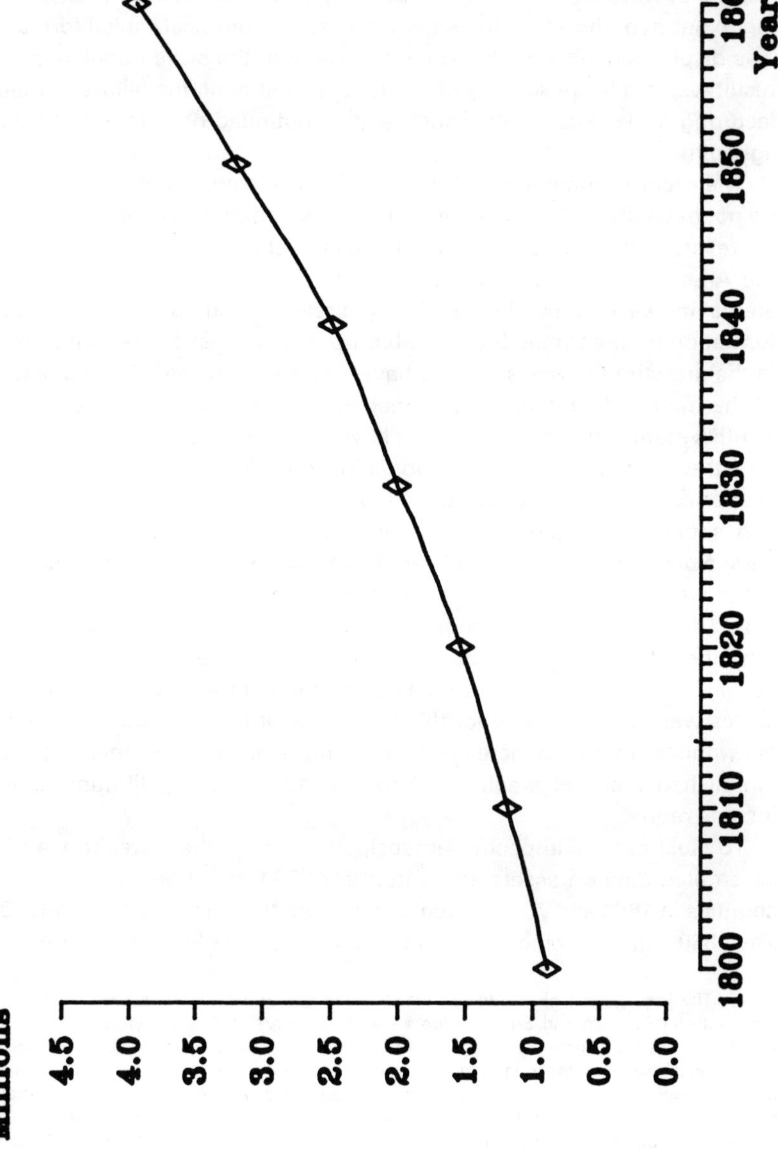

results in the appendix. Figure 3 presents the data graphically. The overall trend is upward and continuously so from 1843 onward. Each year, according to our hypothesis, an increment of potential physical capital formation was displaced by each advance in the value of the slave population. The result was a relative scarcity of funds for investment, the failure of manufacturing to develop in the South, and a continued dependency on slave agriculture.

The order of magnitude of this burden of slavery is not negligible. We do not have data on the flow of savings of southern slaveholders. We can, however, gain some appreciation of the impact which slaveholding had on the American economy as a whole by looking at the estimates of the annual increase in the slave stock together with estimates of total wealth formation for the United States. Table 1 presents these figures. Investment in the growth of slaves seems to have absorbed between 5 and 8 percent of the total additions to the national stock of wealth. The impact on the South's manufacturing sector must have been considerably greater.

If the growth of the slave population retarded capital formation and economic development before the Civil War, then elimination of this burden should have had a positive and permanent impact on the post-emancipation economy. Moreover, the means by which slavery was eliminated would have produced a transitory stimulus to saving that initially would have had an even more powerful impact on the rate of capital formation. With the stroke of a pen emancipation destroyed much of the "capital" the South had accumulated and neither the slaveowners nor the slaves were compensated for this loss of wealth.[25] In response southerners would have had to increase their saving rates to restore their suddenly diminished stock of wealth and to reestablish an equilibrium wealth-income ratio.[26]

To illustrate the impact of emancipation on southern wealth we have assembled data on assets and output for 570 farms from eight southern counties in 1860 and 755 different farms from the same counties in 1870.[27] The 1860 sample included 387 slave farms and 183 free farms. The slave

25. The implications of the way in which emancipation was accomplished is discussed more fully in Ransom and Sutch, "Who Pays for Slavery?" *Working Papers in Economics* Number 88-23 (Department of Economics, University of California, Berkeley, February 1988).

26. An indication of the magnitude of the effect on the wealth-income ratio is suggested by our calculation that the asset-income ratio for the country as a whole fell from approximately five in 1860 to about 3.7 in 1870; see Ransom and Sutch, "Domestic Saving as an Active Constraint on Capital Formation in the American Economy, 1839–1928: A Provisional Theory," *Working Papers in the History of Saving* 1 (December 1984): 56–58, Institute of Business and Economic Research, University of California, Berkeley.

27. A farm is defined as an agricultural unit producing at least $125 worth of crop output. In 1870, only farms that reported the value for personal estate of the farm operator were included, since the 1870 census appears to have underenumerated the data for personal wealth.

Figure 3 Aggregate Value of Slaves

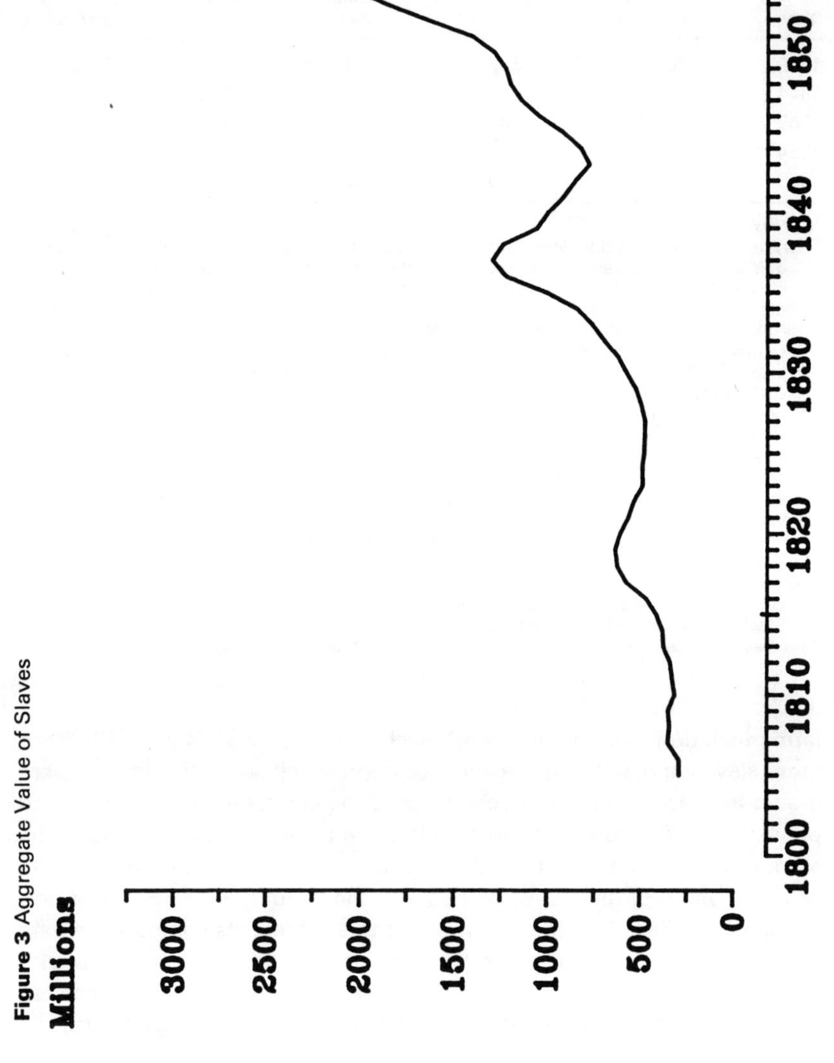

Table 1. Physical Capital Formation and the Value of the Increase in Slaves, 1839–1859, the United States

Year	Physical Capital Formation	*Millions of Current Dollars* Value of Increase In Slaves	Increase In Other Assets	Increase In Total Wealth	Slaves as Percentage of Total Wealth Formation
1839	$218	$22	$57	$275	8.1%
1844	230	21	78	308	6.7
1849	332	31	115	446	6.9
1854	631	43	167	798	5.3
1859	600	60	232	832	7.2

Sources:

Physical Capital Formation defined as Gross Private Domestic Capital Formation is the sum, of Manufactures Durables and New Construction taken from Robert Gallman's GNP figure plus the Change in Business Inventories estimated as one tenth of the decade change in the Stock of Inventories reported by Gallman and Howle. See Robert E. Gallman, "Gross National Product in the United States, 1834–1909," National Bureau of Economic Research, *Output, Employment, and Productivity in the United States After 1800,* Studies in Income and Wealth, Volume 30, (Princeton University Press, 196), Table A-3, p. 34; Gallman and Edward S. Howle, "The U.S. Capital Stock in the Nineteenth Century" (Unpublished paper, 1979); and Roger L. Ransom and Richard Sutch, "A System of Life-Cycle National Accounts; Provisional Estimates, Tables, and Source Notes to: 'Domestic Saving as an Active Constraint on Capital Formation in the American Economy, 1839–1928'," *Working Papers on the History of Saving,* Number 2 (December 1984), Institute of Business and Economic Research, University of California, Berkeley, Table C-3].

Value of the Increase in Slaves is calculated as the numerical increase in the slave population [column 1 of Appendix Table A-1] multiplied by the moving average value of a slave [column 5 of Table A-1].

Increase in Other Assets includes Net Foreign Investment, Purchases of Consumer Durables, Public Land Sales, and the Increase in Federal, State and Local Debt. See Ransom and Sutch, "A System of Life-Cycle National Accounts," Table E-1].

Increase in Total Wealth is the sum of the preceding three columns.

farms included 125 "plantations," each of which was a farm with 20 or more slaves and 100 or more acres of improved land. Both samples were drawn from the manuscript schedules of the censuses of agriculture and population.[28] The census returns for these years included questions on the amount of personal assets and real estate owned by each individual. In 1860 the value of the slaves owned was included in the category of personal estate. Based on these samples Table 2 presents comparative data on the average value of personal and total wealth. The level of wealth reported per farm in 1870 was slightly over one-tenth what it had been ten years earlier. As expected, the great decline was in personal estate; the

28. The farms selected were located in the following counties: Attala, Mississippi; Coweta, Georgia; Dallas, Alabama; Georgetown, South Carolina; Halifax, Virginia; Madison, Louisiana; Red River and Robertson, Texas. For a discussion of the sample collection procedure see Ransom and Sutch, *One Kind of Freedom,* 294–98.

Table 2. Wealth and Farm Output on Southern Farms, Current Dollars, 1860 and 1870, Eight Countries

	Number Farms Sampled	Average Value per Farm			Wealth-Output Ratio	
		Personal Estate	Total Wealth	Value of Output	Mean	Median
1860:						
All Farms	570	14,576	24,881	2,478	8.9	7.3
Plantations	125	45,394	81,609	7,905	9.0	9.7
Other Slave Farms	262	9,114	13,345	1,215	9.7	9.0
Non-Slave Farms	183	1,346	2,648	577	4.6	3.9
1870:						
All Farms	755	772	2,766	1,470	2.5	1.4
White Farms	555	976	3,682	1,702	3.4	2.1
Black Farms	200	203	226	827	0.2	0.2

Source: Sample of farms from the 1860 and 1870 Censuses of Agriculture and Population. See text for details.

Note: Total Wealth is defined as the sum of Personal Estate and Real Estate as reported in the population census. The Value of Output is defined as the sum of the value of field crops produced in the previous year plus an estimate of the value of the increase in the stock of slaves in 1860. The physical outputs of cotton, rice, tobacco, corn, wheat, rye, oats, cowpeas, Irish and sweet potatoes, barley, hay, molasses, and hemp were multiplied by estimates of the farmgate prices of these crops in 1859 and 1869 respectively. For 1859 the prices were as reported in Roger L. Ransom and Richard Sutch, *One Kind of Freedom: The Economic Consequences of Emancipation* (Cambridge University Press, 1977), Table F.4, p. 263, except for the prices of hay, hemp, and cowpeas which were reported in Marvin W. Towne and Wayne D. Rasmussen, "Farm Gross Product and Gross Investment in the Nineteenth Century," in *Trends in the American Economy in the Nineteenth Century,* National Bureau of Economic Research, Studies in Income and Wealth, Volume 24 (Princeton University Press, 1960), pp. 299, 305, and 309, and the price of Molasses which came from Charles E. Seagave, "The Southern Negro Agricultural Worker: 1850–1870," Ph.D. Dissertation, Economics, Stanford University, 1973, p. 112. Prices in 1869 were taken from Arden R. Hall, "The Efficiency of Postbellum Southern Agriculture," Ph.D. Dissertation, Economics, University of California, Berkeley, 1977, Appendix C, except for hay, hemp, and cowpeas which were from Towne and Rasmussen as noted above and tobacco and barley which were from U.S. Department of Agriculture, *Report of the Commissioner of Agriculture for the Year 1869,* pp. 26–28. The value of the increase in the slave stock was estimated as 2.12 percent of the number of slaves owned by the farm operator times the average value of a slave in 1859 ($801) as reported in the Appendix. The Wealth-Output ratio is calculated both as the average of the individual farm ratios and as the median for the sampled farms.

emancipation of slaves produced a decline from $14,576 for all operators in 1860 to a mere $772 in 1870.

We do not have income data for the farms in the sample, however we have made a rough estimate of the average value of output per farm in the two censuses including, in 1860, the value of the natural increase of the slave population. Using these estimates, we present two wealth-output ratios for the sample of farms in Table 2; one calculated as the the average of the ratios of wealth to output for each farm; the other as a median value. These estimates show that the wealth-output ratio for southern farms fell from over 7 to less than 3.5 during the decade. Allowing for the fact that these wealth-output ratios probably exaggerate the change in the wealth-income ratio, it is still clear that emancipation must have seriously disturbed the relationship between wealth and income.[29]

The loss of slave capital was not the only way emancipation reduced the value of assets. One of the more dramatic consequences of freedom from slavery was the partial withdrawal of labor services on the part of the black population.[30] This withdrawal had the effect of creating a labor shortage which made land a redundant resource. Consequently, the price of land dropped considerably. Table 3 presents data on the average value of farm land in our sample counties. Values declined dramatically. Improved acreage was worth only about one-half its value of ten years earlier.

Total wealth fell by the value of the slave stock and also because of the fall in the value of land. Nevertheless, the basis for continued income generation—the potential labor of the former slave population and the natural fertility of the soil—remained unharmed. Thus the wealth-income ratio fell far from the desired level. This undoubtedly stimulated saving by southerners although we do not have data to substantiate this directly. We can, however, see that the prediction is confirmed at the national level by statistics on aggregate capital formation estimated by Robert Gallman.[31] In Table 4 we reproduce two variants of his estimates, both are presented as shares of gross national product. The second estimate is generally preferred, but, by either measure there was a substantial increase in the rate of physical capital formation. In the same table we present the rate of growth of real gross national product. A clear acceleration after 1860 is evident.

29. The wealth-output ratios will tend to be higher than the true wealth-income ratio because the output measure fails to capture all income of the farm operator, or even all of the output generated by the farm. There is a further reason to suspect that the ratios in 1860 may not reflect an equilibrium relationship between wealth and income. Slave prices in 1860 were at their highest level and had been rising for over a decade. The resulting capital gains may have sent wealth-income ratios to abnormally high levels. We note in this regard that the wealth-output ratios for slave farms are in the neighborhood of eight or nine at a time when the average wealth-income ratio for the United States as a whole was around five.

30. Ransom and Sutch, "The Impact of the Civil War and of Emancipation on Southern Agriculture," *Explorations in Economic History* 12 (January 1975): 1–28; and *One Kind of Freedom,* 44–47

31. At the national level capital formation can be viewed as approximately the same as national saving since international capital flows were negligible. We should also note that Gallman has used his new estimates of the real capital stock to make rough esimates of the proportion of net product going to capital formation on a constant dollar basis. These estimates confirm that there was a substantial jump in the proportion coinciding with the Civil War decade (from 14.8 to 22.3 percent). Robert E. Gallman, "The United States Capital Stock in the Nineteenth Century," in Stanley L. Engerman and Robert E. Gallman, editors, *Long-Term Factors in American Economic Growth,* National Bureau of Economic Research Studies in Income and Wealth, Volume 51 (University of Chicago Press, 1986), Table 4.9, page 201.

Table 3. Average Size of Farm and Average Values of Improved and Unimproved Land, 1860 and 1870, Eight Counties in the South

Year	Number of Farms Sampled	Average Number of Acres per Farm	Value per Acre Current Dollars Improved Acre	Unimproved Acre
1860	624	217.1	$29.61	$5.81
1870	917	107.9	14.32	1.36
percentage change	+47.0	−50.3	−51.6	−76.6

Source: Sample of Farms, 1860 and 1870.

Note: All holdings with 10 or more acres of improved land and a reported farm value were included. The average value of improved land and unimproved land per acre was estimated separately for each county. The values reported here are a weighted average across the eight counties, using as weights the total number of improved acres in each county. The estimates are derived from census data on the value of the farm and the number of acres that were improved and unimproved on each farm. Our hypothesis is that the value per acre of improved land differed from the value per acre of unimproved land on each holding by an amount equal to the cost of clearing land. The cost of clearing land was assumed to be a constant for each county. To estimate the price of improved land per acre and the cost of clearing land we used least-square estimation methods. For a more complete explanation of the procedure and the estimating technique, see Roger L. Ransom and Richard Sutch, "Tenancy, Farm Size, Self Sufficiency and Racism: Four Problems in the Economic History of Southern Agriculture, 1865–1880," *Southern Economic History Project Working Paper Series* Number 8 (April 1970), p. 86.

We cannot be certain that the significant expansion of capital formation generated by emancipation took place in the southern states. Indeed, it probably did not. Certainly the South's economy did not experience economic growth comparable to that in other regions of the United States. In fact, the southern economy fell into a prolonged period of stagnation and painfully slow growth lasting well into the twentieth century.[32] Perhaps part of the explanation for this stagnation was the failure of capitalism to revive in the South following the war.

Emancipation destroyed the assets upon which the pre-war "capitalists without capital" had built their economic system. After the war they seemed unable to reestablish a capitalist system based on physical capital located in the South. As a consequence, postbellum southern saving financed capital formation outside of the region. Much of the explanation for this redirection, no doubt, can be attributed to the post-war economic expansion of northern industry and the opening of new channels facilitating interregional capital mobility. But, as we argued in *One Kind of Freedom,* the reconstruction of southern capitalism was blocked by a contradictory institutional structure erected during Reconstruction. The South failed

32. We discuss the post-war stagnation of the South in Ransom and Sutch, "Growth and Welfare in the American South."

Table 4. Share of Gross Capital Formation In Gross National Product, Current Dollars, and The Rate of Economic Growth, Decade Averages, 1839–1888, The United States

		Share of *Capital Formation*	
Decade	*Gallman*	*Davis &* *Gallman*	*Rate* *of* *Growth*
1839–1849	11.5%	12.1%	
1844–1854	12.9	13.9	5.0%
1849–1859	13.3	14.2	5.1
1869–1878	17.4	18.4	
1874–1883	17.3	18.2	5.7
1879–1888	18.9	18.7	6.0

Sources: Computed from the underlying sources using the methods described in Robert E. Gallman, "Gross National Product in the United States, 1834–1909," National Bureau of Economic Research, *Output, Employment, and Productivity in the United States After 1800,* Studies in Income and Wealth, Volume 30, (Princeton University Press, 1966): 3–90; and Lance E. Davis and Robert Gallman, "The Share of Savings and Investment in Gross National Product During the 19th Century in the U.S.A.," *Fourth International Conference of Economic History, Bloomington, 1968* (Mouton La Haye, 1973): 437–466.

to reestablish an effective banking system, white landowners created a system of agricultural sharecropping and mercantile finance that exploited black workers and sapped the initiative of whites as well as blacks, and southerners replaced slavery with an ugly and repressive social regime based on racial intolerance and discrimination. Faced with this inhospitable environment southern savings flowed North. Without a new base of physical capital, southern capitalism simply died.

APPENDIX
THE VALUE OF THE SLAVE POPULATION, 1805–1860

Our estimate of the value of the stock of slaves in current prices for each year from 1805 to 1860 is presented in Table A.1. The estimation procedure follows six steps corresponding to the six columns of the table.

(1) The total slave population each year, given in column 1, is estimated by interpolation between decennial census figures.

(2) The average selling price of prime-aged male field hands in New Orleans is reproduced in column 2.

(3) An index relating the average value of all American slaves to the price of prime-aged males in New Orleans is presented in column 3. The index adjusts for age, sex, location, and skill differences and is set with a base equal to 100 for the price of prime male in New Orleans each year.

Table A.1. Estimation of the Value of the Slave Stock, Millions of Dollars, 1805–1860, The United States

Year	Slave Population	Price of Prime Male in New Orleans	Index of Relative Value	Price of Average Slave	3-Year Moving Average	Value of Slaves
1804	1,002,545	700	44.0	308	—	—
1805	1,031,796	504	44.1	222	282	291
1806	1,061,901	719	44.1	317	275	292
1807	1,092,883	647	44.2	286	321	351
1808	1,124,770	813	44.2	360	306	344
1809	1,157,587	615	44.3	272	303	351
1810	**1,191,362**	624	44.3	277	265	316
1811	1,222,181	555	44.4	246	270	330
1812	1,253,798	643	44.5	286	272	341
1813	1,286,232	638	44.5	284	293	377
1814	1,319,506	694	44.6	309	289	381
1815	1,353,640	610	44.6	272	306	414
1816	1,388,657	753	44.7	337	337	468
1817	1,424,580	900	44.7	403	405	578
1818	1,461,433	1,065	44.8	477	429	627
1819	1,499,238	908	44.9	407	426	638
1820	**1,538,022**	875	44.9	393	397	610
1821	1,579,666	864	45.0	389	359	566
1822	1,622,437	650	45.2	294	331	536
1823	1,666,367	683	45.3	309	293	488
1824	1,711,485	606	45.4	275	287	491
1825	1,757,826	608	45.5	277	273	481
1826	1,805,421	588	45.7	268	264	477
1827	1,854,305	542	45.8	248	257	476
1828	1,904,513	551	45.9	253	261	497
1829	1,956,080	611	46.0	281	269	526
1830	**2,009,043**	591	46.2	273	287	577
1831	2,052,410	663	46.4	308	303	623
1832	2,096,713	707	46.7	330	332	697
1833	2,141,972	765	46.9	359	356	762
1834	2,188,208	800	47.2	378	387	847
1835	2,235,442	893	47.5	424	449	1,005
1836	2,283,696	1,146	47.7	547	535	1,222
1837	2,332,992	1,322	48.0	634	555	1,295
1838	2,383,351	1,002	48.3	484	519	1,237
1839	2,434,798	906	48.5	440	433	1,055
1840	**2,487,355**	773	48.8	377	401	997
1841	2,551,159	788	48.9	385	359	915

(Table A.1. continued)

(1)	(2)	(3)	(4)	(5)	(6)
1842	2,616,599	640	49.0	314	326	853
1843	2,683,718	569	49.1	280	290	778
1844	2,752,559	561	49.2	276	299	823
1845	2,823,166	692	49.4	342	325	918
1846	2,895,583	723	49.5	358	361	1,044
1847	2,969,859	771	49.6	382	384	1,141
1848	3,046,039	830	49.7	413	394	1,200
1849	3,124,174	776	49.8	387	392	1,225
1850	**3,204,313**	756	49.9	377	401	1,286
1851	3,272,371	878	50.1	440	429	1,405
1852	3,341,873	937	50.2	471	492	1,644
1853	3,412,853	1,122	50.4	565	545	1,862
1854	3,485,339	1,189	50.5	601	589	2,052
1855	3,559,366	1,185	50.7	600	619	2,203
1856	3,634,963	1,291	50.8	656	631	2,293
1857	3,712,169	1,249	51.0	636	646	2,397
1858	3,791,013	1,262	51.1	645	694	2,632
1859	3,871,531	1,564	51.2	801	741	2,870
1860	**3,953,760**	1,513	51.4	778	774	3,059
1861	4,037,735	1,440	51.5	742	—	—
	(1)	(2)	(3)	(4)	(5)	(6)

Sources: See text of Appendix.

(4) The New Orleans prices in column 2 are multiplied by the index in column 3 to produce the estimate of the average value of slave given in column 4.

(5) The price series in column 4 is smoothed by taking a three-year moving average. This series is displayed in column 5.

(6) The slave population in column 1 is evaluated using the price series in column 5. The result (in millions of dollars) is given in column 6.

The estimation procedures and their rationale are discussed in more detail below.

Column 1. The U.S. Census Office enumerated the slave population at each of the decennial censuses taken before the Civil War. The bold-face numbers for 1810, 1820, . . . , and 1860 in column 1 report the official figures. The slave population of the intercensus years is estimated assuming a constant rate of population growth between each pair of census dates. The growth rates used are given in Table A.2.

Column 2. The best-known and perhaps the most exhaustive study of

Table A.2. Growth Rate of the Slave Population, 1790–1860, The United States

Year	Enumerated Slave Population	Annual Rate of Growth
1790	697,624	
1800	893,602	2.51%
1810	1,191,362	2.92
1820	1,538,022	2.59
1830	2,009,043	2.71
1840	2,487,355	2.16
1850	3,204,313	2.57
1860	3,953,760	2.12

Source: U.S. Bureau of Census, *Negro Population, 1790–1915* (U.S. Government Printing Office, 1918), Table 6, p. 57.

slave prices was conducted by U. B. Phillips.[33] Phillips presented separate annual series for four geographical regions: Virginia, Charleston, middle Georgia, and New Orleans. Each series gave "approximate prices" of "young male prime field hands."[34] Phillips reported that his method was

> to select in the group of bills [of sale] for any time and place such maximum quotations for males as occur with any notable degree of frequency. Artisans, foremen and the like are thereby generally excluded by the infrequency of their sales, while the middle-aged, the old and the defective are eliminated by leaving aside the quotations of lower range.[35]

The sampling of numbers in Table A.3 has been visually estimated from the chart Phillips presented.[36] The years selected between 1810 and 1843 are the peaks and troughs in the New Orleans series. From 1844 until 1860, according to Phillips, slave prices rose continuously. An examination of the price series reveals a reasonably stable relationship between the four markets.

Robert Fogel and Stanley Engerman have used estate appraisal records to examine the relationship between slave values in different regions. They report for the period 1846–1855 that the average price of male

33. Ulrich Bonnell Phillips, "The Economic Cost of Slaveholding in the Cotton Belt," *Political Science Quarterly* 20 (June 1905): 257–75; "The Economics of Slave Labor in the South," in *The South in the Building of the Nation,* Volume 5 (Southern Publication Society, 1909): 121–24; *American Negro Slavery;* and *Life and Labor in the Old South* (Little Brown, 1929).

34. Phillips, *Life and Labor in the Old South,* 177, and *American Negro Slavery,* 370.

35. Phillips, *American Negro Slavery,* 370.

36. Phillips, *Life and Labor in the Old South,* 177.

Table A.3. Approximate Prices of Prime Field Hands in Four Markets, Selected
Years, 1800–1860
(Dollars per Slave)

Year	Peak or Trough	Virginia	Charleston	Middle Georgia	New Orleans
1800	—	380	500	480	520
1810	P	500	540	600	900
1813	T	400	450	500	600
1819	P	800	900	950	1100
1828	T	400	500	700	770
1837	P	1100	1200	1300	1300
1843	T	500	550	660	700
1848	—	650	720	900	950
1853	—	830	950	1200	1250
1859	—	1100	1200	1650	1690
1860	—	1200	1240	1800	1800

Source: Estimated visually from the chart in Ulrich Bonnell Phillips, *Life and Labor in the Old
South* (Little, Brown, 1929), p. 177.

slaves, aged 18–30, in Maryland, Virginia, North Carolina, and South Carolina was 73 percent of the value in Louisiana and indicate that this relationship held stable for the entire period from 1838 to 1860.[37] Because of the
apparent stability of the relationship between slave prices in different regions of the South, we have chosen to use a series on the price of slaves in
New Orleans to indicate the trend in the average price of all slaves.

Phillips' series of New Orleans slave prices has been widely used from
the time it was published by Conrad and Meyer.[38] Phillips based his estimates on bills of sale that are now in the New Orleans notarial archives.
Since his methods of sampling and averaging are questionable, Fogel and
Engerman drew a new sample from the same archives of approximately
5,800 sales covering the period 1804–1862. This sample has been used by
Engerman and Laurence Kotlikoff to estimate separate price series on
prime males.[39] Engerman's series covers all males, aged 18–30, fully guaranteed and without skill or handicap. Kotlikoff's series averages all males,
aged 21–38. As expected, these averages give figures substantially below
Phillips' maximum values. However, all three series move together. We
have used Engerman's estimates since his excludes slaves with skills and

37. Robert Fogel and Stanley L. Engerman, "The Market Evaluation of Human Capital: The
Case of Slavery," Cliometrics Conference, Madison, April 1972, pp. 8–9.
38. "The Economics of Slavery," Table 17, p. 76.
39. Laurence J. Kotlikoff, "The Structure of Slave Prices in New Orleans, 1804 to 1862,"
Economic Inquiry 17 (October 1979): 496–518.

Figure 4. Price of a Prime Male Slave New Orleans

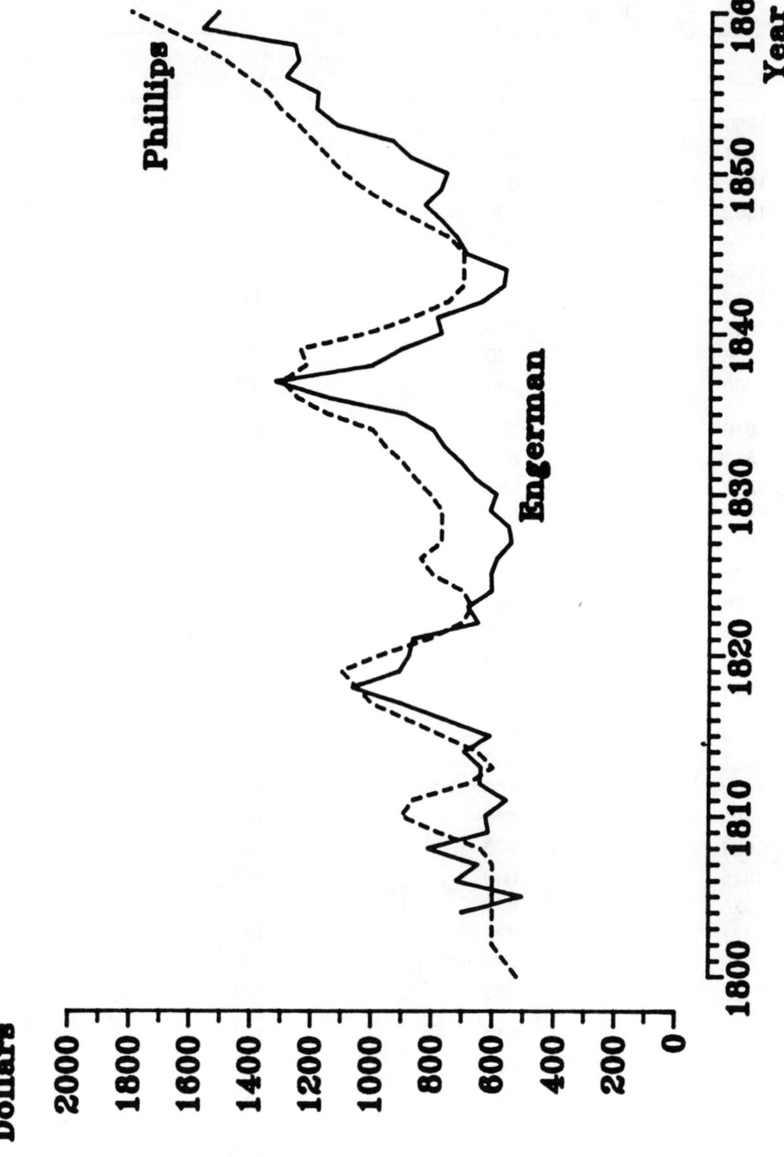

Table A.4. Prices of Prime Male Slaves, New Orleans, 1800–1862
(Dollars per Slave)

Year	Phillips	Price as Estimated by Engerman	Kotlikoff
1800	520		
1801	560		
1802	600		
1803	600		
1804	600	700	
1805	600	504	
1806	600	719	
1807	600	647	
1808	640	813	
1809	780	615	
1810	900	624	
1811	860	555	
1812	680	643	
1813	600	638	
1814	650	694	
1815	765	610	
1816	880	753	
1817	1,000	900	
1818	1,050	1,065	
1819	1,100	908	
1820	970	875	875
1821	810	864	762
1822	700	650	579
1823	670	683	618
1824	700	606	498
1825	800	608	603
1826	840	588	587
1827	770	542	568
1828	770	551	479
1829	770	611	596
1830	810	591	579
1831	860	663	652
1832	900	707	701
1833	960	765	797
1834	1,000	800	714
1835	1,150	893	881
1836	1,250	1,146	1,069
1837	1,300	1,322	1,263

(Table A.4. continued)

1838	1,220	1,002	897
1839	1,240	906	823
1840	1,020	773	800
1841	870	788	746
1842	750	640	608
1843	700	569	547
1844	700	561	547
1845	700	692	608
1846	750	723	709
1847	850	771	656
1848	950	830	797
1849	1,030	776	680
1850	1,100	756	697
1851	1,150	878	831
1852	1,200	937	878
1853	1,250	1,122	1,048
1854	1,310	1,189	1,130
1855	1,350	1,185	1,058
1856	1,420	1,291	1,085
1857	1,490	1,249	1,126
1858	1,580	1,262	1,175
1859	1,690	1,564	1,431
1860	1,800	1,513	1,451
1861		1,440	1,381
1862			1,116

Sources:

Phillips: Prices for 1800, 1801, and 1812 are estimated visually from Ulrich Bonnell Phillips, *Life and Labor in the Old South* (Little Brown, 1929), p. 177. All other figures are from Alfred H. Conrad and John R. Meyer, "The Economics of Slavery in the Ante Bellum South," *Journal of Political Economy* 66 (April 1958), reprinted in Alfred H. Conrad and John R. Meyer, *The Economics of Slavery and Other Studies in Econometric History* (Aldine, 1964), Table 17, column 6, p. 76.

Engerman: Data were supplied by Stanley Engerman. They are mean values of the prices included in a sample of invoices of slave sales held in New Orleans. The sample size for each year ranged between 2.5 and 5 percent. The prices averaged refer to "males ages 18 to 30, without skills, fully guaranteed as without physical or other infirmity." Engerman "utilized only those cases in which there was an individual price listed for a separate slave." For most years there were about 15 to 20 observations used in preparing the averages given.

Kotlikoff: The data are described in Laurence J. Kotlikoff, "The Structure of Slave Prices in New Orleans, 1804 to 1862," *Economic Inquiry* 17 (October 1979), Chart I, p. 498. The numbers on which the Chart was based are taken from Laurence J. Kotlikoff, "A Quantitative Description of the New Orleans Slave Market, 1804–1862," Robert W. Fogel and Stanley L. Engerman, editors, *Without Consent or Contract: Technical Papers on Slavery* (W.W. Norton, forthcoming). Kotlikoff included all males regardless of condition aged 21 to 38.

Table A.5. Age-Sex Profile of Slave Values, Index Numbers Louisiana Males Aged 18–30 = 100

	Old South		New South	
Age Cohort	*Male*	*Female*	*Male*	*Female*
Under 5	10.19	10.75	15.29	15.54
5–9	31.19	31.11	40.88	40.47
10–13	48.27	45.11	62.71	58.79
14	56.58	51.07	73.84	67.08
15–19	64.43	55.86	85.05	74.37
20–23	72.55	59.56	97.85	81.24
24–25	75.21	59.64	103.17	82.91
26–29	75.47	57.86	105.49	82.18
30–35	71.75	52.00	103.78	77.00
36–39	64.16	43.80	96.41	68.29
40–44	55.10	35.35	85.68	58.42
45–49	43.50	26.00	71.39	46.79
50–54	32.16	17.57	56.31	35.70
55–59	21.87	10.66	42.30	26.16
60 plus	7.29	2.45	24.16	14.46

Source: Data provided by Stanley Engerman. See Robert W. Fogel and Stanley L. Engerman, *Time on the Cross* (Little Brown, 1974), Volume I, figures 15, 16, and 18, pp. 72 and 76; Volume II, 79–82, for a description of the method used to derive the estimates. Also see Robert W. Fogel and Stanley L. Engerman, "The Market Evaluation of Human Capital: The Case of Slavery," Twelfth Cliometrics Conference, Madison, Wisconsin, April 1972.

defects, while Kotlikoff's does not. Furthermore, Engerman's series permits us to extend our estimates back to 1804.

Column 3. To convert the price of prime male field hands in New Orleans to an average price for all slaves, corrections must be made for geographical location, age, sex, skills, and handicaps. This was accomplished by estimating the number of prime-aged, New Orleans, male equivalents in the United States slave population at each census using age, sex and location specific price information to weight the enumerated population. The ratio of the prime-equivalent population to the actual slave population equals the ratio of the price of an average slave to a prime New Orleans male.

Fogel and Engerman have used their sample of estate appraisal records to estimate age-sex price profiles for slaves in the "Old South" (defined to include Maryland, Virginia, North and South Carolina) and in Louisiana for 1846–1855.[40] Fogel and Engerman indicate that the age-sex and geographi-

40. Fogel and Engerman, *Time on the Cross,* Volume I, figures 15, 16, and 18, pp. 72 and 76; Volume II, pp. 79–82, and "The Market Evaluation of Human Capital."

Table A.6. The Value of an Average Slave
by State, 1850
(Current Dollars)

States	Value
Delaware, Maryland, and District of Columbia	$300
Virginia, Kentucky, and Missouri	310
North Carolina and Tennessee	330
South Carolina and Arkansas	350
Georgia, Florida, Alabama Mississippi, Louisiana, and Texas	400

Source: Ezra C. Seaman, *Essays on the Progress of Nations* (Scribner, 1852), p. 619.

Table A.7. The Slave Population of the "Old" and "New" South, 1790–1860

Year	Population		Percentage of Total	
	Old South	New South	Old South	New South
1790	668,360	29,264	95.8	4.2
1800	830,707	62,895	93.0	7.0
1810	1,031,385	159,977	86.6	13.4
1820	1,232,770	305,252	80.2	19.8
1830	1,453,548	555,495	72.4	27.6
1840	1,485,308	1,002,047	59.7	40.3
1850	1,693,081	1,511,232	52.8	47.2
1860	1,817,722	2,136,038	46.0	54.0

Source: U.S. Bureau of the Census, *Negro Population, 1790–1915* (U.S. Government Printing Office, 1918), Table 6, p. 57. The "New South" is defined as Georgia, Florida, Alabama, Mississippi, Arkansas, Louisiana, and Texas plus the West North Central (includes Missouri), Mountain, and Pacific Census Regions. The "Old South" is defined as the United States less the New South.

cal patterns remained stable over the period from 1787 to 1860.[41] Accordingly, we have converted the age-sex profiles to relatives using the average appraised price of a Louisiana male between 18.5 and 29.5 years old as a index base equal to 100. The price relatives were calculated for age cohorts consistent with those used to report the ages of the slave population in pre-Civil War censuses. The results are displayed in Table A.5.

On the basis of the slave price data collected by Phillips (see Table A.3) and the estimates of the average value of slaves by state in 1850 made by Ezra Seaman in 1852 and reproduced in Table A.6, it seemed reasonable to

41. Fogel and Engerman, "The Market Evaluation of Human Capital," pp. 8–9, and *Time on the Cross,* Volume II, p. 79.

Table A.8. Comparison of the Value of an Average Slave Presented in Table A.1
with Estimates made by Different Methods
(Dollars per Slave)

| | *Value of Slave* | | |
Year	*Estimate from Table A.1*	*Alternative Estimate*	*Authority*
1805	$222	$220	Blodget
1850	378	357	Seaman
1860	778	771	Phillips
1860	778	778–782	Sholtow

Sources:

Samuel Blodget, Economica: A Statistical Manual for the United States of America (privately
printed, Washington D.C., 1806). In a social table "improved on the plan of Sir William
Petty" Blodget valued slaves employed by planters at $200 and slaves "variously em-
ployed" at $300 [p. 89]. He judged there were 800,000 slaves in agriculture and 200,000 in
other employment, thus giving $220 as an overall average.

Ezra C. Seaman, Essays on the Progress of Nations, (Scribner, 1952). Seaman gives averages for each state (see Table A.6) for 1850. The census of slaves in 1850 was used to
calculate the total value of all slaves and the average price given here; U.S. Bureau of the
Census, *Negro Population, 1790–1915,* (U.S. Government Printing Office, 1918), Table 6,
p. 57.

Ulrich Bonnell Phillips, Life and Labor in the Old South, (Little Brown, 1929). Phillips gives
the prices of prime field hands in Charleston and New Orleans (see Table A.3). Elsewhere,
American Negro Slavery (D. Appleton, 1918), Phillips stated that the "average price for
slaves of all ages and both sexes" was about one-half for slaves of all ages and both
sexes" was about one-half the price of prime field hands [p. 370]. Using the geographical
weights of the Old and New Souths from Table A.7 and the conversion factor of fifty
percent, the weighted average of Phillips' prices for Charleston and New Orleans gives a
figure of $771 for 1860.

Lee Soltow, Men and Wealth in the United States, 1850–1870 (Yale University Press, 1975).
Soltow shows that the value of personal estate returned to the Census Office in 1860
increased on average by $911 for each slave owned [p. 137]. Soltow's estimation tech-
nique will include in the $911 the value of the slave's personal effects and work tools,
including clothing, work animals, and the value of slave cabins. Roger Ransom and Richard Sutch, *One Kind of Freedom: The Economic Consequences of Emancipation* (Cambridge University Press, 1977), report census data to show that the average value of implements, machinery, and livestock totaled $111.71 per slave [Table A.3, p. 209] and estimate
the value of housing at $10 [Table A.2, p. 208]. The value of the slave's personal belong-
ings was probably not much more than the slave owner's annual expenditures for clothing and other semi-durable items provided slaves which ranged from $7.70 to $11.00 per
slave [Table A.5, p. 211]. Thus the total value of slave-related personal capital, according
to these estimates, ranged between $129.41 and $132.71. Subtracting this from $911 gives
an estimate for the value of an average slave of $778 to $782.

weight the Louisiana price relatives in Table A.5 by the slave population of
Georgia, Florida, Alabama, Mississippi, Louisiana, Arkansas, Missouri,
and Texas and the "Old South" price relatives by the remaining slave
population at each census. These definitions yield the weights displayed in
Table A.7. These weights were used to combine the "Old South" and
Louisiana price relatives in Table A.5 thus producing the indexes of relative value displayed for each census date in Table A.1.

The procedures just described yield estimates of the price of an average slave as a percentage of the price of a New Orleans prime-aged male which rise from 45 percent in 1820 to 51 percent in 1860. These proportions seem unreasonable. U. B. Phillips stated that the average slave was worth about 50 percent of the prime male price.[42] For the intercensus years relative value indexes were interpolated along a straight line between the census dates. The Census of 1820 was the first to report an age distribution of the slave population. The relative adjustment factor for 1804 was arbitrarily assumed to be 44 percent. The series used is displayed in column 3 of Table A.1.

Column 4. The adjustment factors in column 3 of Table A.1 were used to convert the New Orleans price series in column 2 into the estimated market value of a slave given in column 4 of that table. No adjustment was thought to be necessary for slave skills or handicaps since the two factors would work in opposing directions. In any case the number of skilled slaves in the total population seems to have been quite small.[43]

Our estimates of the value of an average slave are compared with estimates made by others using different methods of evaluation in Table A.8. In each of the three years for which such comparisons are made our estimate is above the alternative. In 1805 and 1860, however, the difference is less than 1 percent. For 1850 our estimate is about 5 percent above the figure suggested by Seaman.

Column 5. Column 5 displays a centered three-year moving average of the average price series in column 4. We used this smoothed series to evaluate the slave stock because it is likely that individuals discounted somewhat the year-to-year fluctuations when making evaluations of their slave wealth.

Column 6. Column 6 displays the estimated value of the U.S. slave population using the values given in column 5 and the population estimates in column 1.

42. Phillips, *American Negro Slavery,* 370.
43. Sutch, "Treatment of Slaves," 345–53; Herbert Gutman and Richard Sutch, "Sambo Makes Good, or Were Slaves Imbued with the Protestant Work Ethic?" in Paul David, et al., *Reckoning with Slavery,* 77–93.

Farmer Movements and Organizations:
Numbers, Gains, Losses

MORTON ROTHSTEIN

Much scholarly effort in the United States has been expended on describing and evaluating the scattered episodes of farmer protest, some of them marked by violence, that have from time to time challenged and disrupted our prevailing legal and political order. These flashes of violence and intimidation brought much attention, and often fear, but little redress of grievances. In some instances they marked shameful deviations from proclaimed national ideals of group or individual behavior. Yet dissent from authority is also part of the celebrated story of American pioneering, which not only carved farms out of the wilderness from one end of the continent to the other, but simultaneously laid a firm foundation for the nation's extraordinarily productive farm sector. The outbursts in which farmers used or threatened force seem in many respects aberrations from a generally peaceful, orderly path of development. They contrast sharply with the other kinds of organized efforts that absorbed the energies of most American farmers for the period since the mid-nineteenth century.

To be sure, some protests rocked several localities in the new nation at its very beginning, among them, the Carolina "Regulators," Shays's Rebellion in western Massachusetts, the less well-known disturbances in several Maryland rural districts, or the "Whiskey Rebellion," involving the refusal of settlers west of the Appalachians to pay the federal excise taxes on the one form of produce it paid them to ship to market. It was the impending loss of their land, and all that would mean to their livelihood, their security, and their families, that seems to have been the main cause

MORTON ROTHSTEIN is Professor of History, University of California, Davis and Editor of *Agricultural History.*

of protests in these and most other similar episodes both before and since the American Revolution.[1]

Yet we really know relatively little about the leaders or the rank and file in many of these episodes, the dynamics of the movements, or even whether or not they were truly agrarian in nature. Scholars have lavished attention on many of these movements, from Bacon's Rebellion in Virginia during the 1670s to the disruptive tactics of the NFO and the recent American Agricultural Movement, without giving us until fairly recently an account of the numbers involved in these movements or of the social origins of the leaders.[2] Some studies have seen even the late nineteenth century efforts at direct political action as simply isolated, irrational challenges to the larger society. Others, such as the sociologist Carl C. Taylor, have stressed the common elements in the diverse agrarian protests and see in them a continuous "Farmers' Movement."[3] This rather simplistic interpretation lumps to-

1. For a brief discussion of the "Regulator" movement, see Jack M. Sosin, *The Revolutionary Frontier, 1763–1783* (New York: Holt, Rinehart and Winston, 1967), 66–72. Richard M. Brown, *The South Carolina Regulators* (Cambridge: Harvard University Press, 1963) broke new ground in the study of such movements; his prospographic approach has also been utilized and extended in Rachel Klein, "The Rise of the Planters in the South Carolina Back Country, 1767–1808," (unpublished Ph.D. dissertation, Yale University, 1979). A. Roger Ekirch, "The North Carolina Regulators on Liberty and Corruption, 1766–1771," *Perspectives in American History* 9 (1977–1978): 199–256, has an extensive bibliography on the subject. No study of Shays's Rebellion yet provides a good estimate of the number of participants, the status of their leaders, etc. Captain Daniel Shays was a Revolutionary War veteran who led only one of the several groups of farmers participating in the armed struggle to postpone foreclosures on farms. A brief account of the incident is Alden T. Vaughn, "The 'Horrid and Unnatural' Rebellion of Daniel Shays," in Robert Land and John J. Turner, Jr., *Riot, Rout, and Tumult:Readings in American Social and Political Violence* (Westport, Conn.: Greenwood Press, 1978), 56–69. On the alleged connection between the episode and the adoption of the new constitution in 1787, see Robert A. Feer, "Shays's Rebellion and the Constitution: A Study in Causation," *New England Quarterly* 42 (Sept. 1969):388–410. For the revolt by debtor farmers in Maryland in 1786, and its effect of inducing the older political leaders in that state to support the new constitution, see William A. O'Brien, "Challenge to Consensus: Social, Political and Economic Implications of Maryland Sectionalism, 1776–1789," (unpublished Ph.D. dissertation, University of Wisconsin, 1979). Two recent books on the events in western Pennsylvania reveal that the resistance to federal tax agents was broader and lasted longer than earlier studies had indicated: Steven R. Boyd, ed., *The Whiskey Rebellion: Past and Present Perspectives* (Westport, Conn.: Greenwood Press, 1985) and Thomas P. Slaughter, *The Whiskey Rebellion: Frontier Epilogue to the American Revolution* (New York: Oxford University Press, 1986). The deep roots of such direct resistance are traced back three decades in Dorothy E. Fennell, "From Rebelliousness to Insurrection: A Social History of the Whiskey Rebellion, 1765–1802," (unpublished Ph.D. dissertation, University of Pittsburgh, 1981). For the concurrent, though longer and less violent, protest in Kentucky, see Mary K. Bonsteel Tachau, "The Whiskey Rebellion in Kentucky: A Forgotten Episode of Civil Disobedience," *Journal of the Early Republic* 2 (Fall 1982):239–59.

2. I have traced the self-help tradition in my "Technological Change and American Farm Movements," in Joel Colton and Stuart Bruchey, eds., *Technology, The Economy, and Society: The American Experience* (New York: Columbia University Press, 1987), 186–222.

3. Carl C. Taylor, *The Farmers' Movement* (New York: The American Book Co., 1953). The efforts of several rural sociologists to systematize the current knowledge about such move

gether the violent uprisings with concerted political upheavals or the more enduring programs for self-help and economic improvement that mark much of the farmers' own achievements, mostly on the local and state level.

Neither scholarly approach reveals a deep or widespread radical tradition among rural Americans. There is nothing in the national experience to compare in scope or intensity with the peasant revolts of the seventeenth and eighteenth centuries in China, Russia, France, and Central Europe, or the successive uprisings in the "Mother Country" of Great Britain, from Wat Tyler's outbreak to the era of Captain Swing.[4]

On the contrary, in our flexible, increasingly open and accessible political and social system, farmers obtained more and more leverage to help shape local and national policies, not to mention the rich factor endowments they could exploit in a broadening public domain. In a system that provided them with greater freedom as entrepreneurs and increasingly liberal access to resources, farmers had unique opportunities, and most seized them. From the outset they could quickly cast off the remnants of any peasant mentality they may have inherited or brought with them to the "New World," although admittedly some religious settler communities clung to the old ways. Anxious to accumulate a patrimony, most farmers responded to the incentives for commercial production with their own initiatives when it was possible, or in a frontier spirit of "association" with neighbors when it was needed or more convenient.[5]

---

ments, and to establish an agenda for further research is discussed in Denton E.Morrison, ed., *Farmers' Organizations and Movements: Research Needs and A Bibliography of the United States and Canada* (Research Bulletin 24, Agricultural Experiment Station, Michigan State University, 1970). For recent attempts by several thoughtful sociologists to revive and broaden this approach, see Carolyn Howe, "Farmers' Movements and the Changing Structure of Agriculture," in Eugene Havens, et al., eds., *Studies in the Transformation of U.S. Agriculture* (Boulder, Colo.: Westview Press, 1986), 104–49; and John W. Bennett, "Research on Farmer Behavior and Social Organization," in Kenneth A. Dahlberg, ed., *New Directions for Agriculture and Agricultural Research: Neglected Dimensions and Emerging Alternatives* (Totowa, N.J.: Rowman and Allanheld, 1986):367–402.

4. The most convenient overview of this subject is Yves-Marie Berce, "Rural Unrest," in Jerome Blum, ed., *Our Forgotten Past: Seven Centuries of Life on the Land* (London & New York: Thames and Hudson, Ltd., 1982):134–56. See also Roland Mousnier, *Peasant Uprisings in Seventeenth Century France, Russia, and China* (New York: Harper & Row, 1970); Paul Avrich, *Russian Rebels, 1600–1800* (New York: Schocken Books, 1972); Emmanuel Le Roy Ladurie, "Revoltes et contestations rurale en France de 1675 a' 1788," *Annales:Economies, Societies, Civilisations* (Jan.–Feb. 1974), 29th Annee' (No.1), 6–22; and Eric Hobsbawm and Georges Rudé, *Captain Swing: A Social History of the Great English Agricultural Uprising of 1830* (New York: Pantheon Books, 1968). For a study of a recent peasant movement in Latin America, and the contrast with those in both Europe and the United States during earlier stages of development, see Leon Zamosc, *The Agrarian Question and the Peasant Movement in Colombia: Struggles of the National Peasant Association, 1967–1981* (New York and London: Cambridge University Press, 1987).

5. The classic commentary on associations is in Alexis de Tocqueville, *Democracy in America,* 2 vols. (New York: Vintage Books, 1954) which makes a distinction between "political

Most farmers did not trust government, but as they gained personal influence over it at virtually every level through the widening use of the suffrage, the hallmark of the "Second Party System" in the Jacksonian era, they looked to officials for more liberal provisions for buying public land, for help in making it more productive, and for improved means of reaching markets. They usually got it. Between 1800 and the Civil War, federal land disposal policies grew more liberal, while many state and county governments and courts increasingly protected the rights of settlers to the assets they created. Local and state authorities throughout the nation supported schools, agricultural societies, fairs, and farmer clubs, all of which helped them share freely the fund of common knowledge accumulated over generations of clearing and tilling the land. Equally important, officials at all levels promoted "internal improvements," the building of an infrastruc-

---

associations" (I, 198–205) and "public associations," (II, 114–32). There is a continuing debate over the extent of commercialism among farmers in colonial Anglo-America and in the early national period. Influenced by James Scott's work on the "moral economy" of Asian peasants, and aspects of the so-called Brenner debate conducted in various issues of *Past and Present,* the case for resistance among colonial farmers to involvement in the cash nexus of markets has been championed by James Henretta in his "Families and Farms: *Mentalité* in Pre-industrial America," *William and Mary Quarterly,* 3d series, 35 (1978):3–32, and is supported by two recent doctoral dissertations, Bettye H. Pruitt, "Agriculture and Society in the Towns of Massachusetts, 1771: A Statistical Analysis," (Boston University, 1981) and Michael D. Merrill, "Self-Sufficiency and Exchange in Early America: Theory, Structure, Ideology," (Columbia University, 1983). Among the best arguments on the other side are James T. Lemon, "Early Americans and their Social Environment," *Journal of Historical Geography* 6 (1980):115–31 and his "Agriculture and Society in Early America," *The Agricultural History Review* 35 (1987)1:76–94; Carole Shammas, "How Self-Sufficient was Early America?" *Journal of Interdisciplinary History,* 13 (Autumn 1982)2:247–72; Winifred B. Rothenberg, "The Market and Massachusetts Farmers, 1750–1855," *Journal of Economic History* 41 (June 1981):283–14, with an ensuing exchange between her and Rona S. Weiss, ibid. 43 (June 1983):475–80; Bettye Hobbs Pruitt, "Self-Sufficiency and the Agricultural Economy of Eighteenth-Century Massachusetts," *The William and Mary Quarterly,* 3d Series, 41 (July 1984):333–64; and the imaginative use of new materials on motivations of migrants to the colonies in Bernard Bailyn, *Voyagers to the West: A Passage on the Peopling of America on the Eve of the Revolution* (New York: A.A. Knopf, 1986). See also the judicious discussion of tobacco farming in the intensive, quantitative study by Darrett B. Rutman and Anita H. Rutman, *A Place in Time: Middlesex County, Virginia, 1650–1750* (New York: W.W. Norton, 1984) I, 38–44, 69–93, with its interpretation of Bacon's Rebellion as an expression of frustrated hopes for greater material success. For a demonstration that farmer protests stemmed from highly differentiated and complex causes, even in a smaller colony, see Thomas L. Purvis, "Origins and Patterns of Agrarian Unrest in New Jersey, 1735 to 1754," *William and Mary Quarterly* 39 (Oct. 1982):600–27. In New Jersey the main source of felt grievance was confusion over land titles and surveys, which bred insecurity. Darrett Rutman addresses the problem of drawing conclusions about behavior and attitudes of all the social groups in colonial America, and of a prevalence of a "moral economy" in his "Assessing the Little Communities of Early America," *William and Mary Quarterly,* 3d Series, 43 (April 1986):169–78. For a later era, John Mack Farragher, *Sugar Creek: Life on the Illinois Prairie* (New Haven: Yale University Press, 1986) a study in the mode of the *Annales* school, with some quantitative analysis, examines the development of a small central Illinois village community from its condition as a "howling wilderness" to the transformation of its citizens' lives as they eagerly move into production for the market. A more general treatment of commercial orientation is in William N. Parker, "The American Farmer," in Blum, *Forgotten Past,* 182–96.

ture of roads, canals, and then railroads. Along with the telegraph, the transport revolution reduced the barrier of distance that had blocked access to markets. These projects were essential for the commerical growth on which settlers depended as they penetrated deeper into the great continental expanse.[6]

Both the number and average size of pioneer farms began to grow rapidly just as the nation also spurred its industrial growth. Consequently, the relative political strength of rural America declined only gradually. Farmers retained an important if no longer dominant place in most political calculations in terms of their numbers if not of their wealth. By 1850 the value of national manufactures surpassed the value of farm output, yet there was less of the "developmental squeeze" on agriculture, at least directly, than in the policies of most other societies. On the majority of American farms, whose numbers increased from about 300,000 in 1790 to roughly two million in 1860, owners held land in fee simple, used their own and their family's labor, and for the most part came close to fulfilling the national ideal of an independent yeomanry. Thus, for decades the nation avoided the kind of bitter class tensions inherent elsewhere in the modernization process, however we might define that term. In fact, the continued voting strength of farmers through much of the nineteenth century led one economist to wonder why they had not organized themselves with greater success as a separate interest group in the American polity long before the New Deal.[7]

Surely the reason is that until late in the nineteenth century there was

6. For the literature on antebellum political changes, and the new methods of measuring shifts from one "party system" to another, see Sean Wilentz, "On Class and Politics in Jacksonian America," in Stanley I. Kutler and Stanley N. Katz, eds., *The Promise of American History: Progress and Prospects,* the special issue of *Reviews in American History* 10 (Dec. 1982)4:45–63 and Allan G. Bogue, "The New Political History in the 1970's," in Michael Kammen, ed., *The Past Before Us: Contemporary Historical Writing in the United States* (Ithaca: Cornell University Press, 1980). On the spreading use of statistical techniques in various fields of American history, see J. Morgan Kousser, "Quantitative Social-Scientific History," in the same volume, 433–56. The protection by state and local courts of settlers' rights to the fruits of labor in putting up buildings and fences, and making other improvements on the land, is traced in Paul W. Gates, "Tenants of the Log Cabin," *Mississippi Valley Historical Review* 49 (June 1962):3–31. On the prevalence of farmers' clubs, county and state fairs, etc., see Lowell K. Dyson, *Farmers' Organizations* (in The Greenwood Encyclopedia of American Institutions; Westport, Ct.: Greenwood Press, 1986):1–2, 99–101; and Wayne C. Neely, *The Agricultural Fair* (New York: Columbia University Press, 1935). For insights into the manner in which a state farm organization operated, see Rodney H. True, "The Viriginia Board of Agriculture, 1841–43," *Agricultural History* 14 (July 1940):97–103.

7. Paul W. Gates, *The Farmers' Age: Agriculture, 1915–1960* (New York: Holt, Rinehart and Winston, 1960); Mancur Olson, Jr. *The Logic of Collective Action: Public Goods and the Theory of Groups* (New York: Schocken Books Paperback, 1968):148–59. For a recent study with quantitative techniques that verify the implicit thesis about relatively equal opportunity on the farming frontier, see Jeremy Atack and Fred Bateman, *To Their Own Soil: Agriculture in the Antebellum North* (Ames: Iowa State University Press, 1987).

little felt need among most rural Americans. The state had responded well to major agrarian pressures. From the beginning, the federal government opened frontiers to new settlement by "extinguishing" the titles of native Americans to the land coveted by invading whites (while almost extinguishing the native Americans themselves in the process), yet the same authority was easily ignored or thwarted in most of its attempts to protect the remaining property and the rights of the indigenous tribes. In 1832, many agrarian spokesmen from both the South and the West put such pressure on the Jackson administration for lower tariffs that they provoked a major constitutional crisis, but won their basic point. During Europe's "hungry forties," a Congress sympathetic to farm interests approved further tariff reductions to encourage the opening of overseas markets to American food as well as its fiber, a triumph that the coalition of sectional farm interests would not repeat until the work of the Farm Bloc in the early 1920s.[8]

Still, there were serious divisions within the agrarian ranks, and they got worse. On much of the South's best lands slaves provided the farm labor in a system becoming more and more repugnant to the rest of the nation and to many other peoples. By the 1850s, sectional conflict over the spread of slavery to the territories broke the farmer coalition. Northerners who wanted further expansion of opportunities for farmers clashed with southerners increasingly alarmed that such growth would end the delicate sectional balance of power that secured their "peculiar institution." It was no coincidence that during some of the worst fighting of the ensuing Civil War, a Congress no longer thwarted by southern representatives took time to launch the basic programs on their agenda in behalf of the nation's farmers. The Homestead Act, with all its flaws, lowered further the initial barriers to entry by making public land virtually free to settlers. A Department of Agriculture went into operation as the first special agency in the federal government to serve separately a major economic sector. The Morrill Land Grant Act endowed each state with land from the Public Domain in proportion to its representation in Congress for the support of colleges to teach and conduct research in agricultural and mechanical arts. These social inventions, like the subsidies to the transcontinental railroads, were in large part intended to spur the extension of commercial farming wherever feasible in the nation's interior.[9]

8. For an influential traditional historian's summary of these political currents, see the discussion of transport, tariffs, and land policies in Frederick Merk, *History of the Westward Movement* (New York: A.A. Knopf, 1978):214–39. Useful insights are also offered by a noted agricultural economist, and ex-official of the USDA, in Willard W. Cochrane, *The Development of American Agriculture: A Historical Analysis* (Minneapolis: University of Minnesota Press, 1979):57–76.

9. One of the best discussions of these achievements, followed by a chapter on the "agrarian groundswell" already under way at the time, is in Paul W. Gates, *Agriculture and the Civil*

Some of the results can be seen in the greater numbers of those farms, which reached almost 6 million by 1920. Such rapid expansion, indeed over-expansion, of farms and acreage had few parallels, even in an age that opened new "empty lands" on several continents, and world prices for staple commodities responded accordingly. No wonder that in this period spokesmen for farmers who felt distressed began their "agrarian crusade" against a system that then seemed unable to address the problems generated in a more competitive market economy. This series of organized protests by some farmers and their supporters after the Civil War expressed many grievances and proposed a number of solutions, though few dealt with the basic problem of surplus production. In the standard accounts, they also were the main support of the challengers in the presidential election of 1896, in which agrarian hopes, by then largely fused with those of other reform groups, were dashed. After this defeat, the story goes, farmers on the whole turned their political efforts to narrower goals of economic self-interest. As a pressure group overrepresented at many levels in the political system, the "farmer movement" no longer seemed to be in the vanguard of progressive reform. For a generation or more, however, historians have disagreed over whether the Alliance and Populist movements in particular represented a last chance for the true victory of democratic ideals over industrial capitalism, or simply laid some groundwork for the rather moderate reforms of the Progressive and New Deal eras.[10]

Differences over interpreting the Populists, in particular, have wracked a segment of the historical profession in the United States since the early

War (New York: A.A. Knopf, 1965):251–355. A sophisticated recent study of the various alignments in the Senate for these and other measures, and the reasons for the votes, including those on major amendments, is Allan G. Bogue, *The Earnest Men: Republicans of the Civil War Senate* (Ithaca: Cornell University Press, 1981).

10. The first general scholarly study on a national scale of the late nineteenth century farmer unrest was Solon J. Buck, *The Agrarian Crusade* (New Haven, Conn.: Yale University Press, 1980), in which he built on his standard monograph, *The Granger Movement: A Study of Agricultural Organization and Its Political, Economic and Social Manifestations, 1870–1880* (Cambridge, Mass.: Harvard University Press, 1913). The first major scholarly work on the Populist Party, and its predecessor farmer organizations, is John D. Hicks, *The Populist Revolt: A History of the Farmers' Alliance and the People's Party* (Minneapolis: University of Minnesota Press, 1931). For a recent summary of the subject in the Hicks tradition, based on his own considerable research, see Theodore Saloutos, "Farmers' Movements," in Glenn Porter, ed., *Encyclopedia of American Economic History: Studies of the Principal Movements and Ideas* (3 vols.; New York: Charles Scribner's Sons, 1980) II:562–74. A brief, but very useful collection of documents, with some excerpts from interpretations, is Vernon Carstensen, ed., *Farmer Discontent, 1865–1900* (New York: John Wiley & Sons, 1974). There is a judicious discussion of the 1896 election in Paul W. Glad, *McKinley, Bryan, and the People* (Philadelphia: J.B. Lippincott Co., 1964). A good introduction to the differences between the many assessments that appeared in the 1950s and 1960s is Sheldon Hackney, *Populism: The Critical Issues* (Boston: Little, Brown & Co., 1971).

1950s. The debates shed considerable heat but not much light on the nature of the movement, its successes and failures, its relationship to the various forms of racism, antisemitism, imperialism, proto-fascism, and demagoguery that have appeared from time to time in America. Over the last two decades, ideological differences have persisted. Much of the recent debate is as sharp as ever, but also informed by many younger scholars who have been trained in the new, more rigorous quantitative methods.[11]

Our understanding of the National Grange of the Patrons of Husbandry, known simply as the Grange, has been altered over the last generation by this research. It was the first general, national farmer organization, closely associated in the 1870s with two new efforts to address specific economic grievances. One was the set of so-called "Granger Laws" that several states of the upper Midwest used to regulate the railroads and warehouses within their borders. Second was the major effort in many states and localities to build cooperatives to supplant the middlemen with whom farmers had to deal as they relied more on markets. There were some interesting precedents for both efforts in the fact-finding railroad commissions of Massachusetts and some other states, and in the "associations" of dairymen in New York.[12] It also now seems clear that some of the early midwestern laws dealing with railroads and grain elevators, notably in Illinois, were actually passed at the behest of those merchants in the terminal markets whose functions were being displaced by the consolidation of

11. Theodore Saloutos, "The Professors and the Populists," *Agricultural History* 40 (Oct. 1966):235–54 is a thorough survey of the critical literature while defending the Hicks tradition. An analytical review of more recent work, from a sociologist's perspective, is in Donna A. Barnes, *Farmers in Rebellion: The Rise and Fall of the Southern Farmers Alliance and People's Party in Texas* (Austin: The University of Texas Press, 1984):8–50. Its regional bias should be balanced with the historiographic survey by William C. Pratt, "Radicals, Farmers, and Historians: Some Recent Scholarship about Agrarian Radicalism in the Upper Midwest," *North Dakota History* 52 (Fall 1985):12–26.

12. The classic article on earlier efforts to regulate the railroads is Frederick Merk, "Eastern Antecedents of the Grangers," *Agricultural History* 23 (Jan. 1949):1–8. A more detailed study, which set new standards in meticulous revisionism, is Lee Benson, *Merchants, Farmers, and Railroads: Railroad Regulation and New York Politics, 1850–1877* (Cambridge, Mass.: Harvard University Press, 1955). An important survey of scholarship on the broader subject of business regulation in the period is Thomas K. McCraw, "Regulation in America: A Review Article," *Business History Review* 49 (Summer 1975):159–83. An interesting effort to place the demand for action by government in theoretical terms is Thomas S. Ulen, "The Market for Regulation: The ICC from 1887 to 1920," *American Economic Review* 70 (May 1980):306–10. The emergence of producers' cooperatives in New York that seem to have provided models for later organizations in California is traced in H.E. Erdman, "The 'Associated Dairies' of New York as Precursors of American Agricultural Cooperatives," *Agricultural History* 36 (April 1962):82–90. For a more favorable view of Granger coops than Buck's, see George Cerny, "Cooperation in the Midwest in the Granger Era, 1869–1875," *Agricultural History* 37 (Oct. 1963):187–205. The detailed, authorized history of farmer cooperatives is Joseph G. Knapp, *The Rise of American Cooperative Enterprise: 1620–1920* (Danville, IL: Interstate-Printers and Publishers, 1969).

railroads and shipping firms. Only after passage of the Illinois law in 1870 did the Grange suddenly attract members. Farmers obviously connected the Patrons of Husbandry with visible success in curbing the power of railroad companies.[13]

The pattern of rapid growth and equally rapid decline of the "first Grange," as its most recent historian has named the movement of the 1870s, seems plain enough. Its membership soared in the early 1870s, much of it being concentrated in the Midwest. It reached a peak about 1875, then virtually collapsed. Most of its farmer members had become disillusioned with regulatory efforts, since the railroads avoided effective state action by various devices. In addition, virtually all of the Granger cooperatives had quickly failed. Most members simply abandoned the movement. By the 1880s the Order's remaining lodges, located for the most part in the northeastern quadrant of the nation, worked primarily for social and educational improvements in rural life. These goals had been the original ones that had concerned several of the order's founders, who had also opposed involvement in politics as too divisive. Thereafter, except for reducing some of the cultural isolation of farmers and their families, the Grange did little to address pressing economic concerns, related primarily to the felt decline in commodity prices and the burdens of higher real interest rates, that most farmers confronted at the end of the nineteenth century.[14]

Recent estimates of membership in the major farm organizations verify parts of this story. From a mere handful of local lodges in 1870, the number of Grangers increased to 141,000 families in 1874, then to almost 452,000 in 1875, then fell to about 63,000 in 1880. One study shows that almost half the membership in 1875 were in the Midwest. But it defined the region broadly enough to include Missouri, the leading Granger state by far with over 2,000 lodges and 80,000 members, accounting for more than 25 percent of that state's farm families. Illinois, supposedly the archetypal Granger state, had in 1875 a State Farmers' Association which had recently united the many

13. A pioneering revisionist work on the background of the Illinois law regulating railroads and warehouses is Harold D. Woodman, "Chicago Businessmen and the Granger Laws," *Agricultural History* 36 (Jan. 1962):16–24. "Granger" regulation had much the same origin in Wisconsin, as shown in Dale Treleven, "Railroads, Elevators, and Grain Dealers: The Genesis of Antimonopolism in Milwaukee," *Wisconsin Magazine of History* 52 (Spring 1969):205–22. A meticulous, thorough legislative history of these laws in four upper Mid-west states (showing the same business sources) is George H. Miller, *Railroads and the Granger Laws* (Madison: The University of Wisconsin Press, 1971).

14. The only recent scholarly full-length book on the organization is D. Sven Nordin, *Rich Harvest: A History of the Grange, 1867–1900* (Jackson: University Press of Mississippi, 1974). On the divisions over goals among the founders of the organization, which was patterned in its structure and rituals on such fraternal orders as the Masons (though it always admitted women to full standing), see William D. Barns, "Oliver Hudson Kelley and the Genesis of the Grange: A Reappraisal," *Agricultural History* 41 (July 1967):229–42.

politically unaffiliated "farmer clubs" and claimed a membership of 80 to 90,000, more than double the state's Grange adherents. The true geographical center of the movement, therefore, was in the Ohio River Valley and in Missouri, places where the adjustment problems for farmers may have been more severe than on the newly settled prairies and plains.[15]

It is possible that the figures of a membership so ephemeral and scattered across so many states do not in any case measure the real support for or influence of the organization. By one count, the Far West states, including California, Colorado, Montana, Idaho, Oregon, Washington and Wyoming, together did not muster more than 15,000 members in 1875, or a mere 5,000 after 1880. Yet in California, the Grange was the leading agrarian reform organization in the 1870s, challenging with some success the monopolistic wheat marketing practices of the time and launching several cooperatives that endured. In Washington, too, the Grange stood fast as the major spokesman for discontented farmers in the state's politics and served during the 1880s as the direct antecedents of the Populist Party in the rural districts. In such southern states as Texas and Mississippi, on the other hand, Grange leaders were either hopelessly divided about involvement in partisan politics, or too weak politically to seek anything but limited economic goals.[16]

In only a few instances has there been much quantitative analysis of the social origins and economic status of rank and file members or leaders in the later nineteenth century farmer movements. Unique insights into the kind of people who participated in these movements, why they joined and

15. The first effort to gather data on the numbers of those who belonged to this and other national farm groups is Robert L. Tontz, "Memberships of General Farmers' Organizations, United States, 1874–1960," *Agricultural History* 38 (July 1964):143–53. A new set of estimates of selected farm organizations, many of them at variance with Tontz and including several groups he did not cover,is in Lowell K. Dyson, *Farmers' Organizations* (The Greenwood Encyclopedia of American Institutions; Westport, Ct.: Greenwood Press 1986), Appendix 3, 367–69. More detailed figures on the 1875 Grange by each state is available in Nordin, *Rich Harvest,* 29–31, 41–43. For the later history of Illinois' divided farmer politics, see Roy V. Scott, *The Agrarian Movement in Illinois, 1880–1896* (Illinois Studies in the Social Sciences, vol. 521; Urbana: The University of Illinois Press, 1962). For the context of the farmer movement in the largest Grange state, see Nick Adzick, "Agrarian Discontent in Missouri, 1865–1880: The Political and Economic Manifestations of Agrarian Unrest," (unpublished Ph.D. dissertation, St. Louis University, 1977).

16. Rodman W. Paul, "The Great California Grain War: The Grangers Challenge the Wheat King," *Pacific Historical Review* 27 (Nov. 1958):330–49; Gordon B. Ridgeway, "Populism in Washington," *Pacific Northwest Quarterly* 39 (Oct. 1948):284–311; Robert A. Calvert, "A.J. Rose and the Granger Concept of Reform," *Agricultural History* 51 (Jan. 1977):181–96; James S. Ferguson, "Co-operative Activity of the Grange in Mississippi," *Journal of Mississippi History* 4 (Jan. 1942):3–19. For recent studies of state Granges, see Olin B. Adams, "The Negro and the Agrarian Movement in Georgia, 1874–1908," (unpublished Ph.D. dissertation, Florida State University, 1973); Robert E. Culbertson, "Florida's Patrons of Husbandry," (unpublished Ph.D. dissertation, University of South Carolina, 1975); Rexford B. Sherman, "The Grange in Maine and New Hampshire, 1870–1940," (unpublished Ph.D. dissertation, Boston University, 1973).

why they left, as well as into their behavior while they were in it, depends on study at the county or township level. For example, an older study of the Grange in a single Illinois county turned up enough data to indicate that the county membership and the number of its locals were well above the state average in the mid-1870s. Journalistic sources showed a strong sense of relative economic deprivation among the farmers, although the county's agricultural output was on the whole relatively high. Its tillers and livestock owners were still struggling with difficult adjustments to more capital-intensive operations. Consequently, they also bore a heavier average load of mortgage debt on their land than their counterparts in almost any county in the state. The leaders, farmers who founded the Grange lodges, or joined them early, generally owned more assets, and showed more stability in their length of residence, than the average rank and file member. Such leaders were eager to start cooperative stores; but the rank and file who flocked to support them also deserted the movement quickly when many of the stores failed early in 1875.[17]

This pattern was to a large extent characteristic of many other Granger strongholds. An extensive sample study of Wisconsin shows that its Grange leaders and members were also fairly well-to-do, middle-class types. Members on average were about 40 years old, mostly native-born, and still largely relied on frontier-style wheat production for their main income. The officers of the local granges were about ten years older, owned somewhat larger farms, raised more wheat and owned more livestock. Yet as a group, all Grangers had assets and incomes considerably lower in value than that of the average member in the two other farmer organizations in the state, the Wisconsin State Agricultural Society and the Dairymen' Association. These two groups had no interest in dissent or in overt political activity; they worked closely instead with the state's business and political leadership to help their members move more effectively into new, more profitable specialities, mostly butter and cheese production. This upper strata of farmers sustained an informal network between their own leaders, the influential members of the state legislature (and

17. Roy V. Scott, "Grangerism in Champaign County, Illinois, 1873–1877," *Mid-America* 43 (July 1961):139–63. The minutes of an unaffiliated farmers' club in central Texas, covering virtually a century, offers similar insights into the concerns and activities of a strictly nonpolitical group, in Cat Spring Agricultural Society, *The Cat Spring Story* (San Antonio, Texas: Lone Star Printing Co., 1956). The existence of such active bodies at the local level helps account for the speed with which the Grange grew. Additional insights into the social concerns that contributed to the revival of the Grange can be found in Donald B. Marti, "Woman's Work in the Grange: Mary Ann Mayo of Michigan, 1882–1903," *Agricultural History* 56 (April 1982):439–52. On the other hand, in some states the Grange inspired much political dissent; in Arkansas, helped promote the Greenback party and gave it much support for a decade, before turning to rather more radical new movements. See Judith Barsenbruch, "The Greenback Political Movement: An Arkansas View," *Arkansas Historical Quarterly* 36 (Summer 1977):107–22.

with its congressmen), and with the more important figures in the College of Agriculture. In several respects, this coalition of interests was effective in obtaining continued support for the College and the experiment station. They were confident that this emphasis on working within the system would directly help them help themselves.[18]

Such distinctions between groups of farmers, based on the matching of membership rolls, manuscript census returns (both for population and for agricultural production), and land and tax records, in addition to intensive use of local newspapers and other traditional sources, also yield interesting and similar results for larger states. The best-known Grange leader in California, John Bidwell, owned a 10,000-acre ranch in the Sacramento valley, and was taking pains to diversify his grain crops by growing fruit and nuts, even as he led efforts to out-maneuver grain merchants in San Francisco. By contrast, further south in San Luis Obispo County, a large proportion of the immigrant farmers who brought with them from Scandinavia experience with successful producers' cooperatives, used that tradition to sustain themselves on their relatively marginal, isolated hill land, first under the Grange and then the aegis of the Alliance.[19]

It is in studies of the two major Farmers' Alliances, one southern and the other in the high plains and adjoining prairie states, that new research at the local level has had it greatest impact. There have been more such studies than of the Grange, since these movements gathered more followers for a longer time, and served as a major base for the Populist Party, in both the rank and file and in its leadership. As in the Grange, women played an important part in the Alliance organizations, a factor that may have later led the Populists readily to welcome an alliance with the leaders of the women's suffrage movement. There has been some renewed controversy about the proper interpretation of the Alliance movements, and over the role of ideology in its success, and failure. But it has generally been less rancorous than earlier debates.[20]

18. Gerald Prescott, "Wisconsin Farm Leaders in the Gilded Age," *Agricultural History* 44 (April 1970):183–200. On the network of commercial farm interests and its achievements in support of scientific research, see Charles E. Rosenberg, "The Adams Act: Politics and the Cause of Scientific Research," *Agricultural History* 38 (Jan. 1964): 3–12.

19. Gerald L. Prescott, "Farm Gentry vs. the Grangers: Conflict in Rural California," *California Historical Quarterly* 56 (Winter 1977–1978):328–45. The information on Scandinavian immigrants comes from Michael Magliari, doctoral candidate in history at University of California, Davis, who is writing a doctoral dissertation on the Alliance and Populist movement in that county. The information on Bidwell is from Paul, "California Grain War," and the Bidwell papers, Bancroft Library, University of California, Berkeley.

20. One historian has reviewed much of the debate and tried to reconcile the polarities by combining sociological and psycho-history approaches to the entire late nineteenth and early twentieth century series of agrarian movements, with mixed results. James M. Youngdale. *Populism: A Psychological Perspective,* (Port Washington, N.Y.: Kennikat Press, 1975). For the role of women, see Jack S. Blocker, Jr., "The Politics of Reform: Populists, Prohibition, and

About thirty years ago economic historians, persuaded by studies that confirmed the basic rationality of Third World peasants, and confident that their tools of analysis on markets had wide application, turned anew to the study of American farmers. They examined the expressed grievances of the 1880s and 1890s as an empirical matter, subject to the same rigorous testing of evidence as any economic or social question. In any case, they had at hand the massive study by Milton Friedman and Anna Schwartz, based on an exceptional accumulation of data about the nation's money supply, which concluded, among many other things, that there was much foundation to the complaints that it was inadequate in the late nineteenth century. The old charges that the Populist assault on orthodox monetary policy was the quintessence of mindless barbarity began to fall away. Admittedly, some macro-level studies of real prices for farm products, railroad charges, relative interest rates, and land values produced ambiguous results and made it difficult to understand fully the basis of agrarian discontent. Increasingly, however, these economic studies are becoming more subtle, and to some extent confirm the validity of many "radical" charges by the Alliance and Populist leaders. For example, a recent article finds much support for the point that fear of foreclosure on mortgages, with all it could entail in terms of real losses, was a factor in the agrarian unrest of the Midwest.[21]

Woman Suffrage, 1891–1892," *The Historian* 34 (Aug. 1972):614–32; Marilyn Dell Brady, "Populism and Feminism in a Newspaper by and for Women of the Kansas Farmers' Alliance, 1891–1894," *Kansas History* 7 (Winter 1984/85):280–90; Mary Jo Wagner, "Farms, Families, and Reform: Women in the Farmers' Alliance and Populist Party," (unpublished Ph.D. dissertation, University of Oregon, 1986).

21. There is a brief discussion of the influence of Friedman's book on scholarship in Saloutos, "Professors and Poplists." A careful study demonstrating that there actually was something close to a "conspiracy" among some Treasury Department officials to push through the demonetization of silver in 1873—the famous "Crime of '73" that silverites decried—is Allan Weinstein, *Prelude to Populism: Origins of the Silver Issue, 1867–1878* (New Haven: Yale University Press, 1970). On the puzzling results of some economic research, see Robert Higgs, *The Transformation of the American Economy, 1865–1914* (New York: John Wiley & Sons, 1971), 86–102; John D. Bowman, "An Economic Analysis of Midwestern Farm Land Values and Farm Land Income," *Yale Economic Essays* 5 (Fall 1965):316–52; and John Bowman and Richard Keehn, "Agricultural Terms of Trade in Four Midwestern States, 1870–1900," *Journal of Economic History* 34 (Sept. 1974):592–609. On comparisons between American Populism and movements in other nations, see Richard K. Horner, "Agrarian Movements and their Historical Conditions," *Peasant Studies* 8 (Winter 1979):1–16 and R. A. Craig and K. J. Phillips, "Agrarian Ideology in Australia and the United States," *Rural Sociology* 48 (Fall 1983). A pioneering work that recognized the complexity of relationships behind many of the numbers, and found clear connections between the economic changes of the time and the protests is Robert Klepper, *The Economic Bases for Agrarian Protest Movements in the United States, 1870–1900* (New York: Arno Press, 1978). On the specific issue of railroad rates, vindicating some of the complaints, see Mark Aldrich, "A Note on Railroad Rates and the Populist Uprising," *Agricultural History* 54 (July 1980):424–32. A sophisticated econometric study that also infers the rationality of farmer grievances in the wheat and cotton states is Thomas F. Cooley and Stephen J. DeCanio, "Ra-

The locus of "true" radicalism in the Alliances was in the South rather than the Midwest, according to several of the recent studies. The coalition of a movement that began in a few newly settled communities of central Texas and grew slowly into a state-wide organization, it then joined other farm groups in northern Arkansas, and western Louisiana, to sweep across the rest of the cotton South as a single force, its hallmark the stress it placed on farm cooperatives. An insightful sociological study called the Alliance Exchange system, which stressed centralized marketing cooperatives, the "ultimate counter-institution," because it broke the jute-bagging monopoly that cotton growers found oppressive. In Lawrence Goodwyn's brilliantly evocative but flawed *Democratic Promise,* the Texas cooperatives, with their fusion of southern and western agrarian traditions into a "movement culture," created the most effective institution for transforming the nation's prevailing capitalist order. Yet he also concluded that only in Kansas, Georgia, and Alabama did the Southern Alliance movement truly overcome the racism and other sources of division which their opponents tried to exploit and showed the tough-minded commitment essential for a true agrarian victory.[22]

Unquestionably, the Alliances were the most broad-based farmer movements in the nation's history to that point, in some respects perhaps the broadest ever. Between 1884 and 1890, according to the best recent estimates, membership in the Southern Alliance jumped from less than

---

tional Expectations in American Agriculture, 1867–1914," *The Review of Economic Statistics* 59 (Feb. 1977):9–17. On the issue of debt, see James H. Stock, "Real Estate Mortgages, Foreclosures, and Midwestern Agrarian Unrest, 1865–1920," *Journal of Economic History* 44 (March 1984):89–106. For similar conclusions, drawn from an intense study of sample counties in Kansas, see Walter T. K. Nugent, "Some Parameters of Populism," *Agricultural History* 40 (Oct. 1966):255–70.

22. Michael B. Schwartz, *Radical Protest and Social Structure: The Southern Farmers' Alliance and Cotton Tenancy* (New York: Academic Press, 1976) has both a heavily schematic analysis of the origins and functions of the Alliances at the state levels and a full sociological discussion of the "life of protest movements." It also has a probing discussion of the membership, in terms of both numbers and their social status. This study appeared at almost the same time as the traditional, but equally penetrating and thorough work by Robert C. McMath, Jr., *Populist Vanguard: A History of the Southern Farmers' Alliance* (Chapel Hill: University of North Carolina Press, 1975). Lawrence Goodwyn, *Democratic Promise: The Populist Moment in America* (New York: Oxford University Press, 1976) has also appeared in an abridged paperback version as *The Populist Moment: A Short History of the Agrarian Revolt in America* (New York: Oxford University Press, 1978). Some of the more useful of many articles on the participation of blacks in the southern movement are Robert Saunders, "Southern Populists and the Negro, 1893–95," *Journal of Negro History* 54 (July 1969):240–61; R. Jean Simms-Brown, "Populism and Black Americans: Constructive or Destructive?" *Journal of Negro History* 65 (Fall 1980):349–60; Floyd J. Miller, "Black Protest and White Leadership: A Note on the Colored Farmers Alliance," *Phylon* 33 (Summer 1972):169–83; and William F. Holmes, "The Leflore County Massacre and the Demise of the Colored Farmers' Alliance," *Phylon* 34 (June 1973):267–74. See also the article on the Colored Farmers' National Alliance and Cooperative Union in Dyson, *Farmers' Organization,* 47–52.

50,000 to more than 850,000 farm families. It is harder to estimate the levels of strength in the Northwest Alliance, which may not have exceeded 200,000 in its peak year of 1888, partly due to later defections to the Southern Alliance that took place first in Kansas, then in Nebraska. Historians disagree, however, over whether or not the farmers' movement as such ended with the failure at Cincinnati in 1891 to unite the Southern and Northwestern Alliances. In the subsequent fusion of the Alliances with a myriad of other reform groups to pronounce a People's Party lay the full, and foolish, surrender of agrarian interests·and their own reform agenda, according to purists. The accommodations that ultimately put the money question at the center of Populist reform, and the betrayal of southerners who had opposed at great cost their own Democratic party organization in 1892 by the Populist party fusion with that same party in 1896, account for much of the bitterness after defeat.[23]

For most scholars working on the Northwest Alliances and Populist party, however, the broadening of the movement seems a sensible tactic. Much of the new analytic thrust, however, lumps together the various types of dissent in those states or parts of states that were still in the first stages of development. Careful studies of the Nebraska electorate, and of votes which the men they elected cast in the legislature, indicate that much Populist strength was concentrated in the central part of the state. In the eastern third most farmers were comfortably settled and satisfied with the status quo; in the western counties new settlers were eager to bring investment in railroads, grain elevators, and stockyards, the Populists' favorite targets.[24] In addition, plowmen as such had a minor role in the

23. Differences over this question underly the recent discussions of Goodwyn's book in David Montgomery, "On Goodwyn's Populists," *Marxist Perspectives* 1 (Spring 1978):166–73; James Green, "Populism, Socialism and the Promise of Democracy," *Radical History Review* 24 (Fall 1980):7–40, and Goodwyn's extended reply, "The Cooperative Commonwealth & Other Abstractions: In Search of a Democratic Promise," *Marxist Perspectives* 3 (Summer 1980):8–42. See also Roger D. Launius, "The Nature of the Populists: An Historiographical Essay," *Southern Studies* 22 (Winter 1983):366–85. They also underly such recent studies as N. James Wilson, "The Farmers' Search for Order" (unpublished Ph.D. dissertation, University of Oklahoma, 1974) and John M. Wheeler, "The People's Party in Arkansas" (unpublished Ph.D. dissertation, Tulane University, 1975). The question is placed within the context of broader changes in a major state in Dwight B. Billings, Jr., *Planters and the Making of a "New South" : Class, Politics, and Development in North Carolina, 1865–1900* (Chapel Hill: University of North Carolina Press, 1979). See also the excellent intellectual history of Populist writings in Bruce Palmer, "*Man Over Money: The Southern Populist Critique of American Capitalism* (Chapel Hill: University of North Carolina Press, 1980).

24. Stanley B. Parsons, *The Populist Context: Rural Versus Urban Power on a Great Plains Frontier* (Westport, Ct.: Greenwood Press, 1973); Ronald C. Briel, "Preface to Populism: A Social Analysis of Minor Parties in Nebraska Politics, 1876–1890" (unpublished Ph.D. dissertation, University of Nebraska, 1971); David S. Trask, "The Nebraska Populist Party: A Social and Political Analysis," (unpublished Ph.D. dissertation, University of Nebraska, 1971). An imaginative effort to measure levels of community development on the Plains and in the South, with

Populist movements of Colorado or other mountain states, if for no other reason than that miners, lumbermen, and cattlemen outnumbered them, and had their own reform agendas.[25]

After the turn of the century, when a series of gold strikes muted the money issue and even moderate Progressives supported railroad regulation, agrarian radicalism still manifested great strength in a few limited sections where there were few alternatives to staple production. Some of the cooperatives survived in the grain and cotton states, as did stockmen associations. To be sure, in areas where the growers of new specialty crops were concentrated, such as California's rich coastal valleys, they adopted the kind of business-oriented procedures that enabled them to administer prices in successful capitalist style. By contrast, in much of the Southwest, especially in northern Texas and southern Oklahoma, where farmers, many of them relatively recent migants from older parts of the South, still were trapped in a cycle of poverty that stemmed from reliance on cotton and wheat, the Socialist Party flourished, gathering more votes there relative to population than from urbanites in Chicago, Milwaukee, or New York. It was to the southern plains, too, that support for the Farmers' Union shifted after it lost members among veterans of the Alliance movement in a cotton withholding action that failed in 1907. This new organization, rising from the ashes of the old, eventually became the nation's second general farm organization.[26]

---

results suggesting that the key variable determining participation was social and cultural isolation, is in James Turner, "Understanding the Populists," *Journal of American History* 67 (Sept. 1980):354–73. Goodwyn's rather cavalier dismissal of Nebraska populism as unimportant, and his emphasis on "movement culture," has been re-examined with quantitative skill, and found wanting, in Stanley B. Parsons, Karen Toombs Parsons, Walter Killilace, and Beverly Borgers, "The Role of Cooperatives in the Development of the Movement Culture of Populism," *Journal of American History* 69 (Mar. 1983):866–85. In several studies, Karel Bicha, *Western Populism: Studies in An Ambivalent Conservatism* (Lawrence, Ks.: Coronado Press, 1976) asserted that the Plains states Populists were relatively conservative in their leadership and their goals, a position deftly rebutted in Peter H. Argersinger, "Ideology and Behavior: Legislative Politics and Western Populism," *Agricultural History* 58 (Jan. 1984):43–58. On South Dakota farmers who joined the Populists as propertied men fearful of foreclosures in hard times, see John Dibbern, "Who Were the Populists? : A Study of Grass Roots Alliancemen in Dakota," *Agricultural History* 56 (Oct. 1982):677–91.

25. By far the best work on Colorado is James E. Wright, *The Politics of Populism: Dissent in Colorado* (New Haven: Yale University Press, 1974)—a model of rigor in numerical analysis. A convenient new supplement on the region as a whole is Robert W. Larson, *Populism in the Mountain West* (Albuquerque: University of New Mexico Press, 1986). A still neglected area of Populist strength is in the Pacific Northwest, but see the work of Thomas W. Riddle, especially his "Populism in the Palouse: Old Ideals and New Realities," *Pacific Northwest Quarterly* 65 (July 1974):98–109.

26. On socialist strength among southwestern farmers, see James R. Green, *Grass-roots Socialism: Radical Movements in the Southwest, 1895–1943* (Baton Rouge: Louisiana State University Press, 1978); Garin Burbank, *When Farmers Voted Red: The Gospel of Socialism in*

In the dark tobacco regions of western Kentucky and Tennessee, fearful reactions to new marketing arrangements imposed by the international tobacco trust led first to incidents of violent protests by "night riders" in 1906–1907, then to organizing of nascent effective cooperatives by the eve of World War I.[27] Similar incidents of violent attacks by hooded individuals known as "whitecappers" cropped up across much of the South and Southwest at about the same time, their targets being relatively successful and much resented Chicano and black families.[28]

But it was in the wheat state of North Dakota that the "most successful farmer movement of the twentieth century," and the most radical, emerged; the Non Partisan League. Many Catholic and Lutheran farmers there had held aloof from the Grange because of church proscription of its secret rituals, and gave only luke-warm support to the Alliance, or the splintered Equity groups that followed it. Yet in 1916 they elected a governor and a solid majority of its members to the state legislature. These rough-skinned farmer-representatives needed basic instruction in parliamentary procedures, but they learned enough to establish quickly a state-owned terminal grain elevator, state-owned flour mills, state-owned banks, and a state insurance firm, among other "radical" innovations. Whether this amounted to state capitalism for generally prosperous large-scale farmers (it was, after all, a period of high war-time prices) or was truly a nondoctrinaire form of socialism soon became a moot question. Their open resistance to the mili-

---

the Oklahoma Countryside (Westport, Ct.: Greenwood Press, 1976). Worth R. Miller, *Oklahoma Populism: A History of the People's Party in the Oklahoma Territory* (Norman: University of Oklahoma Press, 1987) is a superior explanation of background for later manifestations of radicalism. Miller has also revised the bibliography on the Farmers' Alliances and Populism published by the Agricultural History Center, University of California, Davis. For a recent set of articles on the Midwest, with virtually no reference to farmers, see Donald T. Critchlow, ed., *Socialism in the Heartland: The Midwestern Experience, 1900–1925* (Notre Dame, Ind.: The University of Notre Dame Press, 1986). For a thoughtful discussion of the barriers that prevented radical farm organizations from making alliances with labor unions in the first half of the twentieth century, see David Brody, "On the Failure of U.S. Radical Politics: A Farmer-Labor Analysis," *Industrial Relations* 22 (Spring 1983):141–63. It should be supplemented with Don F. Hadwiger, "Farmers in Politics," *Agricultural History* 50 (Jan. 1976):156–69.

27. Dewey W. Grantham, "Black Patch War: The Story of the Kentucky and Tennessee Night Riders," *South Atlantic Quarterly* 59 (Spring 1960):225–37; Harry H. Kroll, *Riders in the Night* (Philadelphia: University of Pennsylvania Press, 1965); Christopher Waldrep, "Planters and Planters' Protective Association in Kentucky and Tennessee," *The Journal of Southern History* 52 (Nov. 1986):565–88.

28. Robert M. Brown, *Strain of Violence: Historical Studies of American Violence and Vigilantism* (New York: Oxford University Press, 1975):69–83, places the first appearance of the White Cap movement, a forerunner of the Ku Klux Klan, in southern Indiana in 1887. See also William F. Holmes, "White Capping? Agrarian Violence in Mississippi, 1902–1906," *The Journal of Southern History* 35 (May 1969):165–85, and his "Whitecapping in Late Nineteenth Century Georgia," in Walter J. Fraser, Jr., and Winfred B. Moore, Jr. eds., *From the Old South to the New: Essays on the Transitional South* (Westport, Ct.: Greenwood Press, 1981):121–32. There is no reliable way to provide estimates of membership or numbers of participants in such incidents.

tary draft after the United States entered World War I, along with some attempts by their leaders to spread their program to neighboring states, made them vulnerable to prosecution under war-time laws, and to persecution in the subsequent "red scare." In fact, much support in Iowa, Minnesota, Wisconsin, and Montana for the Farm Bureau, then just emerging as a major farm organization, came from fears that the NPL would spread and infect its neighbors with "red" ideas.[29]

By the 1920s the national "agrarian crusade" was over. Much of its broader program had been pre-empted, and won, by the Progressive movement. The more radical notions were stifled and a prosperous, rapidly urbanizing America had little interest in agrarian concerns. In self-defense, the handful of Senators from farm states who were still committed to intervention on behalf of farmers beset by post-war adjustment problems created a nonpartisan "farm bloc," which did its work well, especially on behalf of those farmers who needed help the least.[30] Perhaps that result of special interest politics simply reflects the true nature of an American political and economic system that is now beginning its third century.

29. I do not mean that the Dakota Alliance was inconsequential. Larry Remele, " 'God Helps Those Who Help Themselves':The Farmers Alliance and Dakota Statehood," *Montana: The Magazine of Western History* 37 (Autumn 1987):22–33 shows the crucial role the organization played in territorial politics, including the carving of two states out of the territory and their admission to the Union in 1890. The classic NPL account is Robert L. Morlan, *Political Prairie Fire: The Nonpartisan League, 1915–1922* (Minneapolis: University of Minnesota Press, 1955), re-issued in 1985 by the Minnesota Historical Society, St. Paul, in paperback with a full historiographical essay by Larry Remele. For a convincing effort to apply Goodwyn's "movement culture" concept to the NPL, see Scott A. Ellsworth, "Origins of the Nonpartisan League," (unpublished Ph.D. dissertation, Duke University, 1982). Kathleen Moum, "The Social Origins of the Nonpartisan League," *North Dakota History* 53 (Spring 1986):18–22 is a recent brief effort to emphasize the role of Norwegian ethnicity, with its tradition of cooperatives, in the success of the movement. James F. Vivian, "The Last Round-up: Theodore Roosevelt Confronts the Nonpartisan League, October, 1918," *Montana: The Magazine of Western History* 36 (Winter 1986):37–48, shows that the purpose of Roosevelt's speeches in the state just weeks before the Armistice, and soon his own death, as "a desire to blunt the challenge posed by the Nonpartisan League and the Industrial Workers of the World. . .during the closing weeks of the off-year state and congressional election campaigns." Even in its heyday, the cooperative efforts of the NPL was not always successful; H. Roger Grant has traced the story of two notable failures in his "Western Utopians and the Farmers' Railroad Movement, 1890–1900," *North Dakota History* 46 (Winter 1979):13–18, about farmer owned feeder lines, and "Captive Corporations: The Farmers' Grain & Shipping Company: 1896–1945," *North Dakota History* 49 (Winter 1982):4–10. See also Douglas Bakken, "NPL in Nebraska- 1917–20," *North Dakota History* 39 (Spring 1972):26–31.

30. For the formation of the Farm Bloc, see Murray R. Benedict, *Farm Policies of the United States: A Study of Their Origins and Development* (New York: Twentieth Century Fund, 1953). On the efforts to use cooperatives for controlling prices, see Grace H. Larsen and Henry E. Erdman, "Aaron Sapiro: Genius of Farm Cooperative Promotion," *Mississippi Valley Historical Review* 49 (Sept. 1962):242–68.

Appendix

Table A.1 Membership of Farm Organizations, 1874–1984

YEAR					
1874					
1880					
1882	BofF				
1885	43,000	FMBA			
1886		3,000	NOV		
1888		40,000	250,000	PofI	CFNA
1890	KofR	100–200,000	*	75,000	500,000
1895	126,000	10,000	*	*	*
1900	NFU	ASE			
1902	*	20–30,000	PPA		
1904	50,000	59,000	*		
1908	135,000	100–200,000	35,000		
1912	116,000	40,000	15,000	NPL	
1917	154,000	*		75,000	
1919	131,000	40–50,000		150,000	
1920	120,000	40,000		250,000	
1921	*	*	NCFCMA	*	NPA
1924	*	*	1 Mil.	PFA	30,000
1927	*	*		20–40,000	
1930	91,000	CAWIU			UFA
1933	78,000	25–50,000		STFU	50–100,000
1936	50,000	UCAPAWA	DFU	30,000	
1939	65,000	100,000	13,000	15,000	
1942	100,000	*	35,000	15,000	
1946	140,000	25,000		8,000	
1950	170,000			5,000	
1956	278,000			*	
1962	300,000				
1966	275,000			AAM	
1972	250,000			*	
1978	250,000			100,000	
1984	300,000			50–100,000	

YEAR	GRANGE	ISFA	NAW	NFA	NFAIU
1874	141,000	80-90,000			
1880	62,000				
1882	72,000		300	50-100,000	
1885	62,000		80,000	50-100,000	
1886	58,000		*	200,000	92,000
1888	58,000	*NFL*	200-250,000	150,000	400,000
1890	71,000	50,000		50,000	1.2-1.5 Mil.
1895	94,000		*AOG*	2,000	20,000
1900	98,000		10,000		
1902	115,000		*		
1904	137,000		*		
1908	178,000		*		
1912	218,000		*		*AWO*
1917	297,000		*	*ECE*	20-30,000
1919	268,000		85,000	40,000	4-5,000
1920	232,000	*AFBF*	*	40,000	*FLU*
1921	309,000	466,000	*	*	*
1924	341,000	301,000	70,000	*	165,000
1927	356,000	272,000	*		*
1930	316,000	321,000	*FNCA*	*FHA*	*
1933	303,000	163,000	50-100,000	500,000	
1936	326,000	357,000	*	10,000	*NFG*
1939	336,000	398,000			10,000
1942	332,000	591,000	*NFCF*		5,000
1946	392,000	1.1 Mil.	*		*
1950	443,000	1.5 Mil.	*	*NFO*	
1956	458,000	1.6 Mil.	*	170,000	*UFW*
1962	421,000	1.6 Mil.	*	*	300
1966	421,000	1.7 Mil	*	*	5,000
1972	316,000	2.2 Mil.	*	200,000	30,000
1978	263,000	3.1 Mil.	*	*	30,000
1984	224,000	3.5 Mil.	2.5 Mil.	*	30,000

(See next page for definition of acronyms.)

GRANGE—National Grange of the Patrons of Husbandry
ISFA—Illinois State Farmers' Association
NFL—National Farmers' League
NAW—National Agricultural Wheel
AOG—Ancient Order of Gleaners
NFA—National Farmers' Alliance
NFAIU—National Farmers' Alliance and Industrial Union
AWO—Agricultural Workers' Order 400
BofF—Brothers of Freedom
KofR—Knights of Reciprocity
NFU—National Farmers' Union
FMBA—Farmers' Mutual Benefit Association
ASE—American Society of Equity
NOV—National Order of Videttes
PPA—Planters' Protective Association
PofI—Patrons of Industry
NPL—Non Partisan League
CFNA—Colored Farmers' National Alliance and Cooperative Union
Source: Dyson, *Farmers' Organizations.*

American Agriculture and the Labor Market: What Happened to Proletarianization?

GAVIN WRIGHT

Americans do not believe that their country's agricultural history has much to do with proletarianization, defined as the rise of a class of farm laborers owning no capital and working only for wages. As William N. Parker summarizes the prevailing view:

> Behind the American record of output and productivity growth have lain many production and investment decisions made by millions of separate economic units. Farmers who made these decisions were *small businessmen* holding from thirty to a few hundred acres: their chief source of labor (outside of slaves and sharecroppers in the South) was *themselves and their own families.*[1]

The family farm was not merely persistent, it was ascendant throughout the nineteenth century and into the twentieth, according to recent historical sociology.[2] Writing in 1913, the bourgeois Russian economist Himmer concluded on the basis of the 1910 U. S. Census that "the great majority of farms in the United States are 'individualistic' farms (i.e., *worked by the owners themselves*)" and that "in a large majority of regions . . . small-scale farming by owner-operators is becoming ever more dominant."[3]

Yet Lenin himself wrote: "All these assertions are monstrous untruths. They are the direct opposite of the facts. They are a real mockery of

GAVIN WRIGHT is Professor of Economics at Stanford University.

1. William N. Parker, "Agriculture," in Lance Davis et al., eds., *American Economic Growth* (New York: Harper and Row, 1972), 393–94. Emphasis added.

2. Harriet Friedmann, "World Market, State and Family Farm," *Comparative Studies in Society and History* 20 (1978): 545–86.

3. Quoted by V. I. Lenin, *Capitalism and Agriculture in the United States* (New York: International Publishers, 1934). I. Lenin's study was originally written in 1915. It is reprinted in V. I. Lenin, *On the United States of America* (Moscow: Progress Publishers, 1967), 115–205.

truth."[4] From an extremely detailed review of the 1910 published census material, he went on to argue that American agriculture was becoming more and more capitalistic, including the use of hired labor, which he called the "principal earmark and index of capitalism in agriculture." Close examination of the evidence suggests that Lenin's diagnosis was substantially correct for the time when it was written, not just in terms of the data but in terms of underlying economic forces. The long-run tendency, however, was more strongly toward mechanization than toward wage labor in agriculture.

This essay attempts to examine these issues by surveying available data on farm labor in the United States before World War I, and considering what economic processes were at work in generating these trends. After an initial review of the literature on broader aspects of farming and markets in American history, the focus is on the labor market, primarily outside the South. Clearly the South had a very different agricultural history; there is a much better developed literature on that subject, which is referred to here for comparison.

The coverage extends roughly to 1920, a date which marks a more logical break with the past than does 1900; again, however, later developments will be referred to for comparison.

It is indisputable that American farmers became increasingly involved in product and capital markets over time. Though we do not have continuous indicators of these developments over long periods, the basic trends go back at least to the eighteenth century: rising specialization and production for the market as opposed to home consumption, reliance on trade credit and established marketing channels, and at least for the latter part of the nineteenth century, rising rates of tenancy and farm mortgages.[5] Rates of tenancy, a measure of participation in the land rental market, were recorded in the federal census only from 1880, but diligent researchers have constructed estimates from the manuscript census for 1850, 1860, and 1870. Table 1 presents figures for the state of Iowa, which show a sustained increase after 1870. Table 2 presents the available data for the country as a whole for 1880–1920. Again the rising trend is evident. An alternative to land rental is land purchase on the basis of a mortgage loan; the incidence of mortgaged farms was first reported only in the 1890

4. Lenin, *Capitalism and Agriculture,* 2.
5. Winifred Rothenberg, "The Market and the Massachusetts Farmer, 1750–1855," *Journal of Economic History* 41 (1981): 283–314; Rothenberg, "The Emergence of a Capital Market in Rural Massachusetts, 1730–1838," *Journal of Economic History* 45 (1985): 781–808; Clarence Danhof, *Change in Agriculture: The Northern United States, 1820–1870* (Cambridge: Harvard University Press, 1969); Robert F. Severson et al., "Mortgage Borrowing as a Frontier Developed," *Journal of Economic History* 36 (1966): 147–68.

census (Table 3). By 1890 the majority of American farm operators either rented their land or farmed on mortgaged land; by 1920 this was true for 65 percent.

The questions considered in this section are what broad forces lay behind the increased involvement in land, capital and product markets, and whether these trends may usefully be considered a form of "proletarianization." There are three prevailing interpretations: (1) American farmers had commercial intentions at all times, but were constrained by high transportation costs and poorly-developed markets, including capital markets; (2) American farmers underwent a gradual change in attitudes and behavior, from the traditional motives and pattern of self-sufficiency to new market-based criteria of costs, prices,and profits; (3) American farmers were pushed unwillingly into product and capital markets by economic and political forces, such as land scarcity, involuntary indebtedness, and the rising costs of farm-making. It would be a mistake to rely exclusively on any one or even any two of these factors over the full sweep of American history. Explanation (1) is most congenial to economists, since neoclassical economics has no integral place for such factors as "attitudes" or even "learning." The distinction between "market-oriented" and "self-sufficient" attitudes or behavior is by no means rigid and precise; in the presence of market risk, calculating farmers may try to obtain the benefits of both worlds, as in "safety-first" models of decision-making. Yet it is inconceivable that major expansion in market opportunities over an extended period of time would not also change the way in which farmers think and calculate, eventually changing the things they came to value and aspire to as well. Thus, most accounts blend factors (1) and (2) in some combination.

There are also, however, elements of pressure and coercion in market relations. Consider Figure 1, which depicts a standard production-possibility curve for a family farm with a given stock of productive resources, showing feasible combinations of market and non-market goods which may be produced. OS is the level of non-market goods the farmer desires to produce and consume. But if he owes a debt OD greater than SS, he is forced to give up this goal and produce for the market: "forced" in the specific sense that the creditor has a legal claim which entitles him to seize property if the debt is not paid. Thus the small southern farmers who had to borrow to get through the growing season also had to give up self-sufficiency in corn, in favor of cotton growing. Likewise the frontier farmowners of Kansas and the Dakotas had no choice but to concentrate on specialized wheat production in order to make their mortgage payments. The cash rent which a tenant farmer owes to his landlord has much the same character and effect. A debt-free, ambitious landowning

Table 1.   Tenancy Rates, Iowa, 1850–1900

	1850	*1860*	*1870*	*1880*	*1890*	*1900*
Springfield Township, Cedar County	13.8	11.2	27.9	25.6	29.5	34.3
Union Township, Davis County	11.6	7.8	5.8	24.5	24.8	28.4
Eastern Iowa	17.6	15.1	19.0	24.3	29.7	34.3
Twelve County Sample	20.1	11.2	13.2	24.3	28.9	35.4
State	—	—	—	23.8	28.1	34.9

Source: Donald L. Winters, *Farmers Without Farms: Agricultural Tenancy in Nineteenth-Century Iowa* (Westport, ct., 1978), p. 14.

farmer *might,* of course, make the very same choices; but he is equally free to choose something else, like self-sufficiency. Economists teach that the market system is one in which individuals are "free to choose"; yet they also teach that markets create strong pressures toward efficient resource allocation. The implication is that the range of true freedom of choice is a function of wealth, and in this sense one may say that American farmers were increasingly constrained by market forces, from at least 1870 onward.

Evaluation of this trend and its social implications, however, is by no means straightforward. Any voluntary contract places *ex post* constraints on the parties to it, but the opportunity to borrow capital or rent land through such a contract may be an important means of social and economic advancement. For early New England and for the yeomen of the post-Civil War South, there is a significant literature which argues that small farmers became "entrapped" into market relations by inadvertance, bad luck or connivance.[6] But for the mainstream of northern and western agriculture, historians have generally portrayed farmers as eager players in the market game, borrowing money not out of desperation but in the hope of making profits and accumulating wealth. Even the agrarian insurgencies of the 1880s and 1890s did not so much deny the legitimacy of credit or production for the market, but instead called for reform and regulation of the *existing* credit markets.[7]

There is a deeper analytical point about this process, however, which is

6. James A. Henretta, "Families and Farms: *Mentalité* in Pre-Industrial America," *William and Mary Quarterly* 35 (1978): 3–32; Christopher Clark, "The Household Economy, Market Exchange, and the Rise of Capitalism in the Connecticut Valley, 1800–1860," *Journal of Social History* 13 (1979): 169–89; Steven Hahn, *The Roots of Southern Populism: Yeoman Farmers and the Transformation of the Georgia Upcountry, 1850–1890* (New York: Oxford University Press, 1980).

7. Bruce Palmer, *"Man Over Money"*: *The Southern Populist Critique of American Capitalism* (Chapel Hill: University of North Carolina Press, 1980).

Table 2. Percent of Farms Operated by Tenants

	1880	1890	1900	1910	1920
NORTH	19.2	22.1	26.2	26.5	28.2
New England	8.5	9.3	9.4	8.0	7.4
Middle Atlantic	19.2	22.1	25.3	22.3	20.7
East North Central	20.5	22.8	26.3	27.0	28.1
West North Central	20.5	24.0	29.6	30.9	34.2
SOUTH	36.2	38.5	47.0	49.6	49.6
South Atlantic	36.1	38.5	44.2	45.9	46.8
East South Central	36.8	38.3	48.1	50.7	49.7
West South Central	35.2	38.6	49.1	52.8	52.9
WEST	14.0	12.1	16.6	14.0	17.7
Mountain	7.4	7.1	12.2	10.7	15.4
Pacific	16.8	14.7	19.7	17.2	20.1
U.S.	25.6	28.4	35.3	37.0	38.1

Source: U.S. Special Committee on Farm Tenancy, *Farm Tenancy* (1937), pp. 39, 96.

that it was in many respects a *collective* choice rather than the aggregation of a set of separate *individual* choices. This is so because land values in a given area have the properties of so-called "public goods," in that the general level is shared (to a greater or lesser degree) by all landowners in the vicinity. Rising land values put strong pressure on farmers to commercialize, and land values in any locality were directly affected by collective choices with respect to transportation and marketing facilities (among others). In Champaign County, Illinois, for example, there was little mortgage borrowing before the early 1850s, when completion of the railroad connection to the county became certain.[8] Such "pressure" need not be onerous: it might take the form of increased land taxes which threaten expropriation, but it might instead be merely aggressive bidding by potential buyers who believe they can extract more value from the land. Either way, the greater the value, the greater the pressure, so that in effect the prevailing culture is being shaped by the development process rather than the other way around. "Collective choice" means that the process was essentially political. It may be that commercialization represented the *majority* preference in most areas, but it was never unanimous and there was often conflict and dissent.

To illustrate this point by example, consider the observations of Tocqueville in the 1830s on the striking contrast in farming north and south of the Ohio River. In Ohio he heard the "confused hum" of men at work, while

8. Severson, "Mortgage Borrowing," 152.

Table 3.   Percentage of Owner-Occupied Farms Mortgaged [of those reporting]

	1890	1900	1910	1920
New England	28.3	34.1	34.9	39.8
Middle Atlantic	37.0	40.3	38.3	41.1
East North Central	37.6	39.4	40.9	46.1
West North Central	48.0	44.3	46.1	56.9
South	5.7	17.2	23.5	29.0
Mountain	14.1	14.4	20.8	44.4
Pacific	28.7	27.6	26.8	52.1
U.S.	28.2	31.0	35.6	41.4
Non-South	39.9	39.1	40.1	49.4

Mortgage Debt/Farm Value on Mortgaged Farms

	1890	1910	1920
New England	40.4	31.8	33.8
Middle Atlantic	43.2	34.5	36.3
East North Central	33.2	28.6	31.2
West North Central	33.6	25.8	26.5
South	41.3	26.6	28.6
Mountain	31.8	23.9	30.5
Pacific	30.1	23.4	29.8
U.S.	35.5	27.3	29.1

Sources: 13th Census (1910), Vol. V, *Agriculture,* pp. 159–160; 14th Census (1920), Vol. V, *Agriculture,* pp. 484–486.

Kentucky seemed to be a society "gone to sleep." The slave South represented a radically different culture, he argued, in which work was degraded and idleness glorified.

How should we interpret these statements? It may of course be true that northern and southern cultural roots really were different. but whatever "initial" difference there may have been, these were clearly magnified by the simple fact that land values in Ohio were rising far more rapidly than they were in Kentucky, so that no one could buy into the Ohio land market unless he planned to pursue an aggressively commercial strategy. An aggressively commercial Kentuckian, by contrast, would be well advised to depart for some other location. This regional contrast, in turn, reflected the very different priorities of slaveowners with respect to infrastructure investment.[9] The sharp contrast between North and South in anticipated

9. Gavin Wright, *Old South, New South: Revolutions in the Southern Economy Since the Civil War* (New York: Basic Books, 1986), chapter 2. The difference in return to land improvements between counties (depending on access to markets) is demonstrated by David Weiman, "Farmers and the Market in Antebellum America: A View from the Georgia Uncountry," *Journal of Economic History* 47 (September 1987): 644–47.

Figure 1. Choice between Market and Non-Market Production

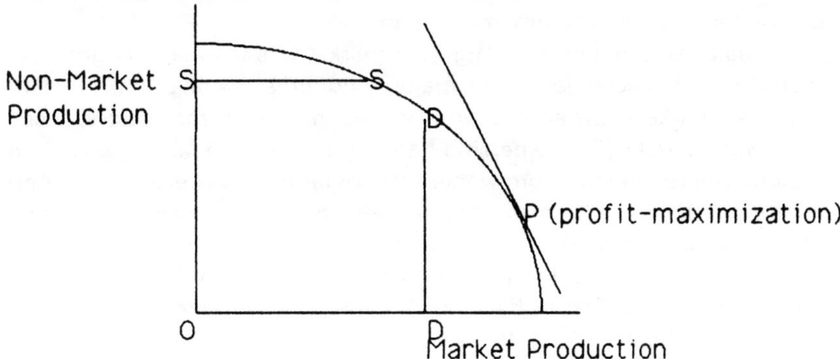

land-values explains the otherwise anomalous observation that many wealthy southerners had large speculative investments in northern lands before the Civil War.[10]

"Commercialization as collective choice" demonstrates the central difficulty with the conflict often posed by social historians between the value of "the market" and the values of "family" and "the community." In rural areas as in small towns, much of the sense of "community" was reinforced if not actually defined by the true shared community interest in local land values. Competition between local communities in the effort to raise property values was a powerful force in spreading railroad lines and commercial agriculture across the northern frontier. The community solidarity which generated these policies certainly had ethnic and kinship and other noneconomic components, but it was fundamentally a solidarity of *landowners,* which often had the effect of squeezing out those who were less fortunate. John Mack Faragher discovered in his study of Sugar Creek, Illinois, that the persistence rates of owners were far higher than those of nonowners. He writes:

In Sugar Creek, community did not work to the benefit of all; rather it was the device that allowed some men and women to succeed and prevail while others, the majority, failed and pushed on. The relative stability of Sugar Creek for the persistent owners was accompanied by high levels of mobility and "shiftlessness" for those whose lacked the means to buy land.[11]

10. Paul W. Gates, *Landlords and Tenants on the Prairies Frontier* (Ithaca: Cornell University Press, 1973), chapter 3.

11. John Mack Faragher, "Open Country Community," in Steven Hahn and Jonathan Prude, eds., *The Countryside in the Age of Capitalist Transformation* (Chapel Hill: The University of North Carolina Press, 1985), 252.

What he does not see quite so clearly is that communities like this were the driving forces behind commercialization.

If wealth accumulation through the market was the majority goal, the market-oriented activities of this majority put pressure on the rest to conform. As markets spread and farm values increased from east to west, those who sought farmownership had to go father and farther west, into regions where the minimum viable scale of farming was ever-larger. And as the continuing efforts of commercial farmers to expand production reshaped farm technology toward higher levels of mechanization and larger scale operation, farm-making costs increased to the point where most new farmers had to begin their careers with a lengthy tenancy or period of mortgage indebtedness.[12]

In all of these ways the deepening of market involvement may be said to imply increased pressure and a reduction in the scope of free choice for a significant number of American farmers. Though American historians often take it as axiomatic that all farmers "chose responses calculated to maximize their incomes," the evidence indicates that capital-market and land-market requirements were often the binding constraints on behavior. In Iowa, for example, tenants practiced significantly more specialized and land-intensive agriculture than did owners.[13] One might conjecture that the overall *variance* in decisions was narrowed over time by the discipline of the market: this is, at any rate, a testable proposition for quantitative research.

Despite this range of considerations, I would not argue that rising tenancy and farm mortgages should be viewed as equivalents to proletarianization. One important reason is that in American history the reduction in the relative equity position of farm operators occurred primarily across generations, rather than for particular individuals through time. Charts of relative tenure proportions (such as Figure 2) convey a sense of steady, inexorable financial loss and demotion. But in fact the absolute numbers of farms and farmowners was growing over time, through World War I for the country as a whole (Table 4). Authorities are agreed that mortgages and tenancies were devices for getting started in farming; cases where former owners "fell into" tenancy were exceptional (though of course not utterly unknown). In a surprisingly large number of cases, renters were in fact related by blood or marriage to landlords: as

12. Clarence H. Danhof, "Farm Making Costs and the Safety Value," *Journal of Political Economy* 49 (1941): 317–59; Jeremy Atack, "Farm and Farm-Making Costs Revisited," *Agricultural History* 56 (1982): 663–76.

13. Donald L. Winters, *Farmers Without Farms: Agricultural Tenancy in Nineteenth-Century Iowa* (Westport: Greenwood Press, 1978), chapters 1 and 2. The previous quotation is also from Winters (p. 8).

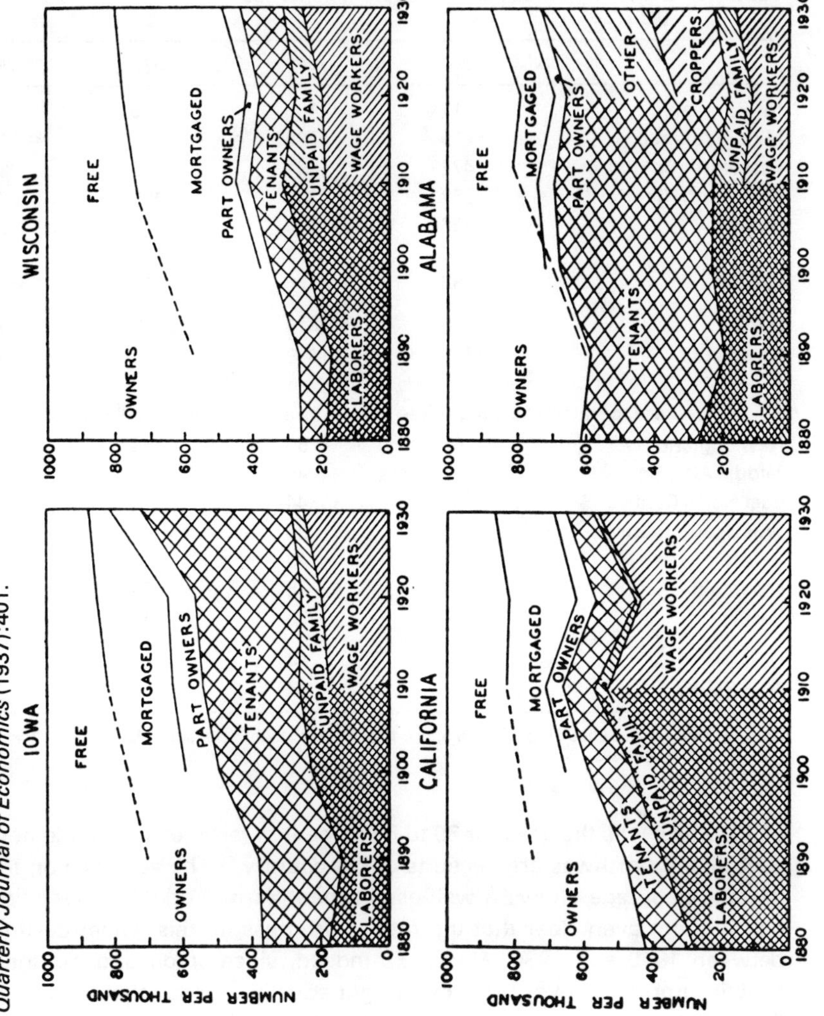

Figure 2. Relative Size of Tenure Categories in Four States, 1880–1930. *Source:* John D. Black and R.H. Allen, "The Growth of Farm Tenancy in the United States," *Quarterly Journal of Economics* (1937):401.

Table 4. Owner-Operated Farms Free From Mortgage
(000)

	1890	1900	1910	1920
New England	117	106	109	80
Middle Atlantic	219	208	217	179
East North Central	472	491	474	383
West North Central	352	395	405	279
South	966	1070	1163	985
Mountain	44	72	126	89
Pacific	58	76	95	79
U.S.	2229	2419	2589	2074
Non-South	1263	1349	1426	1089

Owner-Operated Mortgage-Free Farms, as a Percentage of All Farms

	1890	1900	1910	1920
New England	63	56	58	54
Middle Atlantic	47	44	47	45
East North Central	47	44	43	38
West North Central	39	38	37	28
South	53	43	38	35
Mountain	94	78	69	39
Pacific	62	55	51	37
U.S.	49	44	41	35
Non-South	47	44	44	35

Source: Thirteenth Census (1910), Vol. 5, pp. 150–60; Fourteenth Census (1920), Vol. 5, *Agriculture,* pp. 34, 484–485.

of 1930, this was the case for 35 to 50 percent of farm tenants in the north central and northwestern sections of the country.[14] The evidence on tenure status by age shows a well-established "farm ladder" in operation; and it is not even clear that the rate of progress up this ladder declined between 1890 and 1930 (Figure 3). Indeed, since credit and mortgage facilities improved over time, one might equally well say that the institutions of tenancy and farm mortgages *allowed* increasing numbers to practice family farming even though they lacked the personal resources to purchase the land and necessary equipment.

And this is, I believe, the way most farmers saw the matter. Any farm family would be happy to improve its wealth position and shake off the interference of the landlord or the creditor; but the real fear was loss of farm operator status, being forced into wage labor. American farm move-

14. Jack Temple Kirby, *Rural Worlds Lost: The American South 1920–1960* (Baton Rouge: Louisiana State University Press, 1987), 3.

Figure 3. Rate of Tenancy at Various Ages. 1890–1930. *Source:* Black and Allen, "Growth of Farm Tenancy," 410.

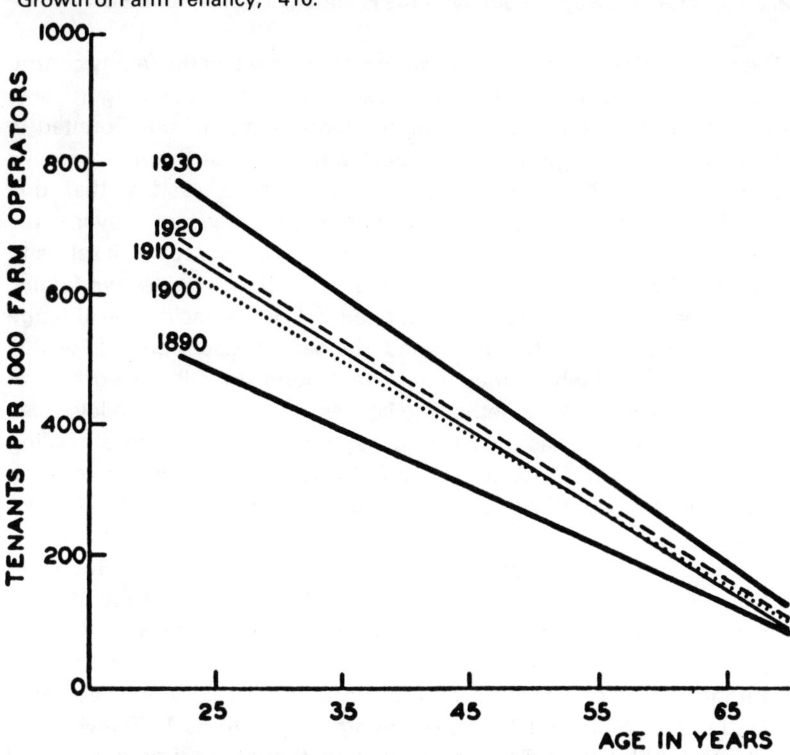

ments before World War I were predominantly composed of farm owners who owed money, and the single most pressing economic motivation was fear of foreclosure.[15] The willingness of small farmers to incur large debts, carry their families over long distances to remote and forbidding locations, and work long hours under conditions that were not uniformly pleasant, is powerful testimony to the desire of members of this class to succeed as family farmers.

They retained control over the detailed allocation of their own work time, supervisory authority over family labor, and a residual claim on the earnings of the enterprise after other legal claims were satisfied. Though they may not have used the term, they were at pains to avoid proletarianization, and most of them succeeded in this struggle, at least until the 1920s. A significant exception was sharecropping in the South, essentially

15. James H. Stock, "Real Estate Mortgages, Foreclosures, and Midwestern Agrarian Unrest, 1865–1920," *Journal of Economic History* 44 (1984): 89–106. See also Allan J. Bogue, "Land Credit for Northern Farmers, 1789–1940," *Agricultural History* 50 (1976): 90–92.

a proletarian though still a family-based status. The comparison between sharecropping and wage-labor systems is considered further below.

The statistical record appears to support the view that the family continued to be the basic work unit in American agriculture. Beginning in 1860, the federal census began to enumerate "farm laborers," and counted in that category 795,679 persons, or about one-fourth of the free persons employed in agriculture. Outside of the South, there was less than one laborer for every two farms. For a number of reasons we will never know precisely what this figure really means. For practical purposes, it refers to males only. Since it includes unpaid members of the farmer's own family, it may be considered a drastic overstatement of the incidence of wage labor in agriculture. On the other hand, the census also reported 969,301 "laborers," many of whom may in fact have done agricultural work for at least part of the year. But whether the figures are too high or too low, they are in a range that makes it clear that wage labor was not the prevailing system in American agriculture. Following the same indicator through time does not suggest that this characterization changed in any fundamental way.

In Tables 5 and 6, the dramatic effects of emancipation in the South are immediately evident and not surprising. Perhaps less well-known is the marked increase in farm labor in the northern states between 1860 and 1870, particularly in the north central states. This increase was not a long-term trend, however, because the incidence of farm labor subsequently declined in the 1870s and 1880s, returning roughly to its 1860 level (outside the South) by 1890. The tables do show that at the time of the 1910 census, when Lenin was analyzing American developments, there did seem to be a more sustained trend toward the use of farm labor. As he argued, the South really has to be separated from the national totals. Outside of the South, the relative importance of farm labor had indeed reached an all-time peak in 1910. And if we focus not on the "wheat factories" of the north central region (which Lenin said were *less* capitalistically developed than elsewhere) but on the truck farms of the northeast, the sugar beets of the rockies, or the fruit orchards of the Pacific coast, the reliance on farm labor was greater than at any previous time. These trends, however, were largely reversed by 1920.

Identifying and counting farm laborers in the decennial census is in any case a risky way to measure proletarianization; much hired farm labor is seasonal and migratory, so that workers may not be present at the time of the count, or they may be assigned to other locations, or they may not identify themselves as "farm laborers" at all. There is an alternative index, however, the responses to the agricultural census question on expendi-

Table 5. Laborers as Percent of Those Engaged in Agriculture
(Home Farm Plus Hiring-Out)

	1860	1870	1880	1890	1900	1910	1920
North Atlantic	29	35	33	33	39	47	39
South Atlantic	19	58	59	49	52	47	38
North Central	24	36	31	25	35	39	32
South Central	17	58	54	31	48	45	39
Western	29	32	27	30	36	48	40
U.S.	25	49	43	35	43	50	39
Non-South	27	35	31	27	36	42	35

Sources: Eight Census (1860), Vol. 1, pp. 656–679; Ninth Census (1870), col. 3, pp. 812–813; George K. Holmes, "Supply of Farm Labor," USDA Bureau of Statistic—Bulletin 94 (1912); Thirteenth Census (1910), Vol. 4, *Population,* Table II;

Table 6. Farm Laborers Per Farm 1860–1920

	1860	1870	1880	1890	1900	1910	1920
North Atlantic	.49	.58	.50	.54	.61	.74	.56
South Atlantic	.35	2.17	1.48	1.08	.97	.72	.50
North Central	.39	.65	.50	.40	.56	.62	.46
South Central	.15	1.10	1.28	.88	.92	.75	.51
Western	.34	.41	.56	.67	.62	.88	.63
U.S.	.39	.97	.83	.66	.81	.90	.51
Non-South	.46	.69	.50	.45	.57	.67	.51

Source: Same as Table 5, plus Fourteenth Census (1920), Vol. 5, *Agriculture.*

tures for hired labor. This question first appeared in 1870, but unfortunately the responses were not published again until 1900. Nonetheless, the figures across this wide spread of historical time demonstrate that the relative importance of hired labor in the production of agricultural *value* was in fact declining (Table 7). From this perspective, the apparent move toward capitalistic wage labor as of 1910 was minor and temporary.

There were many good reasons to think that wage labor would be a temporary phase in regional development, just as it was often a temporary phase in the careers of individuals. As Paul David shows in his analysis of the rise of Chicago, the wage-labor demands from the construction and transport-building components of the westward movement were larger at all times than those of agriculture.[16] The footloose male laborers brought

16. Paul David, "Industrial Labor Market Adjustments in a Region of Recent Settlement: Chicago and the Midwest, 1848–1868," in Peter Kilby, ed., *Quantity and Quiddity: Essays in U.S. Economic History* (Middletown, Connecticut: Wesleyan University Press, 1987).

Table 7. Expenditure for Wage Labor As percentage of Gross Value of Agricultural Output

	1870	1900	1910	1920
North Atlantic	13.7	10.7	11.9	8.6
South Atlantic	19.0	8.0	7.1	3.9
North Central	9.3	6.1	6.0	5.8
South Central	14.4	5.6	5.8	4.6
Western	19.3	16.7	16.9	13.1
U.S.	12.7	7.7	7.7	6.3
Non-South	11.4	8.1	8.3	7.3

Sources: Compendium of Ninth Census (1870), p. 692; Twelfth Census (1900), *Agriculture,* Part 1, p. cxxviii; Fourteenth Census (1920), Vol. V, *Agriculture,* pp. 18, 506.

in to build canals, railroads, or bridges were sometimes available for farm labor as well, but they had no intention of staying in the area unless they could become owners. The farmer's need for outside help was also typically greatest during the land-clearing phase. Many young men came West and got their start doing set-up labor, prairie-breaking, or woodchopping; but this kind of wage labor subsequently declined for both demand-side and supply-side reasons. In many parts of the country, large frontier landholders began operations as large capitalistic estates, only to find that over time wage labor became scarce, unreliable and irresponsible; gradually the holdings were subdivided and assigned to tenants. This scenario was played out in the Indiana-to-Iowa corn belt a generation prior to the better-known example on the "bonanza" wheat farms of the Red River Valley.[17] Far from seeing wage labor as an inexorable tendency of advancing market agriculture, historians and contemporaries have frequently viewed the entire category as transitory and receding into the past. The author of *Hired Hands and Plowboys* writes: "Once an integral though lowly member of rural society, the farm-wage laborer gradually disappeared from sight yet retained a nostalgic place in tradition and American folklore."[18]

There is, however, more here than meets the eye. Though the family farm scale and family labor remained quantitatively dominant throughout the nineteenth century and well beyond, there was an undercurrent of pressure building up over time, toward the use of wage labor on a more

17. John Lee Coulter, "Industrial History of the Valley of the Red River of the North," *Collections,* North Dakota State Historical Society, volume 3 (1910); Allan J. Bogue, *From Prairie to Corn Belt* (Chicago: University of Chicago Press, 1963), 184; Gates, *Landlords and Tenants,* 314–17.

18. David E. Schob, *Hired Hands and Ploughboys: Farm Labor in the Midwest, 1815–1860* (Urbana: University of Illinois Press, 1975), 1.

lasting basis. One clue lies in the evidence that a larger and larger fraction of the enumerated "farm laborers" were not members of the owner's family (Table 8). Table 8 shows that the large number of "farm laborers" in the southern states was mainly an effect of the high regional fertility rate: most laborers were in fact family members. But on the east and west coasts, more than 90 percent of the farm laborers were true wage laborers by 1920.

Other indications of long-term change are more subjective but perhaps no less important. In the first half of the nineteenth century, hired farm labor often had a social status no different from that of farmers. Foreign travelers were amazed to learn that farm help usually took their meals at the family table.[19] The very term "hired man" and "hired girl" were identified in 1820 as Americanisms by the Scottish traveler, James Flynt, who wrote: "*Master* is not a word in the vocabulary of hired people."[20] But by the late nineteenth century, it was clear that these perceptions had changed in many places. In 1891 a Massachusetts farmer noted: "The old-fashioned term, 'help,' has been dropped, and the word 'labor' used, with a peculiar sigificance."[21] The social concerns about what appeared to be small armies of "hobos" "tramps" and "bums" wandering the country-side prompted the hearings of the Industrial Commission into farm labor in 1900. J. R. Dodge of the USDA, who wrote a report for the Industrial Commission on farm labor, explicitly contrasted the hired man with "a class less efficient and desirable, with inferior training in rural occupations, whose rate of compensation is less, and whose condition, from lack of economy and judgement more than deficiencies of compensation, is not so good, and whose lives and aims are not on so high a plain."[22] The allegation of inferiority should certainly not be taken at face value. But there was no mistaking the wide social distance reflected in Dodge's comment, implying that a distinct "farm labor" class had taken shape by 1900.

Thus we seem to have a puzzle. Farm labor was not quantitively large, nor was relative expenditure on farm labor growing. Yet social observers describe the emergence of a new class of proletarianized farm laborers. The next section attempts a resolution.

The crucial consideration is the seasonality of labor requirements in commercial crops. Seasonality is of course a biological fact of life. But a

19. Schob, *Hired Hands and Ploughboys,* 184.
20. Paul Taylor, "The American Hired Man: His Rise and Decline," *Land Policy Review* (1943), 191.
21. Quoted in La Wanda F. Cox, "The American Agricultural Wage Earner, 1865–1900," *Agricultural History* 22 (1948), 100.
22. United States Industrial Commission, *Report,* Volume 11 (1901), 79.

Table 8.   "Home Farm" Workers As Percentage of Farm Labor

	1900	1910	1920
New England	25	19	10
Middle Atlantic	34	24	17
East North Central	49	38	30
West North Central	55	49	35
South Atlantic	58	56	50
South Central	62	59	51
Mountain	35	18	18
Pacific	25	15	8

Sources: Twelfth Census (1900), *Special Reports: Occupations,* pp. 94–95, 104–105; Thirteenth
  Census (1910), Volume IV, *Population: Occupation Statistics,* pp. 96–97, 110–111, 124–125,
  138–139; Fourteenth Census (1920), Volume IV, *Population: Occupations,* pp. 56–57, 74–75,
  92–93, 110–111. In 1900, the phrase used was "Farm laborers (members of family); in 1910
  and 1920, the distinction is between "homefarm" and "working out."

diversified self-sufficient farm has a wide range of activities with different
seasonalities, so that the aggregate fluctuations are moderated. Stanley
Lebergott has argued that land-clearing was ideal for a family farm with a
given quantity of labor, because it provided a profitable activity which
could fill in the work year, repaying the labor at a rate nearly double the
annual wage of a farm laborer.[23] Students of slavery have suggested that
the fixed-cost character of slave labor led the owners to plant large acreages in corn as well as cotton, so as to keep the slaves occupied all year
round.[24] As on family farms, the causation was mutual: mixed and self-
sufficient farming reduced seasonality; and inflexiblity in labor supply led
to a search for remunerative year-round activities, as well as for technologies and varieties that would level off the harvest labor peak for the most
valuable crops.

Pressure gradually built up over time, however, for farmers to extract
the maximum possible value from their improved farm land. We may
attribute this trend to the increasingly commercial attitudes of American
farmers; or to rising land values; or to the financial pressure arising from
the declining percentage of land value owned unencumbered by the
farmer; or we may attribute it to any of the causal forces which may have
lain behind these effects, such as the closing of the frontier, the extension
of the railroad network, and breakthroughs in transport technology which
made it possible to operate specialized commercial farms in distant cor-

23. Stanley Lebergott, "The Demand for Land: The United States, 1820–1860," *Journal of
Economic History* 45 (1985), 184–189.
24. Ralph V. Anderson and Robert E. Gallman, "Slaves as Fixed Capital," *Journal of American History* 64 (1977), 24–46.

ners of the country. However we may enumerate and rank these factors, increasing specialization led to increasing pressure on seasonal labor supplies, especially at the harvest. Figures 4-5 depict the links between market specialization and seasonality on the eve of World War I, for corn, wheat, and fruit farming in different parts of the country. The implication was that a tension or tradeoff existed between the social goal of stable rural communities and the economic goal of extracting maximum value from the land. The saying "wheat farms and hobos go together" became a byword in the plains and prairie states during the late nineteenth century.[25]

The problem on the labor supply side was that the returns to part-time harvest work alone could not cover the cost of maintaining a farm worker and family for an entire year. A more elaborate institutional arrangement whereby labor moves from place to place with the harvest would pay a higher annual wage, but at the expense of stable family life at a standard that most Americans could reasonably aspire to. Thus it was that the wheat harvest drew young unmarried men from neighboring farms or towns on a temporary basis. But these sources became increasingly difficult as whole areas became specialized in single crops and as the national population became increasingly urban. Though a network of more-or-less regularized harvest migration patterns did emerge in the wheat belt, it was inevitable that those attracted to such opportunities would come to be seen as "bummers," "tramp nuisances," and a "low order of farm labor," and inevitable also that reliance on such labor for the precisely-timed demands of the harvest would be seen by farmers as risky and unsatisfactory. There were enough farm laborers in the 1870s to organize militant protests against labor-saving mechanization throughout the Midwest.[26] But such obstreperousness, even if temporarily effective, ultimately only reinforced the farmer's desire to dispense with as much farm labor as he could. Wherever possible, mechanization was a superior choice, and this was the path of the wheat belt.

The crucial point is that outside of the South, the majority of American farmers were (for lack of a better term) middle-class. They were small businessmen indeed, with substantial investment in land and equipment, established credit relationships, and an interest in passing their class status on to the next generation. Despite the oft-repeated statement that Americans have always been highly mobile, American farm *owners* were a stable group, at least by the turn of the century. A census survey on the "Stability of Farm Operators" in 1909 found that the great majority of owners in the northern states had occupied their present farm for more

25. Paul Taylor, *Labor on the Land* (New York: Arno Press, 1981), 62.

26. Peter H. Argersinger and JoAnn E. Argersinger, "The Machine Breakers: Farmworkers and Social Change in the Rural Midwest of the 1870s," *Agricultural History* 58 (1984), 393–410.

Figure 4. Pattern of Seasonality: Corn Belt (Illinois) vs. Winter Wheat Region (Washington). *Source:* Yearbook of the Department Of Agriculture, 1917, 544–55.

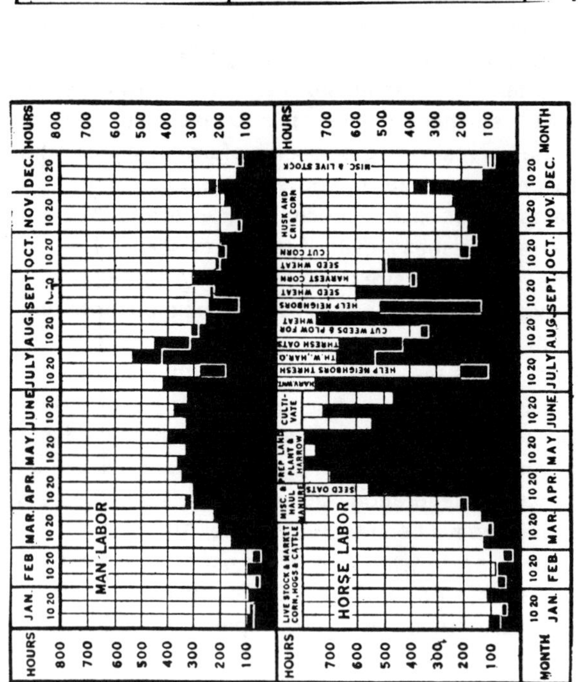

Figure 5. Seasonality in Fruit Farming. *Source:* Yearbook of the Department of Agriculture, 1917, p. 543.

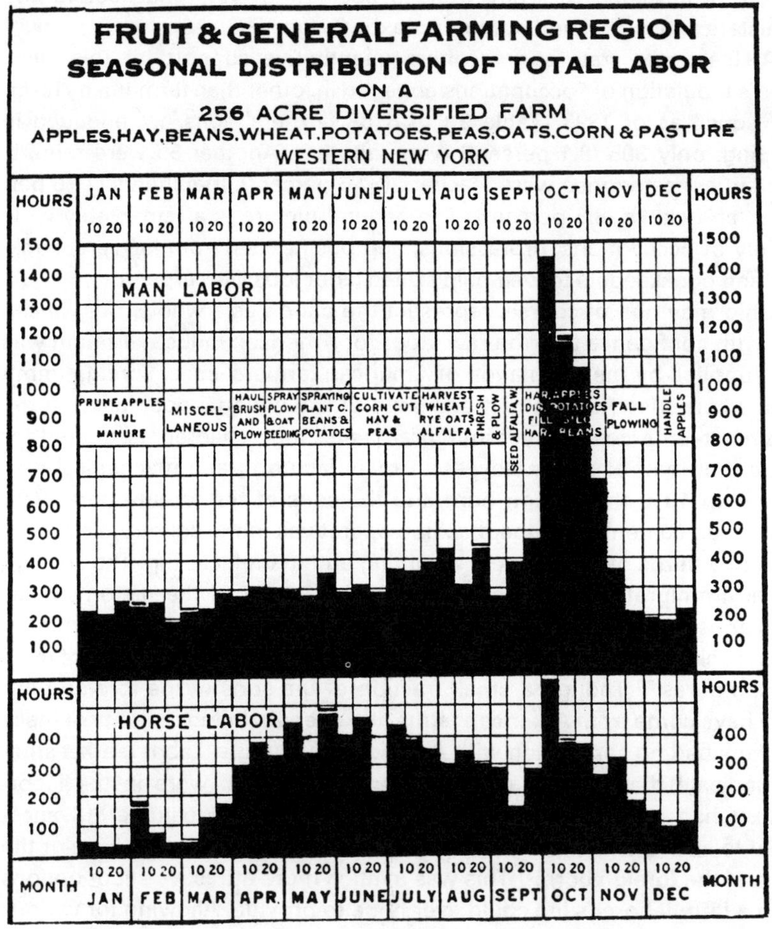

than ten years.[27] Though we do not have good estimates of the rates of off-farm migration for the pre-World War I era, nor the destination of those who departed, there are many indications that it was not a migration of proletarians to unskilled laborers' jobs. When the Industrial Commission inquired in various localities about the tendency of young men to leave the farms, more often than not the responses had to do with *upward* mobility from the status of farmer: "Some enter professions, some mercantile pur-

27. United States Census Bulletin, "Stability of Farm Operators, or Term of Occupancy of Farms" (1910).

suits. . . . They want the higher education. They want to get into and take up the professions. . . . They want to become lawyers and doctors and ministers and all that and not return to the farms."[28]

At least one state survey offers quantitative support for this view, Ohio's tabulation of "occupations engaged in, other than farming, by farmers' sons" as of 1893 (Table 9); of 3758 farmers' sons not engaged in farming, only 305 (8.1 percent) were laborers. Another 55 were miners. The largest categories were teachers (14.6 percent), merchants (11.5 percent), and clerks (8.8 percent). The general picture is a remarkably wide variety of occupations, professions, and businesses, from lawyers, bankers, and bookkeepers to well-paid skilled labor jobs like carpentry.

Ohio may not, of course, represent the nation as a whole. But we can say with confidence that the raw labor for American industrialization was not supplied by the population of American farmers' sons. Foreign immigrants comprised only about one-sixth of the national population in the late nineteenth century, yet immigrants and their children accounted for more than half of all employees in manufacturing and mechanical pursuits.[29] In 1910 the foreign born and the sons of the foreign born were more than 60 percent of the machine operatives in the country, and more than two-thirds of the laborers in mining and manufacturing. The surveys of the immigration commission found that many of the nation's basic industries—agricultural implements, clothing, iron and steel, furniture, meat packing—had foreign-stock labor percentages well in excess of even these figures.[30] Though a small fraction of the sons of the foreign born may have come from American farms, the overwhelming majority of these workers had no connection with American agriculture. Labor market studies indicate that those native-born Americans who were in the labor-market had much better opportunities for quick occupational advancement to skilled and supervisory positions than did the foreign born or the sons of the foreign born.[31] This was a much more attractive set of options than a farm laborer's life could ever hope to provide. Allowing for various local exceptions, we may say that American agriculture did not experience proletarianization as a precursor to industrialization in the classic Marxian fashion.

If there were to be a transition to wage labor in American agriculture, therefore, it would have to involve the recruitment of migrant labor on a

---

28. U.S. Industrial Commission, *Report,* Vol. 10, 133, 321, 848, 868.

29. E. P. Hutchinson, *Immigrants and their Children, 1850–1950* (New York: Wiley and Sons, 1956).

30. United States Senate, *Report of the Immigration Commission,* Vol. 1 (1911), 332, 343.

31. Joan Underhill Hannon, "City Size and Ethnic Discrimination," *Journal of Economic History* 42 (1982), 825–845.

Table 9. Occupations Engaged in, other than Farming, by Farmers' Sons, Ohio, 1893

Occupations	No.	Occupations	No.
Agents	43	Motormen	4
Architects	10	Merchants	434
Blacksmiths	43	Ministers	78
Butchers	35	Miners	55
Broommakers	4	Manufacturers	39
Boilermakers	3	Moulders	13
Bookkeepers	125	Masons	37
Bankers	15	Nurserymen	9
Barbers	9	Opticians	1
Bakers	3	Oil refiner	5
Brick-burners	6	Painters	20
Bee raisers	1	Publishers	3
Brass workers	2	Photographers	26
Commercial men	153	Poultry raisers	5
Cooks	3	Printers	13
Carpenters	183	Postal clerks	8
Clerks	329	Plasterers	14
Civil engineers	30	Plumbers	6
Contractors	9	Potters	4
Chemists	3	Professors	16
Druggists	40	Papermakers	1
Dentists	11	Platers	3
Detectives	2	Physicians	153
Dairymen	18	Railroaders	73
Engineers	72	Restauranters	4
Express messengers	4	Roofers	5
Engravers	1	Ranchmen	2
Electricians	6	Stock dealers	39
Firemen	1	Seamen	5
Glass blowers	4	Stenographers	23
Horticulturists	15	Surveyors	8
Horsemen	4	Shoemakers	8
Hucksters	5	Sawyers	38
Hotel proprietors	3	Saloonists	8
Harness makers	11	Soldiers	2
Iron workers	16	Tinners	5
Insurance agents	20	Teachers	550
Icemen	7	Telegraphers	84
Jewelers	20	Tile-workers	14
Journalists	18	Train dispatchers	2
Lumber dealers	27	Tailors	5
Lawyers	142	Tanners	1
Liverymen	25	Undertakers	8
Laborers	305	Upholsterers	1
Lithographers	2	Veterinary surgeons	14
Musicians	19	Wagonmakers	2
Machinists	75	U.S. weather observers	2
Millers	26	Total	3,758

Source: Seventeenth Annual Report of the Ohio Bureau of Labor Statistics, for the year 1893 (Norwalk, 1894), p. 794

temporary basis from a country with a living standard fundamentally lower than that of the United States. This was the path taken with Mexican labor in Texas and eventually, California and other western states. Mexican labor began to be utilized on a regular basis in both cotton and fruit-vegetable-grain farms in Texas after 1900. Other states began to tap into this market for these crops as well as for sugar beets, and the workers—often traveling as entire families—began to follow a routine which allowed them to fill out a larger and larger percentage of the full working year.[32]

In 1910, an attempt by the Agriculture Department to document a connection between urbanization and farm wages at the county level showed instead that they were largely unrelated in the western states.[33] This finding probably indicates that by 1910 most agricultural wage labor was not local but long-distance, and that the majority of these workers had little access to employment in nearby cities. In California, the use of seasonal wage labor had long been more extensive than elsewhere in the country. Chinese workers were employed during the 1870s and 1880s, partially replaced by the Japanese beginning in the 1890s. Immigration and labor systems had equally long been objects of political dispute. Prior to the war, when substantial numbers of white Americans had been recruited into the farm labor force, employers faced union organizing and political reform campaigns, putting strong pressure on them to improve labor conditions, stabilize their work forces, and adapt their farming practices accordingly. One may think of these pressures as manifestations of a fundamentally political demand that growers accept the standards of the American labor market as the "supply price of labor" in their economic calculations. At the time of the war, this demand was reflected in market forces, as farmers confronted a major exodus of wage labor from agriculture, in California as in the rest of the country. For crops which were unmechanizable at the time, the only alternative to a drastic cutback in production was integration into the low-wage migrant Mexican labor system. This was the direction of California agriculture during and after the war.

Thus the decline in relative expenditure on farm labor reflected two different effects: mechanization and outmigration, which reduced the *number* of farm laborers employed, and the development of a low-wage migrant system, which reduced the *cost* of labor in those states where wage labor was extensively used. Calculating the relative contribution of these two effects would be a useful project. As of 1919, farm wages in California had declined significantly relative to those in other parts of the country

32. George O. Coalson, *The Development of the Migratory Farm Labor System in Texas, 1900–1954* (San Francisco: R&E Research Associates, 1977).

33. George K. Holmes, "Supply of Farm Labor," USDA Bureau of Statistics, Bulletin 94 (1912), 38–40.

(Table 10). For the nation as a whole, the rough equilibration between farm and industry wage levels up to and during the war gave way to a pronounced intersectoral wage "gap" in the 1920s, which has persisted ever since (Figure 6).

It should be stressed that the choice between specialization and diversification is not a choice between a market and a nonmarket policy. If labor is scarce in an economy, or if certain jobs are particularly strenuous or disagreeable, or if seasonal work is a major inconvenience for laborers, the market will force landowners to economize on this scarce factor by choosing land-intensive outputs and diversifying, so as to offer work on an annual basis and attract stable families. But what defines an "economy"? Is it a group of *people* or is it the physical *property* in a given political entity? The concept of "labor scarcity" has no precise meaning if labor markets are continually subject to extension and redefinition across political lines. Yet such decisions are social choices, involving as they do basic issues of citizenship and belonging to a community. There was nothing "natural" about the development of the migrant labor system. Such markets are shot through with problems of risk and incentive compatability. Farmers have a desperate need for labor at the right time, but they don't know in advance exactly when that time will be and how long it will last. The demands of certain areas will be complementary in some years, competitive in others; workers who make a move too early or too late can suffer grievous losses. Transactions are made across long costly distances, yet circumstances are continually changing. A survey of harvest hands in the 1920 wheat belt, for example, found that they had "lost" more than 40 percent of potential worktime during the season, for reasons of communication coordination, distance and weather.[34] It is hardly surprising that Texas soon "realized the necessity for state action in creating a more rational labor market," or that participants pushed for state and federal agencies "to maintain the equilibrium in the labor supply."[35] If we take a snapshot of the migrant labor market at a point in time, it may seem to be a logical set of individual transactions in which all parties gain. Yet it sets up a situation in which employers have an active interest in maintaining barriers between their workers and the surrounding community, through language, information, or (sometimes) brute force. Such systems require legal and political support in innumerable ways, and hence these are fundamentally social choices.

What then determined which crops and regions would go in one direction and which the other? Obviously, technical characteristics of the crops

34. D. D. Lescohier, "Harvest Labor Problems in the Wheat Belt," USDA Bulletin No. 1020 (1922), 10.

35. Coalson, *Development of the Migratory Farm Labor System,* chapter 5.

Table 10. Wages of Farms and Common Labor, 1869–1919
(Daily Wage Without Board)

	1869		1890		1919	
	Farm	*Common*	*Farm*	*Common*	*Farm*	*Common*[a]
Massachusetts	$1.41	$1.60	$1.45	$1.52	$3.40	$3.12
New Jersey	1.20	1.64	1.25	1.44	3.32	3.52
Indiana	1.00	1.56	1.05	1.38	3.21	3.12
Wisconsin	1.15	1.54	1.26	1.37	3.63	3.20
Iowa	1.12	1.69	1.23	1.33	4.24	3.36
Kansas	1.15	1.87	1.10	1.47	4.47	3.44
California	2.13	2.31	1.55	1.96	3.90	3.52
Texas	.85	1.27	.97	1.68	3.15	2.88

[a] Daily wage × 8.
Source: USDA, *Crops and Markets* 19 (1942), pp. 150–151; Stanley Lebergott, *Manpower in Economic Growth,* p. 541.

are basic, such as seasonality, scale economies, and susceptibility to mechanization. The differences by product in the incidence of wage labor are evident at a glance (Table 11). In the case of wheat, the seasonality displayed in Figure 4 is drastically moderated in comparison to the premechanized pattern, when peak labor demands could be as much as 20 to 30 times the rate in the balance of the year.[36] By contrast, mechanizing the cotton harvest was far more challenging, and mechanizing fruits and vegetables was even more so.

But technical determinism is not the whole story. We cannot rule out the possibility that the geography of crop specialization is a function of labor market institutions and other public policies, nor the possibility that technology itself is shaped by labor market conditions. California, for example, was wheat-producing country in the late nineteenth century, under labor-scarce conditions that prompted a degree of mechanization well ahead of the rest of the country. The share of wheat harvest work performed by transient labor varied widely among states: in 1920, from 7 percent in Minnesota to 41 percent in North Dakota and 43 percent in Washington, among the leaders.[37] Or consider the case of cotton. Sharecropping was a way of securing peak-load labor by offering an annualized labor package. Though often portrayed as a quasi-feudal form, sharecropping in the South was a high-turnover system in which labor was highly mobile from year to year, though not from season to season.[38] Perhaps this system was uniquely adapted to cotton, because cotton has

36. Lebergott, "Demand for Land," 186.
37. Monthly Crop Reporter, April 1921, 45.
38. Wright, *Old South, New South,* chapter 4.

Figure 6. Wage Index for Farm Laborers and Factory Laborers, 1914–1939 (1914 =100) Source: Daniel J. Ahearn, The Wages of Farm and Factory Laborers, 1914–1944 (New York: Columbia University Press, 1945), 227.

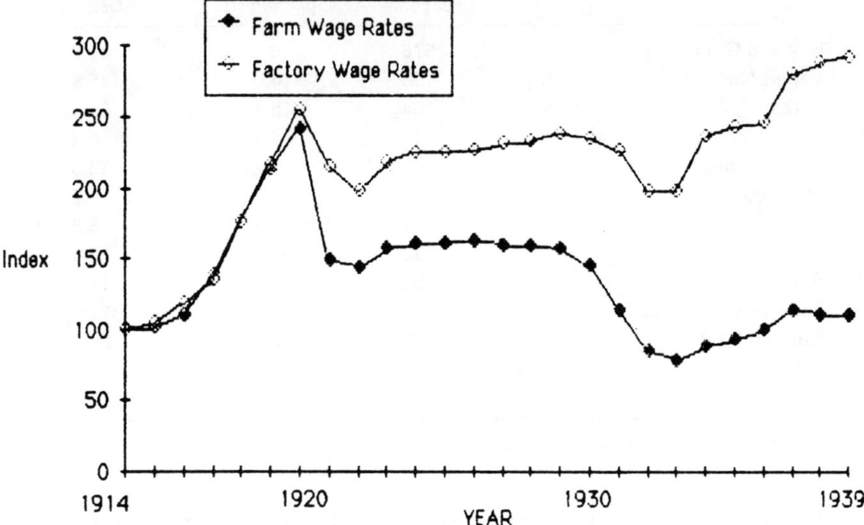

two peak periods, an early one in cultivation and a late one for the harvest. Once in place, however, sharecropping discouraged mechanization.[39] When partial mechanization for the pre-harvest operations became available, the plantation-sharecropping areas were slow to adopt it, but the new growing areas in western Texas rapidly moved to a harvest-only migrant wage-labor system.

It is sometimes argued that the size of the initial land holdings was the key element which shaped the choice of crops, labor systems, and technology. In California holdings of vast size were a legacy of the treaty of Guadalupe Hidalgo, and the average scale of operations in that state was several times larger than elsewhere, even for the same crops.[40] The relationship between wage labor and farm scale is also very striking (Table 12).

But the causation cannot be as simple as that. Initial purchases in the North Central region were also quite large, yet these were subdivided and resold relatively quickly. Is it the case that a set of public policies which

39. Warren C. Whatley, "South Agrarian Labor Contracts as Impediments to Cotton Mechanization," *Journal of Economic History* 47 (March 1987), 45–70.
40. Varden Fuller, "The Supply of Agricultural Labor as a Factor in the Evolution of Farm Organization in California," in U.S. Senate Committee on Education and Labor, *Hearings on Violation of Free Speech and the Rights of Labor,* Part 54 (1940), 19780.

Table 11. Expenditure on Labor by Principal Source of Income, 1899

	Per Farm	Percent of Value All Products	Products not fed
Hay and Grain	$76	8.1	10.0
Vegetables	106	14.0	15.9
Fruits	184	18.4	20.1
Livestock	65	6.1	8.2
Dairy Produce	105	9.7	13.3
Tobacco	51	7.3	8.3
Cotton	25	5.2	5.9
Rice	299	21.8	22.4
Sugar	1985	35.7	37.3
Flowers and Plants	675	22.5	22.6
Nursery Products	1136	22.4	22.9
Taro	51	12.0	12.0
Coffee	360	63.3	63.3

Source: Twelfth Census (1900), *Agriculture,* Part 1, p. cxxviii.

through most of the country's history actively favored family-scale farming, became oriented in a different direction on the West coast, beginning with public support for vast irrigation projects? Public policy debate clearly recognized that the trade-off was between maximum land yield and stable year-round employment.[41] In 1940, Varden Fuller argued that the crossover point came with the capitalization of land values on the basis of large-scale production of maximum value-per-acre crops.[42] At that point the financial incentive to press for supplies of cheap temporary harvest labor was overwhelming. So long as their labor strategies interacted with domestic urban and industrial labor supplies, however, growers found themselves under continuing economic and political pressures. The development of a self-contained migrant labor system at a distinctly lower wage did seem to represent a kind of sociopolitical equilibrium. By the 1920s, Fuller wrote, "the concept of abnormality associated with employment of itinerant and casual workers had largely passed away. California agriculture was declared *by nature* to be such as to demand a permanent supply of itinerant laborers."[43]

The issue of proletarianization in American agriculture, outside of the South, was not a question of dispossessing the existing farm population

41. Cletus E. Daniel, *Bitter Harvest: A History of California Farmworkers, 1870–1941* (Ithaca: Cornell University Press, 1981), chapters 21 and 2.
42. Fuller, "Supply of Agricultural Labor," 19825.
43. Ibid. 19882.

Table 12. Expenditure on Labor by Farm Scale, 1899

Value of Products Not Fed to Livestock ($)	Per Farm	Percent of Value	
		All Products	Products Not Fed
$1–49	$4	8.7	13.6
50–99	4	4.5	6.1
100–249	7	3.2	4.1
250–499	18	3.9	4.9
500–999	52	5.8	7.4
1000–2499	158	8.5	10.9
2500 and over	786	13.6	15.8
Area in Acres	Per Farm	Percent of Value	
		All Products	Products Not Fed
3–9	$18	8.3	9.0
10–19	16	5.9	6.7
20–49	18	4.7	5.6
50–99	33	5.2	6.6
100–174	60	6.5	8.4
175–249	109	7.9	10.3
250–499	166	9.4	12.2
500–999	312	13.0	16.3
1000 and over	1059	17.4	19.9

Source: Twelfth Census (1900), *Agriculture,* Part 1, p. cxxviii.

from their property, so as to rationalize production or technology. The issue instead was whether communities would tolerate an itinerant farm labor force of "outsiders" as a permanent, institutionalized feature of American agriculture. Sharecropping in the South is an example of an institution which developed to accommodate "local" propertyless labor on a permanent basis: in effect an annual contract with incentives suitable for married men with families. Elsewhere in the United States, however, the high seasonality associated with specialized commercial crop production made year-round labor too expensive. But temporary seasonal labor was also expensive, for reasons internal to the labor market (the shiftless, unreliable character of laborers available for such work) and external as well (social objections to wandering "bums" and "hoboes," or demands from farm workers for a stabilized status with opportunities for advancement). The dominant mode of escape from this dilemma was mechanization, a strategy in which the United States was the world leader. But where mechanization was technically difficult, the combined result of these pressures was the emergence of a migrant labor system as insulated as possible from urban and industrial labor markets. This system began in the

Southwest around the turn of the century, spread to California by the 1920s, and continued to expand eastward in the twentieth century, for the most part little noticed or bothered by most Americans, organized labor included. The social invisibility of these migrant workers and their families, their insulation from the mainstream of industrial development (quite in contrast to traditional proletarianization scenarios) is perhaps the major reason why most Americans do not believe that their country's agricultural history has much to do with proletarianization.

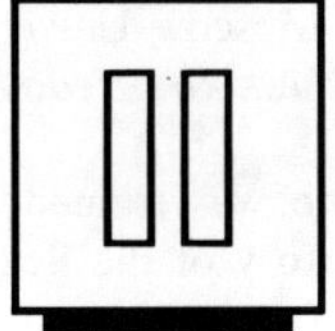

Russian Scholars

Contributors to Part II

Leonid Iosifovich Borodkin, deputy director of the Department of Source Study at the History Department of Moscow University, is author of *Mnogomernyi statisticheskii analiz v istoricheskikh issledovaniiakh* (1986).

Oleg Grigor'evich Bukhovets, who was trained at Moscow University, is senior fellow at the Institute of History of the Belorussian Academy of Sciences in Minsk.

Irina Markovna Garskova, author of *Tendentsii agrarnogo razvitiia Rossii pervoi poloviny XVII stoletiia: Istoriografiia, komp'iuter i metody issledovaniia* (1986) in collaboration with Milov and L. B. Bulgakov, is a research fellow at the Department of Source Study, Moscow University.

Academician **Ivan Dmitrievich Koval'chenko** is director of the Department of Source Study for the History of the USSR, Moscow University, and served for many years as editor in chief of *Istoriia SSSR*. He is the author of many books and articles, including works in collaboration with Milov, Selunskaia, and Borodkin.

Leonid Vasil'evich Milov, a professor at Moscow University, is author of *Issledovanie ob "Ekonomicheskikh primechaniiakh" k General'nomu mezhevaniiu"* (1965) and co-author, with Koval'chenko, of *Vserossiiskii agrarnyi rynok. XVIII—nachalo XX v.* (1974).

Natal'ia Borisovna Selunskaia, assistant professor at Moscow University, is co-author of *Sotsial'no-ekonomicheskii stroi pomeschich'ego khoziaistva Evropeiskoi Rossii v epokhu kapitalizma* (1982).

Daniel Field translated the essays by Soviet historians.

The Political Consciousness of the Russian Peasantry
in the Revolution of 1905-1907:
Sources, Methods, and Some Results

O. G. BUKHOVETS

Introduction

Historians who must assimilate a large quantity of letters, petitions or other similar documents are likely to suffer confusion. They sense patterns, relationships or tendencies, but find it difficult to ascertain and articulate them with precision. In this dilemma, they can rely on intuition and experience, but they may be uneasy about their findings, even if they offer them with outward assurance. Oleg Bukhovets's article, in which he assaults a particularly massive and complicated body of (relatively familiar) documents, shows another way out of the dilemma: the documents are reduced to variables, the variables collapsed into a smaller number of categories, and the categories subjected to statistical analysis. Even historians who are not drawn to quantitative methods will find this way inviting. Yet they may lament, and with some justice, what is lost in the process of reduction (which is just what intuition grasped most surely). To say, "I don't think Monica should have another baby right now," may, in certain circumstances, have the same significance as "I support a woman's right to choose," but the terms of expression suggest very different auspices and influences. By the same token, peasants who maintained that "the land is the gift of God" had the same objective as those who called for the "nationalization" of the land, but the form of expression is as important as the demand itself. It is now feasible, as perhaps it was not when Bukhovets began his research, to enter an entire corpus of historical texts into a computer and to subject every word (if need be) of these texts to analysis—an analysis that would neither preclude nor efface the kind of analysis Bukhovets carries out here. Someone should carry out that kind of analysis on peasant petitions and resolutions from the Revolution of 1905-1907, but it is even more important that Bukhovets and other historians continue along the way he has blazed and, in particular, adhere to the system of variables and categories that he has so carefully worked out.

D.F.

Prior to the Revolution of 1905-1907, the Russian peasant movement did not have a political character. Peasants rose up in behalf of particular interests and demands and were not concerned with altering the socio-economic structure and political system of the nation as a whole; they failed to understand the unbreakable link between the struggle for "land" and the struggle for "liberty." This failure was a direct result of the autocracy's policy of conserving the age-old, patriarchal character of peasant life by circumscribing the rights of the peasant estate and regulating all aspects of the peasants' lives, even their family relationships. Government and church propaganda aimed at cultivating religiosity in the peasantry and inculcating faith in "the good tsar." Public education was at a low level. As a result, right up until the Revolution of 1905-1907, a considerable portion of the empire's peasants remained totally backward, downtrodden, and indifferent to all issues beyond the limits of their village.

Other factors, however, began increasingly to act upon the consciousness of the peasantry; these factors were linked to the development of capitalism and to the ripening of the preconditions for a bourgeois-democratic revolution. The movement of masses of the rural population, attracted by the rapid growth of seasonal work, made peasants more venturesome[1] and broadened the spectrum of their social interests. The slow, steady growth of education in the country, especially in the larger settlements, expanded the peasants' horizons and awakened in them a feeling of self-worth.[2] The activities of teachers, doctors, agronomists and other members of the rural intelligentsia with democratic leanings had a large impact, as did the widespread propaganda and agitation undertaken by neo-populist organizations, which increased markedly on the eve of the revolution. The struggle of the Russian proletariat and its party had an enormous revolutionary influence on the peasantry. As a result (as Lenin emphasized),[3] for the first time in the history of post-reform Russia, the peasant movement achieved the level of a political movement in 1905-1907.

The manifestations of the politicization of the peasant movement in 1905-1907 took many forms: peasant participation in the activity of the All-Russian Peasant Union, political demonstrations in the countryside, the creation by peasants of fighting political organizations, and so on. The most widespread form of political movement in the countryside, and chronologically the first, was the so-called "resolution movement"—that is, the adoption by village assemblies and *ad hoc* peasant meetings of resolutions (*prigovory*), instructions (*nakazy*),[4] declarations, appeals, and telegrams of a political nature; these were sent to the state duma, to various political parties and organizations, to editors, and to government agencies. Thousands of such documents are preserved in central and regional archives; hundreds appeared in newspapers at the time of the revolution. Nearly 300 have been published in document collections.

The authorities interpreted the adoption of these resolutions and instructions as "criminal agitation"; the documents were confiscated, and their compilers were hunted down and prosecuted. There is a great deal of information in the archives about the arrest and imprisonment of persons suspected of composing such documents. The resolution movement represented a political danger to the old regime because, in assembling to discuss and generate the documents, the peasants organized and unified, and so came into confrontation with the authorities. It was also an ideological danger, because the endorsement by peasants of documents expounding a revolutionary-democratic program attested

---

[1] V. I. Lenin, *Polnoe sobranie sochinenii*, 5th ed. (hereafter *PSS*), vol. III, p. 314.

[2] Lenin, "Proekt programmy nashei partii," *PSS*, vol. IV, p. 228.

[3] Lenin, "K derevenskoi bednote," *PSS*, vol. VII, p. 196; "Doklad o revoliutsii 1905 g.," *PSS*, vol. XXX, p. 315.

[4] A *prigovor* was a resolution, in normal times usually concerning routine issues of the village economy, formally voted by an assembly of heads of households in accord with the legislation governing peasant self-administration. A *nakaz* was an instruction to a delegate or an assembly, like a *cahier de doléance* sent to the French estates-general in 1789.

to the diminishing influence of the patriarchal ideology and, by the same token, to the growing appeal of the revolutionary-democratic ideology.

The circumstances in which these resolutions and instructions were composed and adopted varied considerably. Some were voted at village assemblies that took place relatively peacefully, either because the authorities were not aware of them or because they could not bring sufficient police power to bear. Others were adopted by assemblies that convened under the immediate threat of conflict with the authorities. Documents were adopted at underground assemblies, political meetings, peasant congresses and even in prisons. For example, the instruction to the Second State Duma drawn up by peasants from eighteen villages in Bugul'ma District of Samara Province was voted at an underground assembly on March 28, 1907.[5] Another example: Voronezh *agrarniki*[6] confined in the provincial prison nevertheless managed in March of 1907 to compose, sign, and smuggle out an instruction to the duma.[7]

Research into the contents, the stylistic and orthographic peculiarities of these documents, and the circumstances under which they were adopted reveals that some of them were entirely composed by the peasants themselves. An example is the resolution—very distinctive in both form and content—of peasants from the village of Trostianki, Samara District.[8]

Most often the documents were "co-authored" directly or indirectly by peasants in collaboration with Bolsheviks, with members of various neo-populist parties and organizations, and with the rural intelligentsia. This circumstance in no way casts doubt on their peasant authorship. Consider, for example, a resolution by the village of Kanaevki, Nikolaev District, Samara Province.[9] It is written in "intelligentsia" language. There is no doubt about its authenticity, however, as the circumstances of its adoption are known. The assembly convened to approve it was broken up, but the peasants persevered and, shortly thereafter, attached about 300 signatures to the resolution in secret and sent it to the First State Duma;[10] their conduct attests eloquently to the "peasant" content of the text. Even where peasants met to adopt ready-made draft resolutions, the assembly discussed them, made additions and finally voted for (or sometimes against) them.

Peasants did not feel any need to cast the new content in new forms. They simply used the traditional repertoire of written forms that village assemblies

---

[5] A. L. Sidorov et al., eds., *Revoliutsiia 1905-1907 gg. v Rossii. Dokumenty i Materialy. Vtoroi period revoliutsii. 1906-1907 gg. Ianvar'—iiul' 1907 g.*, book 1 (Moscow, 1963), pp. 379, 554.

[6] Active peasant participants in the revolution.

[7] *Trudovoi narod*, March 29, 1907.

[8] For the text of the resolution, first published in *1905 v Samarskom krae* (Samara, 1925), see O. G. Bukhovets, "Matematika v issledovanii obshchestvennogo soznaniia: Krest'ianskie prigovory i nakazy 1905-1907 gg.," *Chislo i mysl'*, fasc. 9 (1986), p. 35.

[9] *Krest'ianskie nakazy Samarskoi gubernii* (Samara, 1906); Bukhovets, "Matematika v issledovanii," pp. 35-36.

[10] *1905 v Samarskom krae*, appendix, p. 138.

used to record decisions on administrative and economic matters. The resolution was one such form. They were formulaic, following a rather strictly defined order, to which most of the documents under study conform. They can be broken down into four major elements: the opening (place, date, number of participants); the narrative (setting forth the reasons for the poverty of the peasantry or the common people); the resolution as such (a program of reforms of a political and socio-economic order); and the closing (signatures or "x's," the stamp of the village elder or cantonal authorities).[11]

The literature on the study of resolutions and instructions is impressive; since 1906 more than 70 specialized and general studies have addressed these documents.[12] The results achieved are also considerable; more than a thousand of these documents have been studied. So far, however, the informational potential of these sources has by no means been fully exploited. Characteristically, studies of the resolution movement are limited to the study of the peasants' *demands*. It is obvious, however, that, important as these elements are, they are only one of many components of interest. Other elements include: pronouncements; inquiries and requests; appeals, slogans and greetings; warnings and assertions; and undertakings to carry out particular actions. We lose much significant social information if we leave these other elements out of account. We also must note that in consequence of this limited approach to the source material, much "outlying" information is lost, even in the demands. Furthermore, the number of demands discussed somehow remains the same—two or three dozen—in all these works, whatever the territorial and temporal limits of the study or the number of documents analyzed, so that the real variety of the demands is not adequately reflected. Their palate was much richer than historians so far have shown.

It follows that working with two or three dozen variables entails examining the mass of demands from a very high level of generalization. This kind of generalization is a totally appropriate and necessary method of studying mass phenomena, for it facilitates isolating what is most common and typical. The level of generalization must, however, be such that, after singling out what is typical, we can also identify what is distinctive.[13] Unfortunately, recent research has not reckoned with the latter condition.

* * *

Our efforts to accomplish the research design suggested above were undertaken using material from two provinces, Samara and Voronezh. These

---

[11] The telegram was, because of technical limitations, usually just a distillation or edited version of the third of these elements.

[12] For a list of the most important of these studies, see O. G. Bukhovets, "Massovye istochniki po obshchestvennomu soznaniiu krest'ianstva (Opyt primeneniia kontent-analiza pri izuchenii prigovorov i nakazov 1905-1907 gg.)," *Istoriia SSSR*, 1986, no. 4, p. 105.

[13] See *Kolichestvennye metody v istoricheskikh issledovaniiakh* (Moscow, 1984), p. 35.

two were among the most populous provinces in Russia, with more than five million peasants in all, and so provide an entirely adequate "polygon" for our research. To get optimal results, we first developed, on the basis of social, economic and political indicators, a typology of both provinces, dividing the nineteen districts into regions. For Samara Province, these proved to be the northern and southern regions;[14] for Voronezh, the northwestern, the central, and the southeastern.[15]

The documents at our disposal, amounting to more than twice the quantity that researchers have collected, consist of resolutions, instructions, manifestos, and telegrams. In all, there are more than 200 documents: 156 for Samara Province and 44 for Voronezh.[16]

Because the documents of the resolution movement were, as explained above, formulaic, they lend themselves to quantitative analysis in general and to content analysis in particular.[17] In the texts under study, words are repeated and constitute invariant components which can serve as units of analysis. For example: 1.) in the category "pronouncements"—"Peasants and their land are indivisible"; 2.) in the category "demands and slogans"—convening of a constituent assembly, "Land and Freedom!"; 3.) "inquiries and requests"—a request for an allocation of land, an inquiry whether peasants should purchase land or await the resolution of the agrarian question; 4.) "greetings"—to the duma, to the *Trudovik*[18] faction; 5.) "appeals, warnings, promises, declarations"—a summons to the army "not to spill the people's blood," warnings to deputies that they will be punished by the common people if they make concessions to the government, promises to render aid to the duma in its conflict with the government; 6.) "undertakings to carry out particular actions"—driving out the police, refusing to provide conscripts.

Let us now attempt to reckon with these components of content. As an example of the process, let us take the shortest of all our documents, a telegram to the First State Duma from the peasants of Verkhniaia Khava in Voronezh Province: "At an assembly held on May 22, the peasants of the Verkhniaia Khava Village Community in Voronezh District resolved to send a greeting to the duma; struggle on...until you gain land and rights for the people. Be not dismayed! There are 100 million people behind you."[19] Even in so short a text, we find at least six variables directly expressed in the text and reflecting needs,

[14] Bugul'ma, Buguruslan, Samara and Stavropol' Districts in the former, Buzuluk, Nikolaev, and Novouzensk Districts in the latter.

[15] Respectively, Voronezh, Korotoiak, Nizhnedevitsa, Zemliansk and Zadonsk Districts; Bobrovsk, Ostrogozh, Pavlovsk, and Biriuchin Districts; and Bogucharsk and Novokhopersk Districts.

[16] Bukhovets, "Massovye istochniki," p. 105.

[17] Bukhovets, "Massovye istochniki," p. 104.

[18] The *Trudoviki* were a peasant-oriented duma grouping with views similar to those of the Socialist-Revolutionaries.

[19] Central State Historical Archive, Leningrad (TsGIA), *f.* 1278, *op.* 1 (Pervyi sozyv), *d.* 2, *l.* 922.

interests, mood, attitudes, and, finally, the illusions of the peasants who adopted this text: 1.) a judgment that the peasants are "dispossessed;" 2.) greetings to the duma; 3.) a summons to the duma to fight for the interests of the people; 4.) a demand for land; 5.) a demand for political rights; 6.) a declaration of support.

The number of such variables in the entire array of documents is impressive, amounting to 177, and the range of themes is extremely broad, extending from a scrupulous determination of the land-allotment norm for every citizen of Russia ("15 *desiatiny* apiece") to a demand for the establishment of an international court to resolve disputes among states. No less broad is the "amplitude," so to speak, of political vacillation; along with a large quantity of firm demands developed by revolutionary democrats (for example, a constitutent assembly, nationalization of the land), are to be found some with a clear imprint of monarchist illusions ("we need a tsar" and the like).[20]

What can be said about the 177 variables to be found in the 200 documents? Frequencies range from 1 to 105, with an average frequency of 11.4.[21] Of the variables of a socio-economic character, most common are various kinds of attestations of poverty, ruination, land shortage, and high rents and taxes. This is a kind of register of the "cursed problems" of the peasantry of the empire. Next come various solutions of the agrarian problem, ranging from the most decisive (nationalization of land) to the more moderate (various forms of supplementary allotment) to the vague (the slogan "Land and Liberty"). Demands for a replacement of indirect taxes by a uniform income tax and for the cancellation of tax arrears and redemption payments are also distinguished by a high rate of frequency. The frequent demand for communal landownership and equalized access to the land is evidence of the "centripetal" disposition of the peasantry concerning the commune.

Among frequently encountered political demands, those that reflect the peasantry's intense hostility to the bureaucracy and to the squires occupy a prominent place. Among them we also find declarations of adhesion to the All-Russian Peasant Union and of readiness to take up struggle, demands for a constitutent assembly and various kinds of demand for political freedoms, for the abolition of the death penalty and of martial law. There follow demands for a democratic reform of the state structure and for elections with universal, equal, and direct suffrage and a secret ballot. The very widespread insistence on amnesty for fighters for "the people's cause" attest to the extent that a sense of class solidarity had emerged within the peasantry, internal divisions notwithstanding. The frequency of demands for free universal public education is remarkably high, and represents a fundamentally new social phenomenon.

Despite its predisposition to revolutionary sentiments, the peasantry, as a class of petty proprietors, was also fertile soil for various kinds of constitutional illusion. In the period of the revolution, this trait found marked expression in

---

[20] Bukhovets, "Massovye istochniki," appendix 1, pp. 115-18.
[21] Bukhovets, "Massovye istochniki," appendix 2, p. 119.

the hopes placed on the duma by peasants; the peasants who adopted resolutions and instructions were no exception, for 158 (79%) of the two hundred documents were addressed to the First or Second State Duma, and there are several variables that show signs of parliamentary illusions: expressions of confidence in the duma or of solidarity with it, willingness to render aid to it, and so on. This confidence, however, was by no means blind; among the variables relating to the duma is a warning that if the deputies make concessions to the government, they will be punished as "enemies of the people."

Let us turn to the variables of low frequency. Of the 177 variables, 128 are to be found less than ten times in the resolutions and instructions under consideration, and 10 of them are encountered only once. This large body of variables can be divided into several groups. One group includes those that reflect the aspiration of small producers to secure their existence in the face of the development of capitalism in agriculture by means of various rigid norms of land allotment, bans on the hire of labor, and so on. In another group of variables are various "back-burner" or, under the circumstances, particular questions, such as the demand for a compulsory survey of land by the state; here we can include demands and petitions for increased food loans and aid to peasants afflicted with famine or murrain. In yet another group fall variables of a, so to speak, strategic character; these demonstrate the level of political maturity of the peasants and the degree to which they understood issues on which the outcome of the revolution depended. Here we find assertions that all the "toiling people" should be united, demands for a republic and for an eight-hour day, and expressions of distrust for the duma and the Kadet Party. Another group of variables is composed of commitments to undertake particular actions, such as seizure of land or refusal to participate in elections to the duma. A fifth group includes variables that register the peasants' understanding that various institutional survivals from the past, such as land allotment and the commune, could not endure; these took such forms as the demand that allotments be converted into private property and that legislation concerning the commune be repealed. A sixth group is composed of variables that attest to the political naiveté and immaturity that were characteristic of the peasantry and took the form of monarchist, constitutional and reformist illusions. Along with variables of the type "We need a tsar" and "Eliminate the bureaucratic barrier between the tsar and the people" should be included endorsement of a "just" appraisal and redemption of lands alienated from the squires for the peasants' benefit, appeal to peasant deputies and Kadet deputies to unite, expression of hopes for the realization of the "cherished hopes of the people" by peaceful means, and so on.

A final group consists of a large number of variables that are in essence variants of high-frequency demands and assertions. For example, the assertion that land is "a gift of God" and should not be held as private property is a variant of the demand for the nationalization of the land; demands for the contraction of the state apparatus and for popular supervision of the activity of all government organs are variants of the demand for a democratic reconstruction of the state. This last circumstance is important, as it opens the way to the amalgamation of variables that we shall resort to later.

This cursory analysis shows rather plainly how fully the variables we have isolated reflect the socio-economic and political needs, interests and ideals of the insurgent peasants, their moods, feelings and attitudes to a wide range of problems confronting the nation as a whole and the peasantry in particular.

Ascertaining the informational potential of the resolutions depends in large measure on a determination of the extent to which they could reflect what was general and what was particular in the peasants' program. This determination is all the more pressing because various parties and organizations circulated massive quantities of model resolutions in the countryside, producing a remarkable uniformity in the documents of the resolution movement. To what extent did this kind of pressure "smooth out" the content of the resolutions? Of the 177 variables, 70 are found in resolutions from both provinces and the other 107, or 60.5%, are present only in resolutions either from Voronezh or from Samara.[22] Could this be because the non-contiguous provinces of Voronezh and Samara did not constitute a single whole? It would obviously be more revealing to establish the extent of the common and the particular in documents from different regions of a single province. But in an intraprovincial cross section, the proportion of common and particular variables is astonishingly similar to our interprovincial finding. Of the 161 variables encountered in documents from Samara Province, 74, or 46%, are found in both the northern and the southern regions, while of the 84 Voronezh variables, 34 (40.5%) are found in resolutions from both the northwestern and the central regions.[23] This coincidence of common and particular variables clearly reveals the complexity of the information contained within the resolutions and instructions as well as their regional coloration. The latter becomes more striking still in light of the finding that variables to be found in documents from all four regions of the two provinces total only 28, or 15.8% of the total.[24] In short, beneath the seeming uniformity of the resolutions and instructions lurks a wide variety of content. And given the differences among the four regions, we can state with confidence that the balance between the common and the particular in the peasant programs was a function of socio-economic and socio-political differences among the locales that produced the documents.

The discussion so far of the 200 documents and of the variables they exhibit represents only the first stage of our analysis. The next, decisive stage is ascertaining the pattern of correlation among the variables and establishing the extent to which it reflects the tendency and structure of the political consciousness of the peasants who produced resolutions and instructions.

---

[22] Bukhovets, "Massovye istochniki," p. 108.

[23] Bukhovets, "Massovye istochniki," p. 110. The total of resolutions from the southeastern region of Voronezh province is only four, too insignificant for comparison with the others.

[24] Bukhovets, "Massovye istochniki," p. 111.

First of all we amalgamate similar variables into categories; the procedure simplifies ascertaining correlation by reducing the number of variables and increasing their statistical weight. For example, demands for the nationalization of land, for the creation of a reserve of land for the whole people, and the pronouncement that land is a "gift of God" can be amalgamated into the category "abolition of private property in land and transfer of land to the people." Similarly, among the 177 variables are several that reflect the peasants' understanding that the people can improve its situation only by taking up the struggle against autocracy—for example, the pronouncement "The people and the army will sweep away the old order." These can be amalgamated into the category "expression of willingness to fight for their interests."

By such means we obtain 79 categories. Although this total is less than half the number of variables, it is still considerable. To analyze the correlation among the categories, it is necessary to render all the documents in a common form. Hence we assigned a code number to each category and noted the presence or absence of each category in the resolutions and instructions. Obviously, to calculate the correlation of 79 variables in 200 observations by hand is scarcely feasible, so we turned to the computer.

* * *

For the analysis of correlation, we chose Chuprov's coefficient (T), which can take a value between zero and one. A value of zero for the T coefficient means that the variables in a pair are statistically independent, that there is no association between them; the nearer the value of T comes to one, the closer the correlation between the variables. The computer was programmed to "deliver" correlations with a value for T of at least 0.1. Since some of the categories are encountered only one or two times in the entire sample, these were excluded from analysis; only those 60 categories with a frequency of at least three were entered into the computer. Preliminary computer analysis revealed that for an understanding of the political consciousness of the peasant authors, values of T exceeding .15 were significant, giving us 42 categories for further analysis.

Figure 1

It is beyond question that close correlations are most important for revealing the political consciousness of peasants. Let us begin our analysis with them, and then descend by stages through the weaker associations. The highest values for T are the five that fall in the range between 0.4 and 0.8. The structure of association among these categories is easier to understand in graphic form, with circles representing categories and lines—the correlation between them, as in Figure 1. In Figure 1, the lines link up five categories, forming two groups. In one group are (category 7) the demand that the commune retain control of allotment land, (no. 9) the demand for government assistance in expanding petty landholding and credit for petty landowners, and (no. 8) a hostile attitude to the misnamed Peasant Bank. This group, then, is a tight cluster of categories reflecting traditional concerns of the peasantry and its desire for more land. In the second group are two demands, one (no. 13) for the introduction of an income tax, and the other (no. 31) for free universal public education; unlike the categories in the other group, these represent something fundamentally new in peasant consciousness. For even here, on the level of the closest correlations we find a socio-economic demand—a very far-reaching and emphatic one—and a purely political demand.

Figure 2

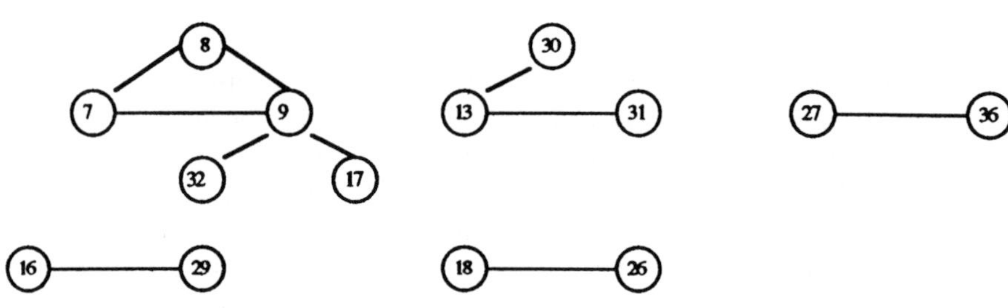

Let us turn now to coefficients with a value between 0.3 and 0.4, which number six. As Figure 2 shows, to the five categories in Figure 1, nine are now added—four economic and five political. Two categories now join those in the first group in Figure 1, and one joins the second; furthermore, three new groups are formed, with two categories in each. In the first group, category 9 (expansion of petty landholding) is now linked, on the one hand, with a demand (no. 32) that the situation of workers be improved and, on the other, with a category (no. 17) that reflects a compromising disposition among the peasantry: redemption of land alienated from the squires. In the second group, there is a natural linkage between category 13, the demand for an income tax, and no. 30, the demand for an end to the disabilities of the peasant estate and for a reform of local government and the courts. Of the new groups, it is remarkable that two of them are purely political and very "harmonious." Thus, one of the most militant categories, no. 18, the demand for a constituent assembly, is logically linked with an expression (category 26) of the peasants' mistrust for the duma on the grounds of its powerlessness. A second group is anti-government and pro-duma and links category 27, a commitment to give the duma every kind of aid in its struggle with the government, and no. 36, an appeal to the duma to mount a firm and determined defense of the people's interests. Then there is a seemingly curious and unexpected grouping; category no. 29, a demand for the introduction of an eight-hour day for rural and urban workers, turns out to be linked with the demand (no. 16) for abolition of the state liquor monopoly. In fact, however, there is nothing surprising in this linkage; peasants opposed the monopoly because they saw it as a instrument to rob the people and also as a means of its moral corruption.

Figure 3

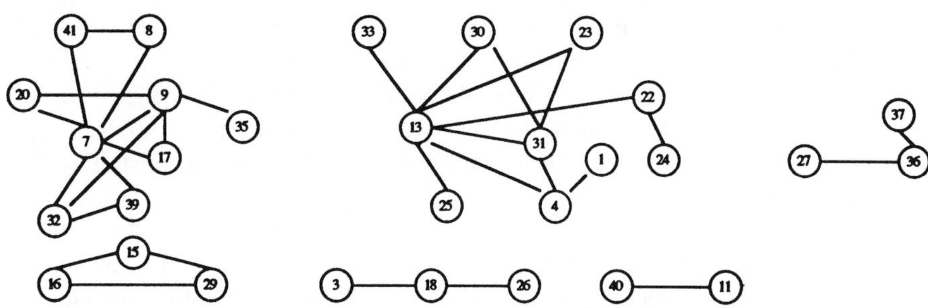

Let us descend to the next, third level and analyze the correlations ranging between 0.3 and 0.2, of which there are 24. Fourteen new categories have insinuated themselves into the five groups already formed, and two of them form a new "anti-squire" group: denunciations of the greed and rapacity of squires (category 40) is linked with a condemnation (category 11) of the system of labor-rental,[25] whereby peasants were compelled to work on estates in return for various kinds of loan in conditions approaching peonage. Each of the three groups that emerged on previous levels takes on a new member. The old revolutionary slogan "Land and Liberty" (category 37) joins the "anti-government, pro-duma" group; to the group concerning a constituent assembly is added category 3, a demand that the assembly decide the land question; and category 15, a demand for increased food loans and aid to peasants afflicted with famine

[25] On labor-rental (*otrabotka*) in this period, see I. D. Koval'chenko and N. B. Selunskaia, "Labor Rental in the Manorial Economy of European Russia at the End of the Nineteenth and Beginning of the Twentieth Centuries," *Explorations in Economic History*, vol. 18, no. 1 (1981), pp. 1-20.

or murrain joins what we can call "the eight-hour day and liquor monopoly group." Each new member corresponds perfectly to the spirit of its group.

Categories mostly adhered to previously formed groups. What emerged as the "traditional-economic" group takes on four new categories: they are a demand to convene the duma (no. 35), to put the church on a government allowance and cease exactions from the populace for its support (no. 39), an appeal for the unity of the entire "toiling people" (no. 20), and hopes for the realization of the "cherished hopes of the people" by peaceful means (no. 41).

At this level, too, we see the same contradiction among the associated categories in the "traditional economic" group, with very resolute categories side by side with relatively accommodating ones. As for the "combined" political-economic group, the new categories are very significantly distinct from those that became part of it on earlier levels. Among the socio-economic categories are moderately militant demands for the nationalization of land (category 1) and for its transfer to those who cultivate it with their own labor (no. 4). Among the political categories are demands for elections with universal equal and direct suffrage with secret ballot (no. 23), the introduction of "democratic freedoms" (no. 24) and the abolition of martial law, the police and the land captains (no. 25). Among the relatively more modest categories is the demand that the duma have "full authority" and that the administration be responsible to it (category 33).

We see that on the first three levels, thirty categories form four small correlational groups and two large ones. The latter embrace 19 categories and—this is beyond doubt—they are fundamentally different in their structures. The demands, slogans, appeals and so on in the "traditional" group are often mutually contradictory in a political sense, while the "combined" group is relatively harmonious. These structural differences will be useful to us.

Figure 4

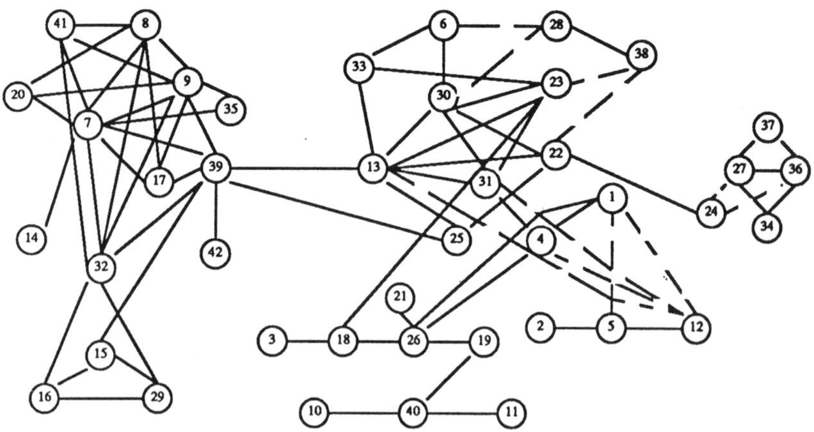

Turning to the fourth and last level, 36 coefficients fall in the range between 0.2 and 1.5. Twenty-five of them reflect associations involving categories that appeared in groups on the first three levels, and so form a tight network of affinity with 30 "old" categories. The other 11 coefficients involve associations among 12 new categories. Figure 4, which depicts the entire network of correlations, displays marked changes compared to Figures 1-3. First of all, two new—and very interesting—groups are formed from new correlations. One of these groups has to be called the "anti-discriminatory" group, for it consists of category 28, a demand for equal rights for all the peoples of Russia, and no. 38, a condemnation of discrimination on the grounds of ethnicity, religion or sex. Another links three socio-economic categories of a revolutionary character: an undertaking to seize land from the squires and the state (no. 2), a demand that peasants hold land on an equal footing and through the commune (no. 5), and a demand for the abolition of taxes and other exactions (no. 12).

A second change is the continued growth of the large groups, but now through the amalgamation of three small groups. The group which we provisionally designated "eight-hour day and abolition of the liquor monopoly" now firmly links all its categories to the "traditional economic" group. The latter group also takes on two new categories which, as before, are strikingly contradictory in their political tendency—in this instance, a firm attestation that taxes are ruinous (no. 14) and "Russia one and undivided," which was a slogan of the bourgeoisie and the squires (no. 42). Several categories now link the "combined" political-economic group to the constituent assembly group, which itself is supplemented by such militant categories as a condemnation of the "gentry party" of the Kadets for its flirtation with the government (no. 21) and an expression of the peasants' willingness to struggle for their interests (no. 19). The "anti-squire" group expands logically by, on the one hand, taking on category 10, a demand for lower land rents, and, on the other, by linking through category 19 to the constituent assembly group, which, in turn, has now become part of the "combined" political-economic group. As a result of this assimilation, the last of these groups now includes 19 categories. Moreover, if we descend further still, to the level where T ranges between 0.15 and 0.1, then—as the dotted lines in Figure 4 show—all the remaining small groups ("anti-government and anti-duma," anti-discriminatory, and the group of revolutionary socio-economic categories) become linked to the "combined" political-economic group.

By following the stages of the crystallization of links, we have brought to light two large groups of correlated categories which are rather firmly demarcated one from the other and linked only by a narrow "isthmus," consisting of the demand for an end to exactions from the populace to support the church. The structure of these groups—their substantive composition and the linking of various categories—reflects the deep structural differences within the political consciousness of the peasants who produced the resolutions and instructions. The categories in the "traditional-economic" group are often strikingly at odds one with another. As we have seen, beginning with the second correlational

level, revolutionary categories and categories manifesting monarchist illusions
and a compromising disposition are introduced in parallel. The "combined"
political-economic group, which consists primarily of new concepts, also
reveals contradictions in the political consciousness of the peasantry, but these
are contradictions of a somewhat different kind. The pattern of association
reveals that within the general political consciousness of the peasants who produced our documents were two, weakly-linked structural subdivisions, one of
which was filled out primarily with traditional demands and perceptions, while
the other consisted primarily of new elements that were politically rather harmonious one with another.

Can we say that one class of categories reflects the distinctive political
consciousness of one group of peasants who wrote resolutions and instructions
and the other—that of another group? Or do both classes represent distinctive
layers in the political consciousness of a single body of peasants? This is a fundamental and important question. We can answer it by ascertaining whether and
to what extent two types of variable are to be found within a single text. One
type, which we will designate the "programmatic," consists of particular pro-
positions in those writings of Lenin and Bolshevik party documents that set
goals and defined strategy in the Revolution of 1905-1907; the other consists of
variables expressive of the parliamentary, monarchist, and reformist illusions of
the peasants or of a disposition towards compromise.[26] We find that the "programmatic" variables occur in 90.5% of our 200 documents, variables reflecting
parliamentary illusions occur in 76%, and variables reflecting monarchist and
reformist illusions in only 16.5%. From these eloquent figures it follows, for
example, that "programmatic" and "parliamentary" variables, which figure
respectively in 9/10 and 3/4 of our sources, must necessarily intersect in many
of them. And indeed they do; the total of resolutions and instructions containing
both is an impressive 137, while monarchist and reformist variables coexist with
programmatic variables in 31 documents. The results appear in graphic form in
Figure 5. It shows that an absolute majority of the documents fall in the cross-
hatched zones of simultaneous incidence. It also draws our attention to two
important circumstances. First, the data show that the admixture of illusion in
"programmatic" documents consists primarily of variables related to the duma,
which, compared to monarchist variables, are a significantly more ephemeral
variety of peasant illusions. Second, despite the obvious linkage of programmatic variables and those reflecting illusions, no less obvious is the significant
degree of statistical independence of the former from the latter. If we take the
documents with variables relating to the duma, 9/10 of them also contain "programmatic" variables, while the inverse relationship is 3/4. Correspondingly,
almost a quarter of all the 181 documents containing "programmatic" variables
are free of duma variables, while less than one-tenth of the 152 documents containing duma variables lack "programmatic" variables. A comparison utilizing

---

[26] Bukhovets, "Massovye istochniki," pp. 112-19.

the monarchist and reformist variables shows the same pattern even more strongly: 82.9% of the documents with ''programmatic'' variables are free of them, while the latter are lacking in only two of the 33 documents that register monarchist illusions.

Figure 5

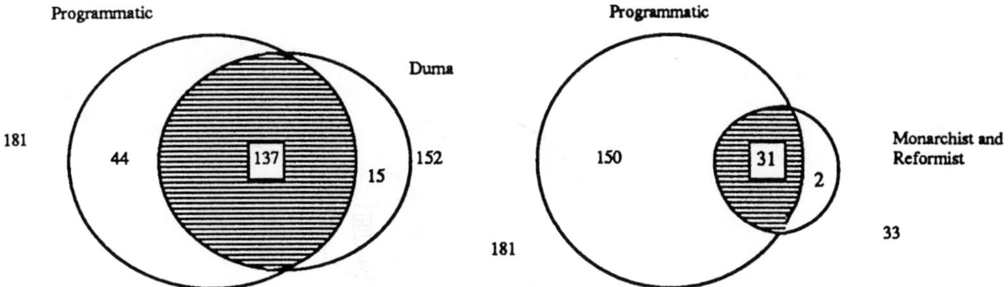

The peripheral role of the variables embodying illusions in the resolutions and instructions—and hence also the consciousness of the peasants who adopted them—also emerges from the fact that there were many more purely ''programmatic'' documents (that is, free of both parliamentary and monarchist illusions) than there were purely parliamentary and especially purely monarchist: respectively 19.5%, 7% and 0.5% of the 200 documents. There were 54 ''pure'' documents of all types; the content of the other 144 reflected the contradictory nature of peasant political consciousness.

To establish the extent of diffusion of various kinds of variable in gross totals and subtotals does not exhaust the question of contradictions in consciousness. It is no less important to establish whether we observe diffusion when we compare only the most significant, crucial variables. Consider in particular the agrarian variables, which both in quantity (about fifty) and in their importance are central to the resolutions and instructions. At one pole are radical and sweeping variables: demands for the abolition of private property in land, for the transfer of all land to those who work it with their own labor, for communal landholding and equal access to the land; resolutions on the seizure of land from the state, the ruling house and non-peasant landowners; and assertions that either a constituent assembly or local peasant committees should resolve the agrarian question.[27] These variables are found in 60 resolutions. The opposite pole

---

[27] Bukhovets, ''Massovye istochniki,'' pp. 112-19. Here and below, see Appendix I (pp. 115-18).

among the agrarian variables is occupied by those that reflect the peasants' reformist illusions concerning the resolution of the agrarian question: allowing the possibility that the state redeem the lands of private owners at a ''just price'' and also various kinds of demands and pronouncements to the effect that the duma should settle the question. To these should be added expressions of the peasants' hopes that they can obtain their ''cherished goals'' (that is, in the first instance, land) by peaceful means. Compromising variables of these kinds are embodied in 27 documents.

Figure 6

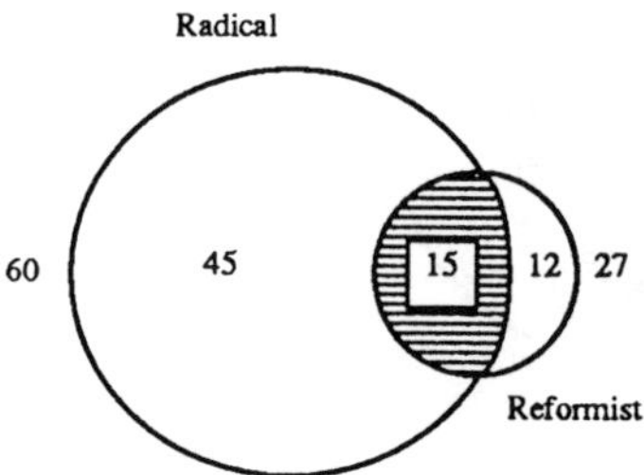

Figure 7

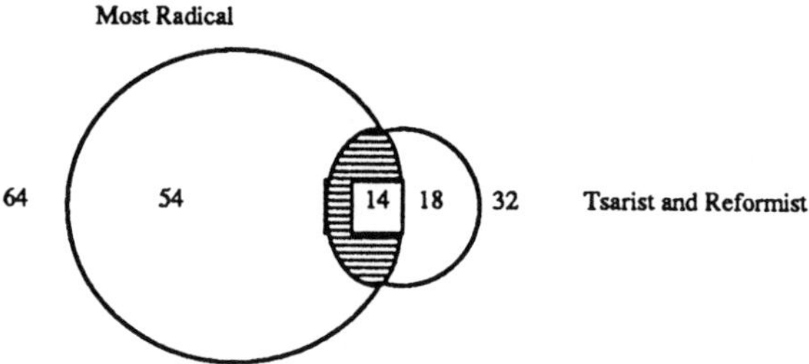

What, then, is the incidence in a single document of variables drawn from opposite poles of the peasants' agrarian program? The left side of Figure 7 shows that the incidence is considerable: a quarter of the 60 documents containing the most revolutionary agrarian variables also contain reformist variables. Once again the contradictory nature of peasant conceptions of means of resolving the agrarian question bespeaks itself, but the pattern is ambiguous and needs examination from another perspective: of the 27 documents containing compromising variables, these are, so to speak, cancelled out by more revolutionary variables in 15 documents. As for political variables, the ideological maturity and determination of the peasants who wrote the resolutions is most accurately conveyed by variables showing their recognition of the leading role of the working class in the revolution and of the necessity of rallying around the All-Russian Peasant Union and the *Trudovik* group, and by several others: the demand for a constituent assembly and a republic; expression of willingness to fight by any means until victory; resolutions undertaking to displace the local authorities, drive out the police, and organize resistance to the army; and finally, declarations of lack of confidence in the duma and in the Kadet Party. As their polar opposites we find expressions of monarchist illusions and of the kind of reformist illusions and compromising disposition that already appeared among the agrarian variables. The pattern that emerges from Figure 7 differs little from what we elicited from Figure 6. More than one-fifth of the 64 documents containing revolutionary variables also embody their opposites. The major difference is that in Figure 6, the radical variables "cancel out" the more moderate in more than half of the documents containing the latter (15 of 27), while in Figure 7 the corresponding relationship is about two-fifths (14 of 32).

Figure 8

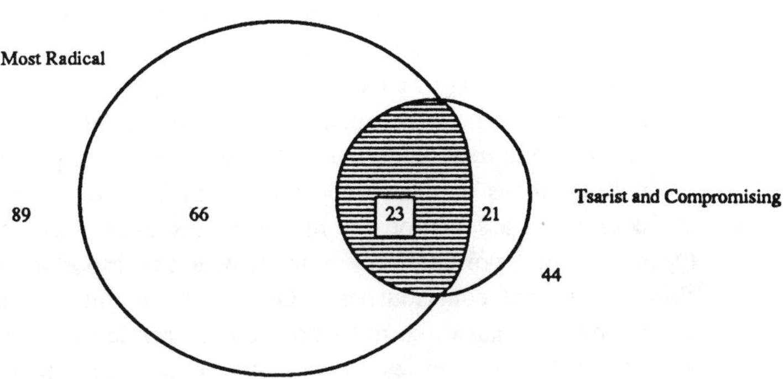

Let us now sum up by conflating agrarian and political variables. Figure 8 shows that the most radical variables appear in 89 documents, the monarchist in 23 and the reformist in 44. Hence a quarter of the documents containing the most radical variables also contain variables that are their political opposites, but the former "cancel out" the latter in more than half of the documents in which the latter appear. Here as before, the revolutionary elements occupy a conspicuously dominant position in peasant consciousness, and the patriarchal and reformist elements are peripheral.

The diffusion that we have brought to light of the components of the resolutions and instructions mirrors the diffusion of the components of peasant consciousness; it follows that the elucidation and analysis of this diffusion is a very promising avenue of research into the richness and the nuances of peasant consciousness. The period of the revolution was revolutionary indeed for the peasants of the empire; the spectrum of their political consciousness expanded dramatically, incorporating a mass of new conceptions. It is, therefore, not surprising that the assimilation of revolutionary ideas even by the most radical-minded peasants often intertwined in bizarre ways with patriarchal, vestigial conceptions. The coexistence in a majority of resolutions of traditional and novel elements derives from a kind of stratification of peasant political consciousness: the revolutionary strata, containing a sound understanding of the peasants' basic interests and objectives as a class, are intermingled with patriarchal-traditionalist strata, reflecting illusions about the means of securing these objectives. This combination is characteristic of peasant consciousness in general. Hence the contradictory nature of the peasant program embodied in the resolutions and instructions composed during the bourgeois-democratic revolution is a completely normal phenomenon, and does not mean that the peasants themselves were not revolutionary.

* * *

Have the results of our study of the resolution movement justified our resort to quantification and the computer? Our methods have enabled us to go beyond a general, abstract conception of the peasants' program to a specific representation of the infrastructure of its component elements. Our understanding of the phenomena under study has become more precise and rigorous. For example, analysis of the strong correlation coefficients reveals that the "crystallization" of association among categories takes place not within but between types (such as "political-economic"), which is important evidence of profound shifts in peasant consciousness. Or, to cite another example, common sense would indicate that there must be a close correlation between such reformist-accommodating categories as peasants' hopes for attaining their "cherished goals" by peaceful means (no. 41) and allowing compensation to squires for land taken from them (no. 17). It turns out that is not the case.

To be sure, the potential for fruitful application of mathematical methods to the study of the social consciousness and the struggle of the peasantry is by no means exhausted in this article, which represents only the first steps in this kind of application.

A Typology of Feudal Estates in Russia in the First Half of the Seventeenth Century (Factor Analysis)

L. V. MILOV
I. M. GARSKOVA

Introduction

Of the papers presented at the Tallinn conference, the contribution by L. V. Milov and I. M. Garskova was the most austere. Densely quantitative and restricted in its immediate subject matter, it is particularly far-reaching in its implications. The authors themselves underline one of these implications when they emphasize their finding that patrimonial estates (*votchiny*) proved to be more vigorous and viable in the 1630s than estates held on service tenure (*pomest'ia*). Historians have long conceived *votchiny* to be losing out to the *pomest'e* form, and particularly in the period in question, during the period of recovery from the Time of Troubles. The supposed superiority of the *pomest'e* form is closely linked to the rise of absolutism, the development of serfdom, and other major issues of Russian political, social and economic life.

On two other counts, Milov's and Garskova's findings confound our expectations. One is the major contribution that *bobyli* made to the restoration of the manorial economy after the devastating calamities of the late sixteenth and early seventeenth centuries. A *bobyl'* was a pauperized, ruined peasant (in the usage of the time, he did not even count as a peasant) who was granted relief from taxes and dues in recognition of his economic incapacity. It would appear—at least, in the area to the southwest of Moscow that the authors studied—that although today the term connotes helplessness and hopelessness, *bobyli* were by no means a burden on the estates where they lived and that the presence on an estate of workers not subject to taxation gave the landowner a crucial advantage.

Even more remarkable is the relative advantage that Milov and Garskova show accrued to smaller estates, especially *votchiny*. The second quarter of the seventeenth century was a period of painfully slow recovery from economic and political collapse. Commerce was seriously disrupted, and the effective authority of the new government in Moscow was limited. To judge by the political documents of the period, the country was dominated by ''the powerful''—politically well-connected holders of large *pomest'ia*. Lesser *pomeshchiki*, *votchinniki*, and townspeople were (or so they complained) unable to resist the depredations of the powerful. The victims beseeched the central government for help, by and large in vain, until 1649. The Law Code issued in that year represented a significant attempt by the government to interpose itself between the powerful and the rest of the population in the interests of securing the flow of tax revenue and maintaining the service system. The form of serfdom and the rigid system of estates of the realm (*sosloviia*) that derive from the code and would dominate Russian life for more than two hundred years represent, so to speak, recognition of the power of the powerful. Yet Milov and Garskova show that the larger estates were comparatively sluggish in their economic recovery and that smaller units were more flexible and more viable—at least, in the circumstances of the 1630s.

The few dense pages that follow, therefore, constitute a restrained but well-grounded appeal for a reappraisal of the Russian economy—and, by extension, of political and social institutions—in the seventeenth century and beyond. Those who dislike quantitative methods in history usually motivate their dislike either by saying that the findings of the quantifiers must be wrong or that they confirm what we already knew. Garskova's and Milov's work will have its critics, but none of them, it is safe to say, will offer the latter objection.

The article also presents a challenge to the humble translator. The variables the authors analyze are derived from cadastral books and often employ a terminology that is distinctive to seventeenth-century Russia. Rather than conjure up archaic and imperfect English equivalents for these terms, I have chosen to transliterate them. The italics-ridden text that results may cause some initial difficulties for readers who do not know Russian, but translating these terms would, I believe, entail a sacrifice of the rigor and precision that is the authors' hallmark. A list of these terms follows.

D.F.

Glossary

bobyl' [pl. *bobyli*]	A householder, often a victim of misfortune, who was relieved of a full-fledged peasant householder's obligations and usually not allotted with plowland.
liudi	Dependent persons living on an estate who were not *bobyli* or full-fledged peasants, such as slaves and others who might once have worked as servants or craftsmen.
"peasant," in Russian *krest'ianin*	In a seventeenth-century village, a full-fledged peasant, the head of an economically viable household economic unit or a member of his family, in contrast to *bobyli* or *liudi*.
tiaglo [pl. *tiagla*]	a.) A bundle of obligations, primarily to the squire. b.) A married couple and their kin capable of meeting these obligations and therefore entitled to an allotment of plowland. A household might consist of one or several *tiagla*.
tiaglosposobnost'	The ability to bear a *tiaglo*, hence a measure of the economic capacity and demographic composition of a household.
perelog	Land left fallow for many years.
pustosh' [pl. *pustoshi*]	Land away from villages and estates, liable to intermittent or casual exploitation.
votchina [pl. *votchiny*]	A hereditary patrimonial estate.
pomest'e [pl. *pomest'ia*]	An estate awarded to a servitor of the state on the condition of his continued service.

The subject of this paper is the agrarian development of Russia during a period when the nation was emerging from decades of economic ruin. The main goal of the paper, however, is methodological: the identification and evaluation of the best methods of forming a typology of a mass of objects. As in our previous publications,[1] we are concerned with types of estates in the first half of the seventeenth century, relying on principal components analysis. In contrast to our earlier work, we have drawn upon a mass of data pertaining to a region, consisting of Vorotynsk and Likhvin Districts,[2] that was distinguished by the complexity of its level of development. Here, according to the cadastral surveys,[3] were 178 secular estates: 91 *votchiny*, or hereditary patrimonies, and 87 *pomest'ia*, or estates held conditionally in return for service to the state. We can

[1] L. V. Milov, M. B. Bulgakov and I. M. Garskova, "Sistemno-strukturnyi podkhod k izucheniiu agrarnogo razvitiia Rossii pervoi poloviny XVII v. i problemy tipologii feodal'nogo khoziaistva (Mnogomernyi analiz pistsovykh knig)," in I. D. Koval'chenko, ed., *Matematicheskie metody i EVM v istoricheskikh issledovaniiakh* (Moscow, 1984); L. V. Milov, M. B. Bulgakov and I. M. Garskova, *Tendentsii agrarnogo razvitiia Rossii pervoi poloviny XVII stoletiia: Istoriografiia, komp'iuter i metody issledovaniia* (Moscow, 1986).

[2] The district towns are, respectively, about 15 and 50 kilometers to the south of the city of Kaluga.

[3] The *pistsovye knigi* are preserved in TsGADA (Central State Archive of Ancient Acts, Moscow), *fond* 1209, *knigi* 84, 1044.

treat the data for these districts as a single whole because they formed an economic and geographical entity in the seventeenth century and because the cadastral surveys pertaining to them were compiled according to a common program and by the same functionaries. Hence we can derive identical sets of variables.[4]

We began our experimentation with factor analysis with a district, Vorotynsk, that was in an extreme situation after the Time of Troubles, in a state of absolute ruin. We chose such a district because the precipitate decline of the economy facilitated making clear distinctions between the characteristic traits of

Table 1: Statistical Characteristics				
	Mean per *pomest'e*		Mean per *votchina*	
Variable	Vorontynsk District	Likhvin District	Vorontynsk District	Likhvin District
Landlord's household	.92	.73	.83	.90
Liudi households	.41	.47	.97	1.16
Peasant households	1.05	1.13	1.67	1.94
Bobyl' households	1.84	3.00	1.77	4.57
Abandoned and destroyed households	8.27	3.32	7.60	4.51
Population of the estate	4.73	8.40	6.03	14.12
Landlord's plowland (in *desiatiny*)	19.26	13.00	23.34	24.07
Peasant plowland (in *desiatiny*)	18.28	19.80	23.07	36.93
Perelog in villages	45.22	27.85	40.60	25.17
Overgrown plowland in villages	60.35	47.13	54.50	30.61
Perelog and overgrown plowland outside village	116.47	79.82	23.00	25.89
Hayfields	10.30	14.75	13.00	16.54
Total land, excluding hayfields	275.79	202.35	181.76	158.22

the two types of estate. Multivariate analysis brought to light striking and significant differences between *pomest'ia* and *votchiny* according to a wide range of social and economic indicators, and we could characterize them as distinct forms of the feudal economy.[5]

The incorporation of data for Likhvin District drastically changes the picture, for in the 1720s and 1730s the estates there were in a much more favorable situation than those in Vorotynsk, as Table 1 shows. In Vorotynsk District,

---

[4] With one exception: the Likhvin survey does not distinguish, as the Vorotynsk survey does, between woodland areas that were used as plowland and those that were not.

[5] Like almost all Soviet historians, we understand "feudalism" to be a historical stage prior to and leading to capitalism and characterized by the economic, political and social dominance of the landowning class. This conception does not require, but does not preclude, the distinctive juridical forms that many historians in the West see as intrinsic to feudalism.

there were almost 2.5 times as many ruined households on *pomest' ia* and twice as many on *votchiny* compared to Likhvin District. There were also striking differences in the number of intact *bobyl'* households and especially in the number of *liudi*, that is, in the density of population of the populated households, which was almost twice as great on the Likhvin *pomest' ia* and more than twice as great on the *votchiny* there. Although the inhabited households of Likhvin District were more populous, the area of demesne plowland was less than on the *pomest' ia* of Vorotynsk District and about the same on the *votchiny*. To be sure, on the Likhvin *pomest' ia*, the peasants' share of plowland was lower per capita than in Vorotynsk, but on the *votchiny* it was one-and-a-half times greater. It is important to emphasize that the amount of abandoned land was much less in Likhvin District than in Vorotynsk, as the variables "*perelog* in villages" and "overgrown plowland in villages" indicate. Finally, we should note a significant difference in the area of exploited lands: the *votchiny* and especially

Table 2: The Class of Absolute Variables	
Factor I: General level of the restoration of the manorial economy	
	33.75% of the dispersion
1. Area of peasants' plowland	.407
2. Population of the property	.388
3. *Tiaglosposobnost'*	.386
4. Number of peasant households	.382
5. Number of *bobyl'* households	.364
6. Area of landlord's plowland	.333
7. Number of *liudi* households	.263
Factor II: Extent of abandonment of plowland	
	24.73% of the dispersion
1. Area of usable land	.471
2. Summary calculation of the decline in the number of *tiagla* and of *tiaglosposobnost'*	.469
3. Area of overgrown plowland in villages	.409
4. Area of *perelog* in villages	.405
5. Number of abandoned and destroyed households	.319
6. Area of *perelog* and overgrown plowland outside villages	.283
Factor III: Number and extent of early manifestations of abandonment	
	10% of the dispersion
1. Area of *perelog* and overgrown plowland outside villages	.650
2. Area of *perelog* in villages	-.444
3. Area of overgrown plowland in villages	-.428
4. Summary calculation of the decline in the number of *tiagla* and of *tiaglosposobnost'*	.253
5. Area of usable land	.238
6. Number of abandoned and destroyed households	-.177

the *pomest' ia* of Likhvin District were significantly smaller than those in Vorotynsk District. It would be natural to expect that the higher level of development of the Likhvin estates will have an impact on the results of factor analysis for the two districts.

We have chosen principal components analysis because this variety of factor analysis performs especially well with the kind of highly specific data to be found in the cadastral surveys. We prefer relative variables, calculated with respect to the area of the peasants' and the squire's plowland and according to the number of peasant and *bobyl'* households, for such variables bring to light nuances of information that absolute variables cannot. Let us begin, however, with an analysis based on absolute variables.

The three factors in Table 2 are determinant, for they account for 68.48% of the dispersion. Furthermore, in their composition they correspond closely to the factors based solely on the data for Vorotynsk District, indicating that the tendencies and patterns shown here were characteristic of an extensive region. Factor I isolates the key components of the development of the squires' and peasants' sectors: the actual area of squires' and peasants' plowland and the populations of full-fledged peasants (who bore the full burden of feudal dues) and of *bobyli* (who were exempt from many of these dues). The variables reflecting the composition of the households are borne out by the data on effective *tiaglosposobnost'*. Finally, among the leading indicators of the restoration of the feudal economy is the number of households composed of *liudi*—slaves, so-called "*delovye liudi*," and the like.

Factor II has the same composition as the corresponding factor derived only from data for Vorotynsk District, but it has the opposite significance. For Vorotynsk, we called this factor the "extent to which abandonment of land was overcome," but taking the region as a whole, the factor primarily reflects the opposite—the extent of abandonment, inasmuch as the variables registering abandonment are positively associated with the factor.

Factor III brings out the importance of land abandonment outside the villages, in the unpopulated enclaves called *pustoshi*; the factor loading of this variable is .650, which makes it predominant for the factor. The positive association of the factor with the variables "area of usable land" and "summary calculation of the decline in the number of *tiagla* and of *tiaglosposobnost'*" reveals that Factor III brings out long-standing elements of abandonment, perhaps dating to the end of the sixteenth century: the larger the estate, the greater was land abandonment outside the villages, and the latter meant a loss in the number of *tiagla*. In Factor III there are indications that land abandonment was being overcome, but they are associated with the more accessible reserves of plowland, close to the villages. When Factor III is less prominent, the reserves were less extensive, and, correspondingly, the greater was the extent of plowland under exploitation. The expansion of plowland goes hand in hand with a diminution in the number of abandoned households—although these components make up only a minor part of Factor III.

Let us turn to an analysis of the factor weights that demonstrate the extent to which each item in the aggregate is associated with the factors. Bear in mind that we conducted this analysis prior to grouping the estates as *votchiny* or *pomest'ia*. We calculated the coordinates of all estates for two factors, in this sequence: 1) the first and second factors; 2) the first and third factors; 3) the second and third factors.

When we plotted the first and second factors, two robust tendencies emerge. The distribution along the axis of Factor I shows that 26 *pomest'ia* and 46 *votchiny* are relatively closely linked to it—that is, their factor weights are above the mean. It follows that 71 *pomest'ia* and 35 *votchiny* are relatively weakly linked. In other words, the *votchina* type of estate is more closely associated with the factor that establishes the extent of the restoration and development of the manorial economy. This pattern is especially prominent for the relatively large *votchiny*, representing 22 of the 46 with factor weights above the mean and only five of the 35 *votchiny* with factor weights below it. As for Factor II, the extent of abandonment of plowland, the converse applies: *pomest'ia*, and especially the larger ones, are more closely linked to this factor. Fifty-three *pomest'ia* and 30 *votchiny* had factor weights above the mean, and among the former are 44 of the 47 large *pomest'ia*. The large *votchiny* exhibit the same tendency, but not so markedly; of the 30 with positive factor weights, 19 were large and another 11 were middling. Moreover, not one small estate had a positive factor weight for Factor II.

If we consider the 49 estates with negative factor weights for the first factor and positive for the second, we find that of the total, 36 are *pomest'ia* and 28 of these were large. In the opposite group, with positive weights for the first factor and negative for the second, *votchiny* dominate, comprising 29 of the total of 38. The *votchiny*, then, were in a relatively favorable position, since most *pomest'ia* stood below the mean level of restoration and development and were more likely to include abandoned lands.

As for Factor III, the essence of which is defined by the presence of long-standing elements of abandonment, 48 *votchiny* and 32 *pomest'ia* had factor weights above the mean, while 49 of each type stood below the mean. Once again we note the predominance of large *pomest'ia*; 33 of the 47 had positive factor weights for this factor, which registers manifestations of long-term stagnation. The sharpest contrast emerges when we plot Factors II and III: of the 37 estates showing signs of stagnation and of land abandonment to a marked degree, 30 were large *pomest'ia*, and not one small estate fell into this category.

Let us now take up scrutiny of the relative variables. Parameters calculated per *desiatina* of demesne plowland permit us to appraise the peculiarities of our region's development from the point of view, so to speak, of the manorial economy.

In Table 3 we see (just as we did when we limited ourselves to Vorotynsk District) a distinctive array of variables making up Factor I and accounting for 34.25% of the dispersion. In first place is the total area of usable land, that is, plowland either under cultivation or abandoned at some time in the past (.405); as a rule, it was land overgrown with trees or brush that predominated. In second place is a variable that also reflects, in the first instance, a tendency towards ruin: the number of persons subject to state taxes makes up only a small

Table 3: The Class of Variables Correlated with the Area
of Demesne Plowland (Calculated per *desiatina*)

Factor I: General conditions of the ruin of the manorial economy

	34.25% of the dispersion
1. Area of usable land	.405
2. Summary calculation of the decline in the number of *tiagla* and of *tiaglosposobnost'*	.394
3. Area of overgrown plowland in villages	.389
4. Area of *perelog* in villages	.381
5. Number of abandoned and destroyed households	.290
6. Area of *perelog* and overgrown plowland outside villages	.260
7. Area of hayfields	.248
8. Density (frequency) of manorial households	.236
9. Number of *bobyl'* households	.186

Factor II: Restoration of the peasant households as the basis of the development of the manorial economy

	21.99% of the dispersion
1. Population of the property	.465
2. Area of peasants' plowland	.435
3. Number of peasant households	.402
4. *Tiaglosposobnost'*	.352
5. Number of *bobyl'* households	.325
6. Area of *perelog* and overgrown plowland outside villages	-.260
7. Summary calculation of the decline in the number of *tiagla* and of *tiaglosposobnost'*	-.213
8. Area of usable land	-.192

Factor III: Basic components of ruin that were overcome by the manorial economy

	12.3% of the dispersion
1. Number of abandoned and destroyed households	.428
2. Area of *perelog* and overgrown plowland outside villages	-.411
3. Density (frequency) of manorial households	-.389
4. Number of *bobyl'* households	-.318
5. Area of *perelog* in villages	.308
6. Area of overgrown plowland in villages	.258
7. Area of hayfields	.222
8. *Tiaglosposobnost'*	-.217

portion of this factor (.394), reflecting a loss of population. Variables 7, 8 and 9 register positive tendencies against the general background of decline. The presence of *bobyli*, however, and the fact that the number of full-fledged peasant households does not figure as a variable reflect the general plight of a region where the *bobyli* (who, on average, were three times as numerous as peasants) were of primary economic significance for the manorial economy.

In Factor II are concentrated all the positive tendencies for the manorial economy: the total population of the estate, the area that peasants had under crops, the number of peasant households and the capacity of these households to bear taxes and dues. It is notable that the factor also includes the population of *bobyli*, who appear to have had considerable importance in the restoration of the manorial economy. These variables are all positively associated with the factor, but other variables are not. For example, a decline of the area of *perelog* and overgrown plowland outside the limits of the village, in *pustoshi*, was a manifestation of the revival of the peasant economy and consequently has a negative value (-.260); the same applies to a decline in *tiaglosposobnost'* and, paradoxically, to the total area of usable land. For, as we have observed elsewhere,[6] the total area of an estate was primarily a function of the extent of *perelog* and overgrown plowland, so that it was small estates that tended to prosper.

Factor III, basic components of ruin that were overcome by the manorial economy, contains variables that are both positively and negatively associated with the factor. Abandoned and destroyed households, land in long fallow, and overgrown plowland in villages were all obstacles to the restoration of the manorial economy. Abandonment of lands away from the villages was not an obstacle to the same degree, which is indirect evidence that the demesne most usually drew on *perelog* and overgrown plowland within the villages for its reserves of plowland. The peasants drew on the same kinds of land, but located in *pustoshi*. Since the squire's primary task was restoring demesne cultivation, the longer the peasants left the *pustoshi* untilled, the better—in the short run, of course—for him. Among the other, negatively associated variables, the number of *bobyl'* households stands out and confronts us again with the special role the *bobyli* played in the manorial economy. It appears that their status was purely formal; they were economically debilitated peasants who had been granted abatements, but in this region the abatement applied in practice only to obligations to the state. For the manorial economy the *bobyli* were an important resource. The role of the full-fledged peasantry in liquidating obstacles to the restoration of the manorial economy is reflected only indirectly in the variable *tiaglosposobnost'*. The relationship is negative; the greater the peasants' *tiaglosposobnost'*, the smoother the course of recovery.

Let us turn now to an examination of factor weights, plotting, as before, the coordinates of all estates for two factors. Plotted against Factors I and II, distribution on the axis of the Factor I shows that 34 *pomest'ia* and 21 *votchiny* were closely associated with this factor and that the association is weaker for 43 *pomest'ia* and 54 *votchiny*, indicating that the conditions that encouraged abandonment of land had more impact on *pomest'ia* (especially the large ones) than on *votchiny*.

With Factor II, by contrast, the distribution of the data points is such that the *votchiny* fit better than do the *pomest'ia*; 25 of the latter and 43 of the former are closely linked to it, while 52 *pomest'ia* and 32 *votchiny*, with factor weights below the mean, are not. Since tendencies towards restoration of the peasant economy as the basis of the landlord's economy are concentrated in Factor II, the *votchiny* come to the fore; in general they were economically more efficient than the *pomest'ia*; note that of the 52 *pomest'ia* with negative factor weights, one-half were large.

---

[6] Milov, Bulgakov, and Garskova, "Sistemno-strukturnyi podkhod," p. 40.

As regards Factors I and II, then, *votchiny* were in a relatively favorable position, with negative factor weights for the first and positive for the second. Conversely, 19 of the 28 estates with positive factor weights for the first factor and negative for the second were *pomest'ia*—and only one of the 19 was a small estate.

Factor III is ambiguous, registering circumstances that impeded the restoration of the manorial economy and also elements linked to its revivial. Thus it reflects the dynamics of development. Plotting the data points shows that 36 *pomest'ia* and 46 *votchiny* were relatively closely associated with Factor III. We conclude that *pomest'ia*, especially large ones, produce this pattern and that the *votchiny* were progressing relatively rapidly along the path of revival of the manorial economy.

When we plot Factors II and III, the estates with either positive or negative factor weights for both factors form polar groups. Eighteen of the 27 estates in the former category were *votchiny*, and 25 of the 29 in the latter category were *pomest'ia*, revealing that the basis for the relative success of the *votchiny* was the restoration of the peasant economy.

In Table 4, the variables are calculated per *desiatina* of peasant plowland. Here we can term Factor I "the level of abandonment from the perspective of the peasant economy." Along with six variables embodying data on abandonment as such, we find such variables as frequency of manorial households and area of landlord's plowland; from the peasants' point of view, these were expressions of their own exploitation and were by no means favorable to the rapid development of their economy. The significance of the variable that ranks seventh, the number of *bobyl'* households, is clear: for the manorial economy, *bobyl'* households were ripe for exploitation and important to the development of the feudal sector, but the more *bobyli* there were, the harder it was for the peasants who bore a full *tiaglo*.

Factor II can be called "the initial restoration of the manorial economy through the ruination of the population at large." The tendency towards ruin is stronger, however, than the tendency towards restoration. The logic of the correspondence among the variables in the factor links the development of demesne plowland primarily with the growth of households populated by *liudi*—that is, the contingent of slaves. The factor is negatively associated with five variables; among them, the role of the number of abandoned and destroyed households is not clear. It would appear that the number of *liudi* households was a function of the number of empty homesteads and that the intensity with which the full-fledged peasants were exploited obliged the squires to rely to a considerable extent on the labor of slaves and day laborers.

Table 4: The Class of Variables Correlated with the Area of Peasant Plowland (Calculated per *desiatina*)	
Factor I: Level of abandonment as regards the peasant economy	
	42.23% of the dispersion
1. Summary calculation of the decline in the number of *tiagla* and of *tiaglosposobnost'*	-.385
2. Area of usable land	.380
3. Density (frequency) of manorial households	.371
4. Area of overgrown plowland in villages	.354
5. Area of *perelog* in villages	.353
6. Area of *perelog* and overgrown plowland outside villages	.329
7. Number of *bobyl'* households	.270
8. Area of landlord's plowland	.220
9. Number of abandoned and destroyed households	.175
Factor II: Initial restoration of the manorial economy through ruination of the population	
	16.01% of the dispersion
1. *Tiaglosposobnost'*	-.577
2. Population of the property	-.508
3. Number of peasant households	-.441
4. Area of hayfields	-.261
5. Number of abandoned and destroyed households	-.222
6. Area of usable land	.141
7. Number of *liudi* households	.115
8. Area of landlord's plowland	.115
Factor III: Conditions of the rehabilitation of the economically enfeebled peasantry	
	10.63% of the dispersion
1. Number of abandoned and destroyed households	-.515
2. Area of hayfields	-.423
3. Number of *bobyl'* households	.363
4. Area of *perelog* in villages	-.267
5. Number of *liudi* households	.264
6. Population of the property	.248
7. Area of landlord's plowland	-.247
8. Number of peasant households	-.242

For Factor III, the conditions of the rehabilitation of the economically enfeebled peasantry, the variable with the greatest factor loading is the decrease in the number of abandoned and destroyed households. The elimination of *perelog* within the village was also important. A decrease in the area of hayfields facilitated the revival of debilitated peasants, that is, the *bobyli*, a revival that found expression in the increase in the *bobyl'* population, which was negatively associated with an increase in the number of peasant households: the fewer peasants, the more *bobyli*. Note also the inverse relationship between the *bobyl'* population and the area of demesne plowland, for it was the full-fledged peasants that predominated in the squire's fields. Hence the more *bobyli*, the more *liudi*, who played a major role in the manorial economy. To be sure, with respect to Factor III they figure as a condition of the increase of the population of *bobyli* rather than as a component of the manorial economy. Note finally that an increase of the population of *bobyli* leads to an increase in the total population.

When we plot these data, we get the following results. Plotting the data points on the axis of Factor I reveals a strong association of 45 *pomest'ia* and 11 *votchiny*, while 51 *pomest'ia* and 69 (mostly small) *votchiny* have factor weights below the mean. Only 13 of the 45 strongly-associated *pomest'ia* counted as small. Hence, from the peasant perspective, the level of abandonment was much higher on *pomest'ia*. As for Factor II, "initial restoration of the manorial economy through ruination of the population," it is strongly associated with 58 *pomest'ia* and 49 *votchiny*, while 38 *pomest'ia* and 31 *votchiny* have factor weights below the mean. The linkage of this factor with *votchiny* and with *pomest'ia* is about the same. When we take size into account, however, it turns out that large *pomest'ia* are primarily associated with this factor. Of the 57 *pomest'ia* with positive factor weights, 48 were large or middling, and 46 of the 59 estates with negative factor weights were middling or small. Finally, plotting the data points on the axis of Factor III shows that a majority of both *pomest'ia* and *votchiny* are closely linked to it. When we take scale into account, it turns out that the association of small and middling *votchiny* is rather strong, whereas of the 44 *pomest'ia* with negative factor weights, 25 were large. And in general, with respect to both Factors II and III, the grouping of estates by scale is striking.

Let us turn now to an analysis of the variables calculated per peasant household. In Table 5, Factor I has a rather general character and registers the conditions of life and work of the peasantry. It is composed primarily of variables indicating the abandonment of productive resources. The key variables reflecting the development of the landlords' sector (area of landlord's plowland and frequency of manorial households) also bear upon the conditions of life and work in the peasant sector. Finally, in last place with respect to the value of the correlation coefficient is the area of the peasants' plowland. It follows that the predominant variables in this factor are those that indicate ruin.[7]

The second factor is of a different character; almost all the positive processes are clustered here. The factor can be termed "conditions favorable to the recovery and development of the peasant economy," inasmuch as the variables are calculated per peasant household. Here population of the property is

---

[7] We should note that in the more delimited territory of Vorotynsk District this factor had a somewhat greater share of positive elements of development.

Table 5: The Class of Variables Correlated with the Number
of Peasant Households (Calculated per Household)

*Factor I: Basic stages of the economic crisis and of
economic development that significantly determined the
situation of a peasant household*

	31.67% of the dispersion
1. Area of usable land	.435
2. Summary calculation of the decline in the number of *tiagla* and of *tiaglosposobnost'*	.426
3. Area of *perelog* in villages	.334
4. Area of overgrown plowland in villages	.330
5. Number of abandoned and destroyed households	.304
6. Area of *perelog* and overgrown plowland outside villages	.253
7. Area of landlord's plowland	.213
8. Density (frequency) of manorial households	.207
9. Area of peasants' plowland	.205

*Factor II: Conditions favorable to the recovery and development
of the peasant economy*

	21.34% of the dispersion
1. Population of the property	.454
2. Number of *bobyl'* households	.453
3. *Tiaglosposobnost'*	.441
4. Area of peasants' plowland	.391
5. Area of landlord's plowland	.228
6. Area of overgrown plowland in villages	-.177
7. Number of *liudi* households	.175
8. Area of *perelog* in villages	-.170

Factor III: Conditions assuring the renewal of the peasant economy

	11.67% of the dispersion
1. Area of landlord's plowland	-.419
2. Area of overgrown plowland in villages	.414
3. Area of *perelog* and overgrown plowland outside villages	-.370
4. Area of *perelog* in villages	.368
5. Number of *liudi* households	-362
6. Density (frequency) of manorial households	-.342
7. Number of *bobyl'* households	.195
8. Area of hayfields	.178

the most important variable, but the number of *bobyl'* households and the *tiaglosposobnost'* of the population follow closely behind. It is interesting to note that although the area of peasants' plowland plays only an ancillary role for this factor, the variable has a higher value than the area of the landlord's plowland. The contribution of the number of *liudi* households is also modest; as this number increases, the burden of *barshchina* on the peasants must be less. As for the two variables with a negative association with the factor, they both register abandoned plowland within the villages. The smaller the area of this kind of land, the better provided the peasants were with plowland and hayfields. In this region, apparently, the peasants undertook to restore these lands to cultivation, although they were primarily associated with the landlords' sector.

As for the factor we have called ''conditions assuring the renewal of the peasant economy,'' its level of dispersion, 11.67%, is typical for third factors.

Because some circumstances favored the development of the peasant economy and others did not, some of the variables have a positive association with this factor, others a negative. Among the former are variables registering the increase of unexploited lands within the village, which is indirectly linked to the factor: the increase of such lands meant a failure of the demesne plowland to increase. Conversely, a decline in the area of abandoned lands in *pustoshi* could increase the peasants' irregular sowing there. *Bobyli* and hayfields constituted, each in their own way, a reserve for the development of the peasant economy; hence their positive association with Factor III. For this factor in general, the components of the manorial sector are in a kind of antagonism with those of the peasant sector; hence the negative associations with the area of landlord's plowland, the number of *liudi* households, and the density of manorial households. We emphasize, however, that here these variables are counterposed not to the peasant economy as such, but to its reserves.

Let us now plot the factor weights, beginning as before with Factors I and II. The distribution of data points along the axis of Factor I reveals a clear tendency towards association with *pomest'ia*, especially the larger ones. Of the 67 *pomest'ia*, 34 (26 of them large) have factor weights above the mean; only 22 of the 72 *votchiny* do so. That is to say, the elements of crisis that influenced the peasant economy were also sharply felt within the manorial sector. Factor II is closely linked with 31 *pomest'ia* and 44 *votchiny*; conversely, the link to 40 *pomest'ia* and 28 *votchiny* is weak. It follows that estates of a patrimonial type, as opposed to *pomest'ia*, strongly tend to be associated with the factor that embodies the conditions facilitating the renewal and development of the peasant economy. This tendency, which emerged from our analysis of the data for Vorotynsk District, finds confirmation in the larger data set. Furthermore, it is the 29 small and middle-sized estates that dominate among the 40 *votchiny* with positive factor weights for Factor II, while 24 of the 40 *pomest'ia* with negative factor weights were large. Taking the first and second factors together, two polarized groups emerge: one is composed of the eighteen large *pomest'ia* with positive weights for Factor I and negative for Factor II, the other of 23 small and medium *votchiny* with negative weights for Factor I and positive for Factor II. Compared to *votchiny* of modest size, the big *pomest'ia* exhibit signs of crisis and of comparatively unfavorable conditions for the renewal of the peasant economy.

A third tendency emerges when we turn to Factor III, with which 31 *pomest'ia* and 27 *votchiny* are strongly associated. Note that of the 45 *votchiny* with negative weights for this factor, it is the small and medium that dominate, while large estates constitute more than half of the 31 *pomest'ia* with positive weights. It follows that on estates of the latter kind, attempts to exploit the reserves within the peasant economy intensified antagonism between peasants and the manorial administration.

Table 6: The Class of Variables Correlated with the Number
of *Bobyl'* Households (Calculated per *Bobyl' Household)*

Factor I: Bobyl' *status as a component of ruin*
and of development

	39.5% of the dispersion
1. Area of usable land	.372
2. Summary calculation of the decline in the number of *tiagla* and of *tiaglosposobnost'*	.366
3. Area of *perelog* in villages	.337
4. Area of overgrown plowland in villages	.323
5. Number of abandoned and destroyed households	.316
6. Area of landlord's plowland	.272
7. Area of peasants' plowland	.270
8. Number of peasant households	.249
9. Area of *perelog* and overgrown plowland outside villages	.232

Factor II: The role of bobyl' *status in the revival of the*
peasant economy and the tiaglosposobnost' *of the population*

	17.28% of the dispersion
1. Population of the property	.421
2. Number of peasant households	.411
3. Area of peasants' plowland	.385
4. Area of *perelog* and overgrown plowland outside villages	-.343
5. Summary calculation of the decline in the number of *tiagla* and of *tiaglosposobnost'*	-.281
6. Number of *liudi* households	.278
7. Area of usable land	-.273
8. *Tiaglosposobnost'*	.245

Factor III: Conditions characterizing the bobyli *as*
a diminished element of the manorial economy

	10.42% of the dispersion
1. Density (frequency) of manorial households	-.635
2. *Tiaglosposobnost'*	-.506
3. Area of landlord's plowland	-.290
4. Area of hayfields	.225
5. Population of the property	.222
6. Area of peasants' plowland	.210
7. Number of *liudi* households	-.210

Let us turn finally to the fifth and last category of variables, those correlated with the number of *bobyl'* households. In Table 6, as before, Factor I embraces both positive and negative tendencies, but here they pass through the prism, so to speak, of the development of *bobyl'* status. It is notable that variables relating to abandonment prevail: the total area of usable land (which usually included an enormous amount of abandoned plowland) and all the varieties of direct abandonment, including both lands within villages and those remote from them. The area of landlords' and peasants' plowland also contributes to the factor, as does the number of peasant households.

Factor II retains the character it had on the basis of Vorotynsk data, and so we have retained the name: "the role of *bobyl'* status in the revival of the peasant economy and the *tiaglosposobnost'* of the population." Here the positive tendencies are concentrated, with the population of the estate in first place and the number of peasant households in second. Note also the positive association between this factor and the number of *liudi* households; the increase in *bobyli*, an economically feeble category that enjoyed some relative advantages, inevitably was accompanied by an increase in the number of craftsmen, other non-agricultural workers, and slaves, all of whom worked directly for the manor. Factor II also contains variables negatively associated with it, reflecting the ongoing liquidation of negative phenomena. Thus, the more powerfully this factor is manifest, the smaller the area of abandoned lands outside the village; so too with variables 5 and 7.

Although the *bobyli* represented, on the whole, a positive element in the manorial economy, their presence did have negative consequences, and these are reflected most prominently in Factor III. Therefore, the greater the impact of the *bobyli* on the initial restoration of the manorial economy, the more painfully were these accompanying negative consequences felt. In essence, the stronger the manifestation of Factor III, the lower the frequency of manorial households and the smaller the area of demesne plowland. This relationship is intelligible, inasmuch as the basis of the manorial economy in that era was the population of the *liudi* households and the *tiaglo* borne by the peasants, so the negative association (-.210) between this factor and the number of *liudi* households is consistent with our expectations. Furthermore, the more prominent the activity of the *bobyli* within the manorial economy, the lower the *tiaglosposobnost'* of the population, since it was peasants who bore the *tiaglo*. The impact of the *bobyli* as an element of the manorial economy finds primary expression in the increase of population and in the area of the peasants' plowland (.222 and .210, respectively).

Let us turn to an analysis of the connection of the estates with the factors. Thirty-seven *pomest'ia* and 25 *votchiny* are closely linked to Factor I, with factor weights above the mean; 56 *pomest'ia* and 52 *votchiny* have weights below the mean. Hence *pomest'ia* (especially 27 large ones) are somewhat more closely associated with the factor than *votchiny* are.

With Factor II, the pattern is very different. The data points are distributed along the axis of this factor in such a way that 31 *pomest'ia* and 58 *votchiny* have factor weights above the mean, while 62 *pomest'ia* and only 19 *votchiny* are weakly linked to this factor. The preponderance of *votchiny* is very clear, which provides further evidence that it was on estates of the *votchina* type that the process of restoration of the economy of the direct producers advanced most swiftly, with the *bobyl'* population playing a considerable role in the process. So it is that, plotted against Factors I and II, 25 estates—of which 19 were large *pomest'ia*—have positive weights for Factor I and negative for the second. By the same token, in the opposite group, with negative weights for the first factor and positive for Factor II, we find mostly *votchiny* (35 out of 52), and only ten of these were large.

As for Factor III, 48 *pomest'ia* and 43 *votchiny* are closely associated with it, while 45 *pomest'ia* and 34 *votchiny* are not. Most of the closely associated estates were large: 20 of the 43 *votchiny*, and 30 of the 48 *pomest'ia*. Among the estates where the situation of the *bobyli* was comparatively difficult and their role was relatively modest—that is, estates with positive factor weights for Factor III and negative for Factor II—it is large *pomest'ia* (20 out of a total of 41) that predominate. Conversely, the situation of the *bobyli* was easier on *votchiny*, which comprise 25 of the 39 estates with negative factor weights for Factor III and positive for Factor II. In other words, it was on the *votchiny* that the *bobyli*, despite their economic weakness, made a contribution.

To sum up, in the 'twenties and 'thirties of the seventeenth century, by almost any yardstick, the traits of the *votchina* type of estate were more favorable than those of the *pomest'e* type. Despite its high level of exploitation of the direct producers, the *votchina* created more favorable conditions for the restoration of the peasant economy. It was able to overcome the manifestations of economic ruin, land abandonment and stagnation comparatively rapidly. In this connection we must emphasize that these findings apply to *votchiny* of all sizes. Conversely, the onerous manifestations of ruin had a pronounced impact on the economy of *pomest'ia* as a group. This distinction, we believe, proves that *pomest'ia* and *votchiny* represented two distinct economic types of estate in the context of a natural economy.

Our findings, based as they are on a large aggregate of 178 estates, permit us to address some more general issues. In particular, our research shows that in the 1630s, when we penetrate to the actual operations of the servile regime, the *votchina* form of feudal property was able significantly to surpass the *pomest'e* in its capacity to create optimal conditions of development. Most historians, however, have regarded and still regard the rise of the *pomest'e* system at the end of the fifteenth century and the beginning of the sixteenth as a step forward in the development of feudalism. They associate the development of the *pomest'e* system with the rise of a *barshchina* (corvée) economy, devoted to the production of grain for the domestic market. And it is to the *barshchina*-based economy of the *pomest'e* that they attribute, in the last analysis, the onset of the era of full-fledged serfdom, which, in turn, they perceive to be the beginning of the so-called "stage of decline" in the history of feudalism. Inasmuch as our findings here are at odds with prevailing conceptions, they encourage reflection and attempts to construct new conceptions of the development of feudalism. One such conception might be a revisionist interpretation of the causes of the emergence of serfdom.[8] Another might be a new approach to the *pomest'e* system and to the causes of its genesis. We hope that our findings here will facilitate the resolution of these major problems in the literature on Russian feudalism.

---

[8] See L. V. Milov, "O prichinakh krepostnichestva v Rossii," *Istoriia SSSR*, 1985, no. 4.

Two Paths of Bourgeois Agrarian Evolution in European Russia: An Essay in Multivariate Analysis

I. D. KOVAL'CHENKO
L. I. BORODKIN

Introduction

Academician Koval'chenko is the doyen of Soviet quantitative historians and the mentor of many of them.* While the sheer volume of Koval'chenko's publications is due to his erudition and energy, some credit for the increasing sophistication of his work should go to his collaborators and first of all to L. I. Borodkin, who was initially trained in information science. The article published here is the culmination of a series, each more sophisticated than the last, devoted to the agrarian typology of prerevolutionary Russian provinces. The categorization itself holds no great surprises. It corresponds to the authors' earlier findings and, in general, to the common usage of social scientists. The importance of the article lies in showing the extent to which a province matches a particular type and, more particularly, the social and economic determinants of the agrarian development of provinces and types of provinces. Like Selunskaia, Koval'chenko and Borodkin emphasize the sway of the peasantry in prerevolutionary agriculture; indeed, what they now call the "manorial-peasant type" they once called simply the "manorial type." Yet they also bring out the viability, in some regions, of the development on a manorial or "Prussian" basis.

In analysis of the kind they undertake, gains in sophistication mean losses in transparency, as more and more operations are performed behind the scenes, usually in the bowels of a computer.† One turns almost with relief to the wearying rows of numbers in their appendix, which represent the absolute values for each province that were the point of departure for their analysis; one-third of the rural population in Iaroslavl' Province was literate in 1897—*there* is a fact to sink your teeth into. Some years ago, Lawrence Stone wondered whether "the sophistication of...mathematical and algebraic formulae are [sic] not ultimately self-defeating, since they baffle most historians." He went on to complain of "the virtual impossibility of checking up on the reliability of final results, since [one] must depend not on published footnotes but on privately owned computer tapes..."‡ There is no reason to share Stone's disdain for what he cannot bother to understand. We can, however, share his nostalgia for an era, extending from the time of Gibbon until very recently, when all historians could adhere to a common standard of verification. The humble footnote was the basis of the comity of scholars. Koval'chenko, Borodkin and their colleagues explain what they are doing with admirable patience, they share their data with rare generosity, but they are increasingly obliged to characterize their data and the operations they perform rather than identify them, as historians used to do. We can directly verify their findings only by retracing step by step the arduous path of data collection and analysis that they travelled. Perhaps the lesson for historians is that humility and caution become ever more important as our methods become more complex.

D.F.

---

* *Tematicheskii bibliograficheskii ukazatel' otechestvennoi literatury po primeneniiu matematicheskikh metodov i EVM v istoricheskikh issledovaniiakh*, published in 1986 by Moscow University's *Kafedra istochnikovedeniia istorii SSSR*, which Koval'chenko heads, lists almost 300 books and articles, of which a solid majority are the work of members of the *kafedra*.

† Some of the operations performed for this article are illuminated in I. D. Koval'chenko, T. L. Moiseenko and N. B. Selunskaia, eds. *Sotsial'no-ekonomicheskii stroi krest'ianskogo khoziaistva Evropeiskoi Rossii v epokhu kapitalizma* (Moscow, 1988).

‡ Lawrence Stone, "The Revival of Narrative," *Past and Present*, no. 85 (1979), p. 11.

Our purpose in this article is to establish a social agrarian typology of the provinces of European Russia at the turn of the twentieth century. We seek first of all to determine the boundaries within which two types of development prevailed. These were the manorial and the peasant types, or the "Prussian" and the "American" paths.[1] We further seek to reveal the economic structure of these types and to show the social and economic results of their functioning. To do so we use a variety of forms of multivariate analysis.

Sources and Methods

Our source base is rich and permits us to use a large number of variables relating to the agrarian profile of the provinces of European Russia.[2] In forming a data base, the first task is selecting variables for analysis from among the hundreds to be found in the sources. We began with about a hundred, reflecting all the basic demographic, economic and social aspects of agrarian structure, and entered them into the computer. These we had to winnow down, for so large a number might serve to ascertain the typology of provinces but would impede the analysis of the structures and of the functioning of the various types. We did so by identifying the factors that characterize the basic social aspects of agrarian relations and hence reflect the differences between the two paths of agrarian evolution. For each factor, we selected from our initial array several variables that convey the essence of the factor. The result was nine factors, which characterize 31 variables (see Table 1).[3] These factors embrace the basic components that determine the specifics of the social aspects of agrarian relations. Working from these 31 variables and using the methods of cluster analysis, probability grouping and pattern recognition, we were able to distinguish types of provinces presenting similar "bourgeois profiles." To analyze the *structure* of the types thus brought to light, we used the nine factors. For each of them we obtained aggregate indices and factor weights (based on the component variables) that

---

[1] These terms were coined by Lenin and first used in his "Agrarian Program of Social Democracy in the First Russian Revolution...," (1908). See V. I. Lenin, *Polnoe sobranie sochinenii*, 5th ed., 55 vols. (Moscow, 1958-65), vol. XVI, p. 216.

[2] On these sources, see I. D. Koval'chenko, ed., *Massovye istochniki po sotsial'no-ekonomicheskoi istorii Rossii perioda kapitalizma* (Moscow, 1981), pp. 219-343.

[3] The variables, which are shown in Appendix A, were calculated from the population census of 1897, the military horse censuses of 1900-1901 and 1912, the landownership surveys of 1877 and 1905. See Tsentral'nyi statisticheskii komitet, *Statistika zemlevladeniia 1905 g.*, 50 fasc. (St. Petersburg, 1906-1907), *Materialy Vysochaishe uchrezhdennoi 16 noiabria 1901 g. Kommissii po issledovaniiu voprosa o dvizhenii s 1861 g. po 1901 g. blagosostoianiia sel'skogo naseleniia...*, 3 vols. (St. Petersburg, 1903) and data taken from three secondary studies: A. N. Anfimov, *Zemel'naia arenda v Rossii v nachale XX v.* (Moscow, 1961); I. D. Koval'chenko and L. V. Milov, *Vserossiiskii agrarnyi rynok. XVIII—nachalo XX v.* (Moscow, 1974); and I. D. Koval'chenko, "Sootnoshenie krest'ianskogo i pomeshchich'ego khoziaistva v zemledel'cheskom proizvodstve kapitalisticheskoi Rossii," in L. M. Ivanov et al., eds., *Problemy sotsial'no-ekonomicheskoi istorii Rossii* (Moscow, 1971), pp. 171-94.

<table>
<tr><td colspan="1" align="center">Table 1: Social Agrarian Structure—Factors and their Component Variables</td></tr>
</table>

Note: Allotments and other lands held by peasant communities were, whatever the prevailing form of tenure, subject to various legal and customary encumbrances. Prerevolutionary Russian usage therefore distinguished between such lands and "privately owned" (*chastnovladel'cheskie*) lands, held as ordinary private property, and this distinction is embodied in the data underlying Table 1. An individual peasant who purchased land would be a private landowner with respect to that land and a peasant landholder with respect to his allotment.

Factor I: Tendencies in landownership by nobles
   1. Ratio of nobles' land to peasant allotment lands in 1905.
   2. Proportion of privately owned lands owned by nobles in 1905.
   3. Proportion of nobles' lands occupied by latifundia.
   4. Ratio of nobles' landholding in 1905 to their holdings in 1877.

Factor II: Tendencies in bourgeois landownership
   5. Proportion of privately owned lands owned by non-nobles in 1905.
   6. Ratio of non-nobles' landholding in 1905 to their holdings in 1877.
   7. Ratio of all privately owned lands sold between 1891 and 1900 to the total area of
      privately owned lands in 1905.
   8. Sale prices of land in 1900 (rubles per *desiatina*).
   9. Rate of growth of land prices, 1885-1910 (rubles per year).

Factor III: Peasant landholding and land use
   10. Mean allotment per peasant, early 20th century.
   11. Ratio of peasant allotment lands to land purchased by peasants in 1905.
   12. Ratio of lands purchased by peasants to all privately owned lands, 1905.
   13. Ratio of lands rented by peasants to peasant allotment lands, early 20th century.
   14. Land cultivated by peasants (*desiatiny* per person).

*Factor IV: The regular use of hired labor and the development of capitalism in non-peasant
   agriculture*
   15. Ratio of regular hired agricultural laborers to all local workers in agriculture.
   16. Proportion of non-peasant agricultural units employing hired labor.

Factor V: Peasants working as hired laborers in agriculture
   17. Ratio of peasants working locally as hired agricultural laborers to
       all agricultural workers.
   18. Proportion of peasants working as migratory agricultural laborers.

Factor VI: The proletarianization of the village
   19. Proportion of peasant household with one horse or with none, 1900-1901.
   20. Ratio of the proportion of peasant household with one horse or with none in 1912
       to the proportion in 1900-1901.
   21. Ratio of peasants working for hire in industry or agriculture as a proportion of
       all workers.

Factor VII: The embourgeoisement of the peasantry
   22. Proportion of peasant households with 4 or more horses, 1900-1901.
   23. Proportion of horses belonging to these households.
   24. Ratio of the proportion of peasant households with many horses in 1912 to the
       proportion in 1900-1901.
   25. Concentration of land purchased by peasants (ratio of peasant-purchased lands in
       units of 50+ *desiatiny* to all such lands in 1905).

Factor VIII: Relationship of the peasant and manorial economies
   26. Ratio of peasants' to nobles' landholding, early 20th century.
   27. Ratio of area sown by peasants to that sown by non-peasants in the capitalist era
       (mid-19th century to 1916).
   28. Ratio of peasants' to non-peasants' work horses, 1888-91 and 1912.

Factor IX: The intensification of agricultural production
   29. Work horses per *desiatina* of sown area, early 20th century.
   30. Cattle per *desiatina* of sown area.
   31. Grain yields (puds per *desiatina*).

indicate each province's level of development according to the particular factor. And these indices open the way to discovering the most important traits of each type of province.

Table 2: The Results of Agrarian Development—Factors and Variables

Factor I: General level and productivity of cereal cultivation
1. Harvest of grains and potatoes, puds per head of total population.
2. Harvest of grains and potatoes, puds per head of population engaged in agriculture.
3. Ratio of harvest, 1909-13 to that of the late nineteenth century.

Factor II: General level and productivity of cattle-raising
4. Cattle per head of total population.
5. Cattle per head of population engaged in agriculture.
6. Ratio of draft animals to all cattle.

Factor III: Situation of the peasants
7. Peasants' harvest (puds per head of peasant population).
8. Peasants' cattle per head of peasant population.
9. Off-farm earnings per head of peasant population.

Factor IV: Situation of agricultural workers
10. Level of wages (kopecks per day in 1900).
11. Rate of increase of wages per day (kopecks per year, 1900-13).
12. Real wages (nominal wages weighted by the price of rye per pud in the autumn).

Factor V: The socio-demographic and cultural development of the village
13. Ratio of locally born villagers living elsewhere to the total rural population.
14. Ratio of local peasants in cities to total rural population.
15. Proportion of literates in total rural population.

Distinguishing among types of provinces and establishing the differences among them represents only the first stage of analysis. The second stage is discovering the social and economic results of the functioning of these types and finding out which of them promised the highest level of agricultural production and secured the most favorable situation for the direct producers, the peasants and agricultural laborers. We selected the variables necessary for the second stage as we had for the first. We identified five factors that characterize the economic and social results of agrarian development. Fifteen variables convey the essence of these factors (see Table 2). For each factor, we used principal-components analysis to determine factor weights that reflect the level of development of each province and then, by means of repeated factoring, to reduce the five factors to two: I. The level and productivity of agriculture and the situation of the peasants; and II. The socio-demographic and cultural development of the village. Finally, using multivariate regression analysis we were able to determine the effect of each of the nine structural factors on the two general factors that register the results of the function of systems of agrarian relations.

Table 3 : Clustering of Social Types of the Provinces
of European Russia at the Turn of the Twentieth Century

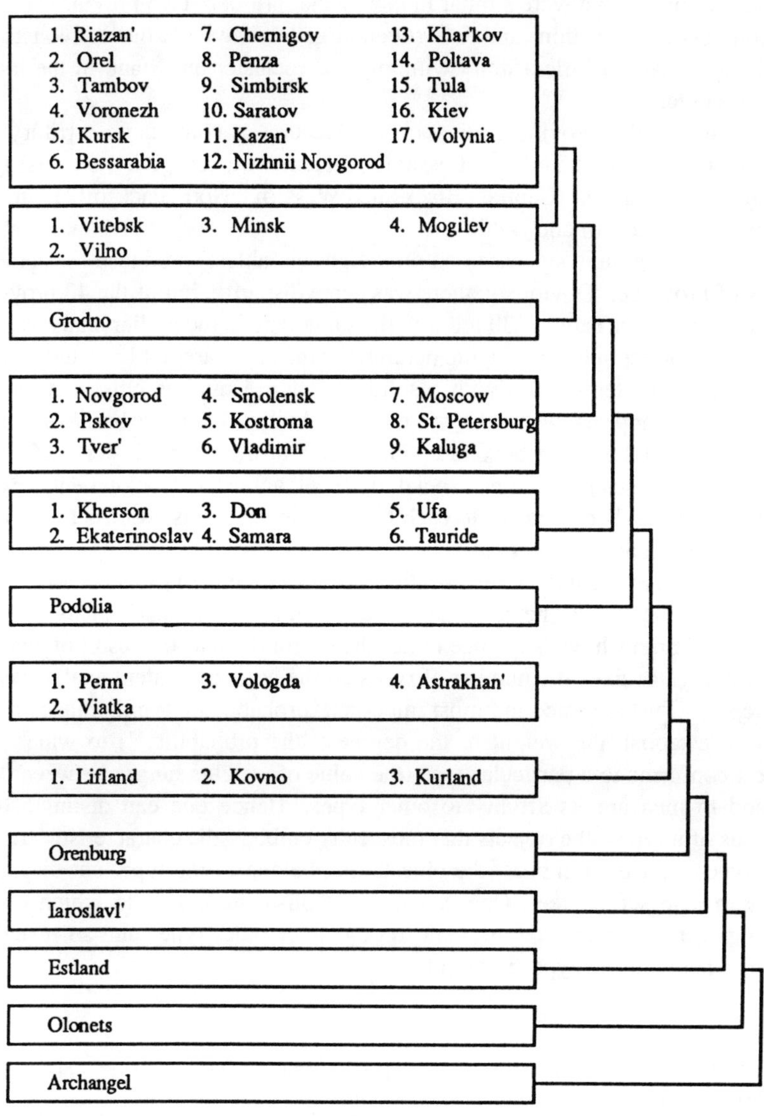

Cluster analysis provides the most flexible multivariate classification of provinces according to their social profiles and, working from our 31 variables, produces the result shown in Table 3. It discloses six macroclusters of provinces (43 in all) that were similar in their social profile.[4] Two provinces occupy an intermediate position, and five (Orenburg, Iaroslavl', Estland, Olonets and Archangel) exhibit little affinity with any macrocluster and stand in the hierarchy as singletons.

Usually the provinces in a cluster occupy a contiguous territory; the exceptions are Astrakhan' and Bessarabia. Hence we can give the social types geographical names: Central Blacksoil, Western, Non-Blacksoil Industrial, Steppe, Northeastern and Baltic.

An earlier analysis, based on only eight variables,[5] produced the same six types of province. Obviously, there was some disparity, but of the 43 provinces that form types in Table 3, 30 fell into the same type in the earlier analysis. This degree of correspondence is quite natural, for the types are not hard and fast, but rather social systems, each with a solid core and a nimbus pulsing around it. This pulsing appears when the systems are analyzed on the basis of various arrays of variables. It is beyond doubt, then, that the macroclusters in Table 3 represent significantly distinct social types of agrarian development. Before examining what is distinctive to each type, however, let us address the issue of typology using another method.

This other method is the so-called "fuzzy typing" of objects based on the theory of fuzzy sets.[6] The advantage of this method over cluster analysis is that one can distinguish types at once rather than form them as the result of analysis; the investigator fixes the number of types on the basis of content analysis of the aggregate. Furthermore, and most important, probabilistic typing makes it possible to establish the weight or the degree ("the probability") to which each object conforms to a particular type (the value of weights ranges between 0 and 1) and to measure its affinity to other types. Hence one can distinguish the nucleus of a type—the objects that most fully embody the characteristic traits of that type—and the nimbus—the objects that belong to the type, but display its traits to a lesser degree. One can also establish the extent to which objects belonging to a given type are akin to objects of other types and so reveal the zones where various types intersect.

---

[4] Within each cluster the provinces are arranged according to their degree of similarity one to another, so that the most similar provinces are adjacent.

[5] I. D. Koval'chenko and L. I. Borodkin, "Agrarnaia tipologiia gubernii Evropeiskoi Rossii na rubezhe XIX-XX vv.," *Istoriia SSSR*, 1979, no. 1, pp. 82 ff.

[6] On the use of this method in historical research, see I. D. Koval'chenko and L. I. Borodkin, "Veroiatnostnaia mnogomernaia klassifikatsiia v istoricheskikh issledovaniiakh," in I. D. Koval'chenko, ed., *Matematicheskie metody i EVM v istoricheskikh issledovaniiakh* (Moscow, 1985) and L. I. Borodkin, *Mnogomernyi statisticheskii analiz v istoricheskikh issledovaniiakh* (Moscow, 1986), pp. 33-35.

Table 4: The Bourgeois Agrarian Typology of the Provinces
of European Russia at the Turn of the Twentieth Century

I. Steppe Type

Tauride	.88	Kherson	.80	Don	.88
Ekaterinoslav	.72	Samara	.83	Ufa	.59

Other provinces akin to the Steppe type

Saratov	.32	Orenburg	.30

II. Baltic Type

Lifland	.85	Kurland	.59	Kovno	.74
Estland	.46				

Other provinces akin to the Baltic type

Grodno	.20	Minsk	.18	Vilno	.19
Petersburg	.16				

III. Western Type

Vitebsk	.74	Minsk	.58	Vilno	.68
Grodno	.53	Mogilev	.60	Podolia	.37

Other provinces akin to the Western type

Volynia	.35	Khar'kov	.23	Kiev	.32
Perm'	.22	Poltava	.27	Smolensk	.21

IV. Non-Blacksoil Industrial Type

Kostroma	.90	Smolensk	.65	Vladimir	.88
Moscow	.64	Tver'	.84	Kaluga	.56
Novgorod	.67	Petersburg	.49	Pskov	.66
Iaroslavl'	.38				

Other provinces akin to the Non-Blacksoil Industrial type

Nizhnii Novgorod	.30	Poltava	.21	Tula	.30
Khar'kov	.18	Vologda	.27		

V. Central Blacksoil Type

Tambov	.94	Chernigov	.59	Kursk	.94
Kazan'	.57	Voronezh	.94	Nizhnii Novgorod	.56
Orlov	.88	Tula	.54	Riazan'	.78
Volynia	.46	Penza	.77	Kiev	.43
Simbirsk	.70	Khar'kov	.41	Bessarabia	.65
Poltava	.37	Saratov	.64		

Other provinces akin to the Central Blacksoil type

Podolia	.35	Astrakhan'	.20	Kaluga	.27
Ekaterinoslav	.19	Ufa	.25	Grodno	.18
Moscow	.25	Mogilev	.18		

VI. Northeastern Type

Viatka	.68	Vologda	.53	Perm'	.65
Astrakhan'	.47				

Other provinces akin to the Northeastern type

Archangel	.22	Pskov	.18	Kazan'	.20
Iaroslavl'	.18	Novgorod	.20	Olonets	.18

VII. Provinces Not of Any Type

Archangel	.67	Orenburg	.67	Olonets	.65

Probabilistic typing of provinces according to their social profile, carried out using 31 variables and seeking to distinguish, as with cluster analysis, six types gave the results shown in Table 4. The same six types were obtained as with cluster analysis, and three of the types (Steppe, Central Blacksoil and Northeastern) are composed of the same provinces. Grodno and Podolia Provinces, which cluster analysis left outside the macroclusters, now belong to the Western type. Iaroslavl' and Estland Provinces, which also stood outside any cluster, now belong to the Central Industrial and Baltic types, respectively, while Archangel, Olonets and Orenburg are still singletons. The virtually total correspondence of typing carried out using various methods attests that it is a suitable and effective method of multivariate typology.

It is also clear that each type has its nucleus and its nimbus. Objects with a weight of .60 or more can be assigned to the nucleus with some confidence. The Steppe type displays the highest average density (.80), while the Non-Blacksoil Industrial, the Central Blacksoil, and the Baltic range between .65 and .78, and the Western and Northeastern stand at .58. On the whole, the density of the types is considerable, which attests to their marked "crystallization." Yet the types are by no means self-contained formations. They intersect and pile up, so to speak. Thirty-one provinces belonging to one of the types or to none (as with Archangel, Olonets and Orenburg) clearly display signs of affinity with other types—sometimes a degree of affinity almost as high as that of provinces belonging to the other type. For example, Podolia Province belongs to the Western type with a weight of .37, while Volynia Province, which belongs to the Central Blacksoil type, has an affinity with the Western type with a weight of .35. Poltava Province belongs to the the Central Blacksoil type with a weight of .37, while Podolia was akin to that type with a weight of .35. Nizhnii Novgorod and Tula display a high degree of affinity (.30) with the Non-Blacksoil Industrial type. In short, the social types are firm only with respect to qualitative definition; as regards their determining variables, they are relatively unstable and blurred.

The Content Structure of Social Types

To characterize the essence of each of the social types we used the factor weights for the nine content factors discussed above. Their mean values are shown in Table 5. This table strikingly reveals the distinctiveness of the Steppe and Baltic regions. The Steppe type has the most extensive employment of peasants as hired agricultural laborers (1.43), the highest level of peasant embourgeoisement (2.44), and the lowest intensity of agricultural production (-1.11). The Baltic type stands out for the extent of landownership by nobles, use of regular hired labor, and capitalist development in the non-peasant economy and also for the lowest level of bourgeois landownership (-1.13) and correlation of the manorial and peasant economies (-.92). Furthermore, the Steppe and Baltic types stand at opposite poles. For five factors they display a contrary tendency and for three others (IV, VII and VIII) they differ considerably one from the other. These data confirm the well-known fact that the Baltic and the

Southern and Southeastern Steppe regions exhibited very different forms of bourgeois agrarian evolution. These differences are expressive of two types of bourgeois agrarian evolution; the manorial variant of that evolution dominated in the Baltic, and the peasant variant on the steppe. To be sure, one can only say

Table 5: The Content Structure of the Social Types of the Provinces of European Russia at the Turn of the Twentieth Century								
Mean Factor Weights According to Social Type								
FACTORS								
I	II	III	IV	V	VI	VII	VIII	IX
Steppe Type								
-.83	.21	.29	.78	1.43	-.69	2.44	-.36	-1.11
Baltic Type								
2.62	-1.13	-.34	3.97	-1.56	-.95	1.04	-.92	1.33
Western Type								
1.86	-.89	-.47	.49	.59	-.99	-1.01	-.78	-.19
Non-Blacksoil Industrial Type								
-.86	1.18	-.54	-1.15	-1.04	1.29	-1.17	-.09	.73
Central Non-Blacksoil Type								
.001	-.98	-.49	-.47	.43	-.04	-.12	-.46	-.76
Northeastern Type								
-.14	.80	1.30	.76	-.94	-.64	.07	3.00	.09
Provinces Not of Any Type								
-2.19	3.39	3.65	-.34	.27	.49	.24	2.41	2.56

Factors
1. Landownership by nobles.
2. Bourgeois landownership.
3. Peasant landownership and land use.
4. Use of regular hired labor and its development in non-peasant agriculture.
5. Peasant agricultural labor for hire.
6. Proletarianization of the village.
7. Embourgeoisement of the peasantry.
8. Correlation of the peasant and manorial economies.
9. Intensification of agricultural production.

that a particular variant prevailed, not that it reigned unchallenged, for neither variant was completely triumphant in Russia or in any particular region; as Lenin noted, tendencies of development along the manorial and the peasant paths of development were to be found wherever peasants and squires were neighbors.[7]

---

[7] Lenin, *Polnoe sobranie sochinenii*, vol. XVI, pp. 217-18.

The other types occupied an intermediate position between the poles, displaying lines of affinity with both and inclining to a greater or lesser extent to one or the other. This inclination is most marked with the Western and Northeastern types. The former has the second highest proportion of landownership by nobles (1.86), a lower correlation between the peasant and manorial economies (-.78), and the lowest level of proletarianization of the village. The Northeast was distinguished by the highest[8] level of peasant landownership (1.30) and land use and the strongest sway of the peasant economy over the manorial (3.00).

The notable traits of the Non-Blacksoil Industrial type were the extent of bourgeois landownership, the highest level of proletarianization of the village and the lowest level of embourgeoisement of the peasantry. For the Central Blacksoil type they are the modest extent of bourgeois landownership, and the low intensity of agricultural production. The provinces not conforming to any type tended to have low levels of landownership by nobles and high levels of bourgeois landownership, intensification of agricultural production, and landownership and land use by peasants.

In sum, factor analysis provides us with a multifaceted perspective, a perspective that is at once specific and generalized, on the characteristic traits of various social types of agrarian structure and so reveals the tendency of their development through the extent of their conformity to the manorial or the peasant path of bourgeois agrarian evolution.

The analysis of the content of the social types of agrarian development permits us to relate each type to the manorial or the peasant path of agrarian evolution, but it does not establish with any precision the territorial demarcation between these paths. We can render that demarcation using the method of pattern recognition.[9] We used the variant recognition ''with a teacher.'' Groups of objects are selected that have to the maximum extent the traits inherent in a particular type. These groups serve as the ''teacher.'' Then the rest of the objects are compared to the ''teacher'' and assigned to a type or, if they are distinct from the types, to a special group.

We took the Baltic Provinces except for Kovno to characterize the manorial (strictly speaking, the ''manorial-peasant'') variant of agrarian evolution and the Steppe provinces except for Ufa to characterize the peasant, or ''peasant-manorial,'' variant. Then, working from 31 variables, we arrayed the provinces into two types and isolated a group of provinces that was significantly different from both types. In this group, bourgeois evolution was so firmly based upon the peasant economy that one can classify it as the peasant type pure and simple.

---

[8] Excluding the provinces that do not conform to any type.

[9] On this method, see Borodkin, *Mnogomernyi statisticheskii analiz*, p. 31.

The results are shown in Table 6. Thirty-one provinces conform to the peasant-manorial type, 14 to the manorial-peasant type and seven to the pure peasant type. St. Petersburg and Mogilev Provinces belong to both types, for their affinity has the same weight (.50) for each. Geographically, the manorial-peasant type prevailed in the West, the pure peasant type on the northern and eastern peripheries, while the peasant-manorial type prevailed in the provinces of the center and the South.

Province	Weight	Province	Weight	Province	Weight
Table 6: Distribution of the Provinces of European Russia According to their Tendency of Bourgeois Agrarian Evolution					
		Peasant-Squire Type			
1. Tauride	1.00	12. Tambov	.71	23. Kazan'	.64
2. Kherson	1.00	13. Khar'kov	.71	24. Novgorod	.64
3. Ekaterinoslav	1.00	14. Ufa	.70	25. Kursk	.64
4. Don	1.00	15. Pskov	.70	26. Tula	.58
5. Samara	1.00	16. Bessarabia	.67	27. Poltava	.57
6. Saratov	.77	17. Tver'	.67	28. Petersburg	.50
7. Chernigov	.77	18. Smolensk	.67	29. Mogilev	.50
8. Orel	.76	19. Vladimir	.66	30. Kaluga	.47
9. Nizhnii Novgorod	.76	20. Voronezh	.66	31. Moscow	.44
10. Riazan'	.75	21. Penza	.66		
11. Simbirsk	.75	22. Kostroma	.65		
		Squire-Peasant Type			
1. Lifland	1.00	6. Vilno	.67	11. Petersburg	.50
2. Kurland	1.00	7. Volynia	.60	12. Mogilev	.50
3. Estland	1.00	8. Kiev	.57	13. Perm'	.40
4. Kovno	.71	9. Grodno	.56	14. Podolia	.38
5. Minsk	.67	10. Vitebsk	.53		
		Peasant Type			
1. Archangel	.82	4. Orenburg	.57	7. Astrakhan'	.36
2. Olonets	.77	5. Viatka	.53		
3. Iaroslavl'	.68	6. Vologda	.40		

The Economic and Social Results of the Functioning of the Agrarian Structure

Let us now examine the economic and social impact of the two types of bourgeois agrarian evolution. To this end, as already explained, we isolated five factors reflecting the state of agricultural production, the situation of the peasants and agricultural workers, and the level of the socio-demographic and cultural development of the countryside. The factor weights for each type are shown in Table 7. From these factor weights we calculated two integral factors. The sum of the index of these factors registers the general level of the economic and social development of the two types of agrarian structure.

Note first of all that two types, the Steppe and the Baltic, are sharply distinct from all other types in Table 7. The summary integral index of economic and social development pent in the agrarian sphere is 3.08 for the Steppe and 1.54 for the Baltic. In other words, bourgeois agriculture was a success in Russia in both its "American" and its "Prussian" variants; attempts to represent the Prussian variant of agrarian development as a significant restraint on socio-economic development cannot succeed. We should emphasize, however, that the sway of the American variant assured a higher overall level of agrarian development, for the combined index of development was much higher in the Steppe than in the Baltic. The high level of agrarian development in these two regions had a number of components. The integral factors (VI and VII) show that progress on the steppe was due primarily to the high level of agricultural development and the relatively auspicious situation of the peasants, while in the Baltic it was due primarily to the level of socio-demographic and cultural develop-

Type of Agrarian Structure	FACTOR WEIGHTS							
	Initial Factors				Integral Factors			
	I	II	III	IV	V	VI	VII	VI+VII
Steppe	2.14	1.11	1.15	2.06	-.63	2.27	.81	3.08
Baltic	.89	1.46	-.05	-.33	2.94	.01	1.53	1.54
Western	-.53	.39	.35	-1.45	-.52	-.13	-1.11	-1.24
Non-Blacksoil Industrial	-1.34	-.73	-1.27	-.12	.86	-1.51	.39	-1.12
Central Blacksoil	.52	-.93	-.04	-.01	-.42	-.62	-.12	-.14
Northeastern	-.98	1.57	1.02	-.67	-.93	.76	-1.23	-.47
Provinces Not of Any Type	-1.75	.64	-.13	.41	-.66	-.16	-.50	-.66

Table 7: Economic and Social Results of the Functioning of the Agrarian Structure

ment of the countryside. More particularly, the Steppe region stands first among the regions of European Russia with respect to the level of development of cereal cultivation and the situation of the peasants and agricultural workers; the Baltic region stands much lower than the Steppe with respect to the latter criterion, but first with respect to factor V and second after the Northeast with respect to cattle-raising. It follows that, since the American path promised higher productivity and a better lot for the direct producers, it was, taking a broad historical perspective, the more progressive form of bourgeois agrarian development.

As for other types, the sway of the Northeast in cattle-raising is striking, albeit well known, but note also the relatively favorable situation of the peasants. In general, according to factor VI, the Northeast stood second after the Steppe—further proof of the advantages of the American path. Note also that, in their socio-demographic and cultural development, the Non-Blacksoil Industrial provinces (along with the Baltic) significantly surpassed the rest of European Russia, attaining a value of .86 for factor V. (Recall that for European Russia the mean of each factor weight is zero.) For progress in this sphere depended not only on the state of agriculture and the situation of the peasantry (on these counts, the Non-Blacksoil Industrial provinces stood very low) but also on non-agrarian factors. The converse appears from the minimal levels of socio-demographic and cultural development on the steppe and in the Northeast, where the level of agricultural production was high and the situation of the peasants was most favorable. Note finally that the Western region is in last place overall (factor VIII = -1.24), while the Central Blacksoil belt was close to the average.

In short, the types of agrarian structure were significantly distinct from one another not only in their own traits but in the economic and social results they produced.

Table 8: The Dependence of the Economic and Social Results
of Agrarian Development of the Provinces of European Russia
at the Turn of the Twentieth Century on Various Factors

The table gives the results of equations of multiple regression, obtained with a stepwise algorithm. It shows how much the value of the explained dispersion of the resultant variable is altered by the introduction of each successive factor into the regression model. A minus-sign in parentheses denotes that the corresponding regression coefficient is negative.

FACTORS	OUTCOME VARIABLES			
	Level of Economic Development		Level of Social Development	
	50 prov.	29 prov.	50 prov.	29 prov.
1. Landownership by nobles	—	0.4	1.5 (-)	—
2. Bourgeois landownership	—	7.5 (-)	—	—
3. Peasant landownership and land use	2.5	7.9 (-)	13.3 (-)	0.3 (-)
4. Development of capitalism in non-peasant agriculture	0.5	4.4	12.0	2.0
5. Peasant agricultural labor for hire	0.8	1.2 (-)	2.9 (-)	28.7 (-)
6. Proletarianization of the village	2.9 (-)	—	18.4	1.7
7. Embourgeoisement of the peasantry	45.9	60.5	9.6	9.3
8. Correlation of peasant and manorial economies	0.4	0.6 (-)	0.5 (-)	5.9 (-)
9. Intensification of agricultural production	5.1 (-)	—	—	28.3 (-)
Total weight of factors (% of explained dispersion)	58.1	82.5	58.2	76.2

It remains to establish which of our nine social-structure factors had the most significant impact on the results of the functioning of the agrarian order. To this end we conducted a multivariate regression analysis. As outcome variables indicating the effectiveness of the function of the agrarian system, we took the two integral variables (VI and VII from Table 7), one economic, the other social. As causal factors we took the nine variables that characterize the basic social traits of the agrarian order. We used a linear model of multivariate regression, which was feasible inasmuch as the factors were not closely correlated one to another; of 36 correlation coefficients, only three surpassed .50.

The highly suggestive results of this analysis are shown in Table 8, both for the 50 provinces of European Russia and for the 29 provinces that followed the peasant-manorial path.[10] Note first of all that the nine factors we used determine more than 50% of the variation in the outcome variables, which means we were able to identify and employ the factors that did have the greatest impact. Among these factors, both for the whole of European Russia and especially for the 29 provinces where the American path of development prevailed, the embourgeoisement of the peasantry played the leading role. In other words, the level and productivity of agriculture and the situation of the peasants depended first of all on the bourgeois-capitalist development of the peasant economy. The role of other factors, including those relating to landownership, was secondary, although the level of the peasants' landownership and land use did exert a positive, if modest, influence.

The impact of the factors on the results of socio-demographic and cultural development of the village was more complicated. Taking all 50 provinces, four factors exerted a significant influence: peasant landownership and land use, the development of capitalism in non-peasant agriculture, the proletarianization of the village, and the embourgeoisement of the peasantry. In other words, the level of the social development of the village depended primarily on the development of agrarian capitalism in all its manifestations—on bourgeois progress in *both* the manorial and the peasant economies and on the proletarianization of the village. This last consideration deserves emphasis. It confirms quantitatively the widespread assumption that the development of capitalism was socially progressive not only for the emerging rural bourgeoisie but also for the strata of the peasantry undergoing proletarianization. The negative correlation between factor 3, peasant landownership and land use, and the level of social development confirms this point.

The picture that emerges from an analysis of all 50 provinces is confirmed when we turn to the 29 where the American path of development prevailed. Here the level of social development was primarily determined by the extent of peasant agricultural labor for hire and the intensification of agricultural production. In the provinces of the southern steppe and the Southeast, where the embourgeoisement of the peasantry proceeded more intensely and the intensification of agriculture was least advanced, social development of the village was at a higher level. In the provinces of the Non-Blacksoil Industrial region, on the other hand, where peasant labor in industry was widespread, there was an inverse relationship between the level of rural social development and peasant agricultural labor for hire. In general, in these 29 provinces, social progress in the village was a function of the level of bourgeois agrarian

---

[10] Unfortunately, the number of provinces that followed the manorial-peasant path is too small to permit a corresponding regression analysis of them.

development.

Multivariate regression analysis also permits us to establish the magnitude of the variation in the outcome variables determined by variation in the causal factors of agrarian development on the eve of the twentieth century. To achieve this we use coefficients of the natural equations of regression.

	Coefficients of Natural Equations of Regression*	
FACTORS	Economic Development	Social Development
1. Landownership by nobles	—	-.123
2. Bourgeois landownership	—	—
3. Peasant landownership and land use	.159	-.336
4. Development of capitalism in non-peasant agriculture	.120	.305
5. Peasant agricultural labor for hire	.166	-.236
6. Proletarianization of the village	-.156	.453
7. Embourgeoisement of the peasantry	.379	.410
8. Correlation of peasant and manorial economies	.112	-.070
9. Intensification of agricultural production	-.320	—
Intercept term of the regression (a_0) equation	.000	.000

Table 9: The Impact of Various Factors on the Dynamics of the Level of the Economic and Social Agrarian Development of the Provinces in European Russia in the Early Twentieth Century

*The coefficient of the natural equations of regression shows the extent to which the value of a variable changes when the factor operating upon it changes by 1.0.

A very clear and persuasive picture emerges from Table 9. It shows that further improvement in the level of economic and social agrarian development could be achieved primarily as a result of bourgeois progress in the village—that is, by the development of capitalism in the peasant economy. As regards economic development and the situation of the peasantry, five factors, with coefficients ranging from .379 to .112, had the biggest impact: the further embourgeoisement of the peasantry, the expansion of peasant agricultural labor for hire, expanded use of hired labor, the level of peasant landownership and land use, and an increased contribution by peasants to agricultural production. By comparison, the role of the development of capitalism in non-peasant agriculture was relatively insignificant.

It is revealing that the basis of agrarian economic progress was extensive rather than intensive. Given the prevailing agrarian structure, the intensification of agricultural production had a negative influence on its *level* of development. By the same token, the increased proletarianization of the peasantry, which first of all took the form of employment in industry outside the village, also had a negative impact on the development of agriculture. In short, there was a conflict between agricultural development on an extensive footing and the progress of capitalism in industry.

The advance of social progress in the village (as expressed, to recapitulate, in the mobility, urbanization and literacy of the rural population) was also closely linked to the development of capitalism in the countryside. Here the proletarianization of the village (enlargement of the poorest stratum of the peasantry and expansion of peasant employment in agriculture and in industry) was most significant, as Table 9 shows, with the embourgeoisement of the peasantry in second place. The development of capitalism in non-peasant agriculture had much more of an impact on social progress than on economic. Note also another conflict. The expansion of peasant landowership and land use and of peasant labor for hire in agriculture, which exerted a positive influence on economic development, correlates negatively with social development; the coefficients are -.336 and -.236, respectively.

All of the factors had a strongly interactive influence on the achieved level of agrarian development and on its potential for further development. For all the complexity of the interaction, it is clear that it was primarily the development of capitalism in the peasant economy that exerted a determining influence on both the economic and the social aspects of agrarian progress.

Let us conclude by drawing some conclusions from the analysis we have carried out. First of all, in European Russia at the turn of the century, there were in reality two paths of bourgeois agrarian evolution, and not just a manorial or Prussian path, as some scholars maintain. Second, as regards territorial extent, agrarian evolution based on the peasant economy was the prevailing tendency in the system of agrarian relations. Third, while both the American and the Prussian paths of development were certainly progressive, on the whole the American path promised a higher level of social and economic progress. Objectively, then, there was a footing for achieving a high level of capitalist agrarian development based on the peasant economy. The ruling classes and the autocracy that defended their interests, however, made every effort to secure the victory of manorial capitalism, despite the lack of the preconditions for that victory. These efforts generated intense contradictions and antagonisms, which influenced the course of the entire nation's development and could not be resolved in a framework of bourgeois social, economic and political relations.

Appendix A: The Social Agrarian Structure of the Provinces of European Russia at the Turn of the Twentieth Century
This appendix shows the values for each of the 31 variables in Table 1 for each province of European Russia.

Province	1.	2.	3.	4.	5.	6.	7.	8.	9.	10.	11.	12.	13.	14.	15.	16.	17.	18.	19.	20.	21.	22.	23.	24.	25.	26.	27.	28.	29.	30.	31.
1. Archangel	.003	.017	.100	.100	.983	1.00	.334	10.0	.20	1.1	.185	.950	.053	1.3	.033	.040	.152	.204	.770	1.113	.462	.026	.128	.171	.011	52.6	65.0	6.5	.667	2.338	56.0
2. Astrakhan'	.057	.674	.946	.400	.326	.149	.334	20.0	1.95	4.7	.011	.132	.295	6.1	.135	.044	.022	.220	.675	1.041	.281	.079	.387	.958	.985	17.7	26.0	4.9	.190	1.849	55.0
3. Bessarabia	.461	.529	.645	.910	.471	.584	.287	137.0	7.05	1.3	.176	.202	.127	1.7	.053	.044	.143	.033	.505	.987	.189	.079	.287	.599	.634	2.5	5.5	4.4	.181	.505	44.0
4. Vilno	.971	.810	.552	.862	.190	.190	.319	53.0	1.45	1.1	.144	.120	.025	1.3	.073	.096	.107	.028	.718	1.206	.155	.023	.084	.726	.589	1.1	2.7	3.1	.201	.848	36.0
5. Vitebsk	.813	.607	.718	.806	.393	.436	.463	46.0	2.24	1.4	.226	.168	.047	1.8	.056	.071	.093	.047	.629	1.194	.152	.031	.100	.355	.653	1.5	7.5	3.2	.289	.930	36.0
6. Vladimir	.203	.306	.644	.542	.694	.661	.293	56.0	2.41	1.6	.153	.231	.111	2.0	.024	.037	.036	.255	.825	1.046	.482	.013	.071	.655	.716	5.6	9.6	22.1	.220	.568	40.0
7. Vologda	.067	.168	.689	.321	.832	.904	.424	14.0	.52	2.9	.198	.494	.021	3.5	.017	.011	.044	.090	.711	1.036	.143	.015	.056	.773	.300	17.8	86.0	27.9	.317	1.208	46.0
8. Volynia	.889	.725	.742	.843	.275	.280	.256	155.0	4.29	1.0	.120	.098	.063	1.2	.033	.068	.074	.019	.561	.789	.108	.046	.421	1.120	.385	1.2	3.5	2.9	.322	.787	59.0
9. Voronezh	.266	.627	.748	.722	.373	.360	.293	105.0	6.33	1.6	.100	.235	.196	2.0	.014	.035	.042	.057	.582	1.110	.110	.109	.353	.580	.702	4.1	9.6	11.8	.174	.538	39.0
10. Viatka	.042	.396	.948	.665	.604	.658	.111	28.0	3.12	2.6	.013	.117	.022	2.7	.018	.048	.027	.056	.554	1.177	.110	.087	.256	.545	.730	23.9	100.0	65.4	.135	.466	43.0
11. Grodno	.901	.840	.536	1.390	.160	.229	.119	58.0	3.85	1.4	.151	.141	.031	1.6	.055	.083	.093	.029	.747	1.028	.170	.016	.067	.972	.435	1.2	3.1	2.9	.184	.957	38.0
12. Don	.102	.434	.336	.511	.566	.615	.372	87.0	6.05	4.4	.097	.412	.087	5.2	.041	.078	.095	.028	.463	.999	.160	.175	.548	.994	.923	10.7	11.4	3.4	.189	.795	33.0
13. Ekaterinoslav	.468	.422	.714	.498	.578	.577	.313	141.0	4.30	1.4	.412	.371	.244	2.3	.043	.096	.118	.036	.408	.949	.266	.178	.488	1.090	.901	3.0	3.9	5.9	.171	.510	41.0
14. Kazan'	.143	.623	.630	.775	.377	.380	.258	69.0	2.32	1.6	.044	.191	.052	1.8	.010	.022	.062	.133	.693	1.098	.219	.033	.127	.598	.739	7.3	19.6	18.1	.193	.332	39.0
15. Kaluga	.302	.385	.532	.556	.615	.588	.275	74.0	1.97	1.4	.236	.301	.109	1.9	.023	.033	.048	.302	.581	.452	.380	.063	.193	.702	.655	4.0	9.6	8.9	.342	.638	38.0
16. Kiev	.725	.730	.690	.833	.270	.282	.318	122.0	6.79	.8	.151	.153	.092	.9	.035	.059	.064	.036	.680	.998	.143	.035	.191	.636	.592	2.1	2.6	2.4	.246	.552	58.0
17. Kovno	.521	.657	.558	.531	.343	.259	.370	72.0	1.28	1.4	.125	.158	.065	1.7	.208	.164	.064	.031	.540	1.135	.108	.143	.393	.610	.509	3.1	2.8	2.0	.285	.751	44.0
18. Kostroma	.456	.316	.737	.487	.684	.707	.356	19.0	1.21	1.7	.422	.292	.092	2.6	.029	.034	.050	.240	.794	1.042	.373	.010	.046	.999	.699	.9	6.0	13.7	.245	.624	43.0
19. Kurland	1.009	.922	.924	1.027	.078	.082	.334	75.0	1.51	1.7	.006	.005	.128	1.9	.298	.241	.094	.048	.334	1.001	.176	.392	.698	.999	.927	.9	4.1	1.6	.305	1.012	58.0
20. Kursk	.350	.617	.440	.737	.383	.373	.296	141.0	6.42	1.1	.153	.269	.209	1.5	.021	.040	.078	.088	.483	1.137	.182	.149	.408	.731	.343	3.2	5.3	9.1	.218	.367	43.0
21. Lifland	1.170	.750	.876	.808	.250	.256	.334	65.0	1.53	1.1	.000	.000	.255	1.4	.305	.237	.075	.042	.411	1.105	.210	.191	.140	.670	.000	.6	3.8	1.5	.285	1.221	63.0
22. Minsk	2.077	.769	.839	1.015	.231	.284	.467	32.0	1.73	1.3	.249	.092	.062	1.7	.050	.076	.118	.021	.625	1.156	.149	.039	.140	.493	.691	.6	5.8	2.4	.250	1.033	38.0
23. Mogilev	.866	.609	.706	.709	.391	.402	.309	53.0	2.21	1.2	.422	.296	.105	1.8	.033	.082	.108	.028	.383	1.242	.150	.148	.077	.530	.645	1.6	5.6	5.3	.410	.846	40.0
24. Moscow	.288	.438	.440	.618	.562	.533	.310	163.0	3.61	.8	.125	.190	.215	.9	.023	.065	.010	.153	.791	1.054	.553	.017	.098	.486	.775	3.9	9.8	2.5	.547	.755	45.0
25. Novgorod	.495	.294	.698	.582	.706	.847	.394	20.0	.89	2.3	.588	.350	.109	3.9	.032	.044	.060	.201	.632	1.185	.300	.041	.119	.359	.645	3.2	17.5	13.2	.373	.985	43.0
26. Nizhnii Novgorod	.341	.378	.738	.575	.622	.783	.316	47.0	3.16	1.3	.215	.239	.102	1.7	.024	.026	.089	.120	.769	1.042	.254	.034	.158	.604	.582	3.5	21.3	12.0	.193	.404	41.0
27. Olonets	.014	.106	.265	.412	.894	1.181	.081	13.0	.13	11.5	.023	.169	.002	11.8	.029	.069	.069	.192	.707	1.002	.298	.030	.116	.763	.960	71.7	100.0	29.5	.421	1.422	58.0
28. Orenburg	.052	.274	.706	.558	.726	1.150	.595	22.0	1.97	7.7	.064	.333	.035	8.5	.045	.064	.133	.050	.359	1.296	.245	.340	.736	.713	.990	20.3	31.0	7.3	.290	.753	34.0
29. Orel	.421	.533	.504	.686	.467	.461	.276	108.0	4.79	1.1	.191	.243	.178	1.5	.025	.036	.078	.099	.571	1.032	.215	.095	.308	.954	.501	2.8	5.2	5.7	.223	.371	42.0
30. Penza	.434	.623	.724	.731	.377	.370	.306	83.0	4.39	1.3	.087	.126	.169	1.5	.019	.034	.116	.053	.592	1.152	.189	.079	.256	.580	.664	2.5	3.9	7.7	.178	.385	43.0
31. Perm'	.806	.759	.999	1.004	.241	.251	.324	30.0	1.36	2.9	.014	.013	.047	3.0	.028	.057	.011	.131	.541	.998	.249	.114	.339	.907	.804	1.2	100.0	40.4	.161	.863	51.0
32. St. Petersburg	.948	.515	.811	.689	.485	.485	.446	64.0	1.51	.7	.325	.176	.149	1.0	.031	.068	.032	.077	.757	1.116	.369	.023	.103	.372	.787	1.3	8.8	1.5	.597	.727	46.0
33. Podolia	.747	.805	.608	.836	.195	.189	.469	62.0	10.13	.7	.126	.135	.043	1.1	.030	.046	.075	.017	.426	1.484	.111	.158	.195	.084	.643	1.5	2.8	3.0	.244	.439	44.0
34. Poltava	.492	.571	.447	.625	.429	.399	.404	143.0	8.60	.9	.245	.284	.340	1.4	.047	.067	.070	.037	.778	.935	.113	.028	.166	1.205	.447	2.5	3.7	3.0	.149	.553	49.0
35. Pskov	.435	.299	.605	.522	.701	.808	.425	41.0	2.78	1.4	.512	.352	.163	2.3	.027	.047	.098	.095	.611	1.198	.201	.041	.128	.412	.545	3.4	5.1	4.7	.350	1.030	41.0
36. Riazan'	.362	.468	.429	.648	.532	.540	.270	99.0	4.26	1.1	.215	.277	.196	1.6	.020	.038	.055	.125	.688	1.024	.209	.055	.227	.954	.538	3.3	6.1	6.0	.193	.454	45.0
37. Samara	.138	.261	.855	.456	.739	.791	.264	45.0	3.16	2.6	.176	.333	.413	4.1	.044	.059	.223	.018	.430	1.008	.247	.251	.604	.887	.957	8.5	10.3	9.0	.257	.452	31.0
38. Saratov	.391	.499	.791	.580	.502	.457	.421	78.0	3.46	1.6	.253	.322	.275	2.4	.031	.016	.187	.036	.545	1.164	.253	.128	.382	.685	.885	3.1	8.6	8.3	.182	.564	39.0
39. Simbirsk	.382	.579	.689	.539	.421	.351	.346	71.0	3.05	1.2	.163	.248	.226	1.6	.026	.023	.148	.058	.653	1.094	.228	.053	.195	.678	.691	3.0	8.4	8.4	.171	.346	41.0
40. Smolensk	.497	.365	.473	.502	.635	.693	.486	55.0	3.08	1.4	.498	.366	.274	2.5	.033	.046	.081	.177	.430	1.264	.273	.119	.279	.519	.583	3.0	8.4	7.2	.416	.934	48.0
41. Tauride	.444	.314	.803	.548	.686	.698	.339	110.0	5.24	1.7	.563	.398	.273	3.2	.097	.125	.197	.038	.311	1.430	.273	.316	.692	.784	.916	5.4	18.1	2.6	.147	.354	36.0
42. Tambov	.391	.518	.630	.699	.482	.507	.293	109.0	6.89	1.1	.138	.183	.196	1.5	.020	.036	.071	.051	.579	1.123	.133	.114	.391	.690	.671	2.9	3.7	8.8	.189	.419	51.0
43. Tver'	.237	.291	.499	.499	.709	.767	.295	44.0	2.82	1.6	.430	.528	.224	2.7	.029	.030	.036	.175	.667	1.156	.242	.026	.089	.461	.525	6.0	13.8	13.7	.387	.921	48.0
44. Tula	.591	.641	.359	.777	.359	.370	.250	115.0	3.15	1.1	.183	.199	.196	1.5	.024	.044	.093	.273	.552	1.045	.406	.080	.250	1.052	.481	2.0	3.7	4.0	.207	.373	46.0
45. Ufa	.221	.437	.849	.747	.563	.750	.401	29.0	3.12	1.0	.196	.387	.067	3.8	.027	.041	.134	.029	.541	1.003	.199	.171	.520	.689	.817	5.4	26.0	17.3	.334	.649	43.0
46. Khar'kov	.303	.503	.618	.601	.497	.482	.443	127.0	7.92	1.2	.192	.318	.083	1.5	.020	.050	.058	.063	.704	1.006	.154	.038	.161	.888	.655	3.9	5.4	5.1	.160	.536	40.0
47. Kherson	.542	.392	.692	.487	.608	.572	.338	142.0	7.12	1.2	.391	.283	.576	1.4	.052	.071	.175	.036	.414	.916	.246	.183	.515	.787	.941	2.5	3.7	2.3	.167	.343	36.0
48. Chernigov	.339	.432	.426	.584	.568	.635	.338	93.0	4.46	1.2	.299	.381	.155	1.7	.041	.053	.102	.054	.495	1.110	.171	.134	.380	.623	.303	3.8	8.9	10.7	.298	.534	34.0
49. Estland	2.805	.901	.944	1.298	.099	.132	.334	70.0	1.50	1.1	.001	.000	1.162	2.3	.178	.165	.065	.040	.543	1.029	.188	.118	.318	.621	.000	.3	1.9	1.6	.307	1.146	62.0
50. Iaroslavl'	.281	.301	.540	.507	.699	.739	.281	39.0	1.47	1.5	.987	1.057	.141	3.1	.040	.051	.031	.367	.834	1.051	.488	.008	.044	.640	.467	7.0	27.8	8.5	.278	.810	55.0

Appendix B: The Results of Agrarian Development in European Russia at the Turn of the Twentieth Century
This appendix shows the values for each of the 15 variables in Table 2 for each province of European Russia.

	1.	2.	3.	4.	5.	6.	7.	8.	9.	10.	11.	12.	13.	14.	15.
1.	9.15	11.48	1.061	.405	.508	3.508	10.12	3.94	24.0	59.0	3.9	.653	.088	.042	.203
2.	8.37	11.36	1.269	.620	.841	9.717	16.19	10.35	12.0	47.0	1.0	.950	.035	.126	.120
3.	37.80	50.00	1.975	.425	.562	2.797	44.29	4.79	14.0	44.0	3.1	.603	.035	.025	.125
4.	25.47	34.84	1.338	.500	.684	4.211	24.82	5.02	10.0	30.0	2.6	.471	.087	.042	.256
5.	22.05	29.90	1.223	.482	.653	3.216	24.88	4.71	8.0	38.0	2.9	.598	.100	.040	.210
6.	22.41	38.44	1.338	.286	.490	2.579	22.49	3.02	28.0	58.0	2.7	.912	.123	.078	.240
7.	20.46	22.77	1.189	.523	.583	3.817	21.20	5.29	5.0	39.0	2.1	.544	.054	.020	.174
8.	26.96	35.98	1.359	.427	.570	2.446	27.95	4.23	7.0	36.0	1.8	.644	.040	.024	.151
9.	34.41	40.41	1.568	.449	.527	3.089	32.72	4.34	9.0	51.0	3.0	1.079	.076	.041	.142
10.	41.83	46.98	1.053	.447	.502	2.521	42.62	4.52	11.0	34.0	2.1	.711	.072	.015	.148
11.	24.38	35.50	1.323	.495	.721	5.192	25.49	5.16	8.0	32.0	2.5	.481	.064	.043	.255
12.	45.87	60.84	1.899	1.079	1.432	4.207	48.08	9.51	16.0	80.0	3.6	1.333	.028	.074	.195
13.	47.60	64.33	1.868	.573	.774	2.976	43.30	5.62	27.0	78.0	3.9	1.147	.058	.052	.188
14.	31.58	36.70	1.339	.263	.306	1.719	31.88	2.64	7.0	34.0	1.1	.551	.068	.047	.150
15.	23.55	30.69	1.090	.352	.459	1.864	23.62	3.47	17.0	53.0	.7	.833	.202	.029	.164
16.	27.93	39.83	1.503	.252	.359	2.246	26.01	2.28	11.0	40.0	2.2	.753	.072	.046	.140
17.	26.15	38.22	1.579	.401	.586	2.633	27.09	3.75	11.0	37.0	3.9	.566	.096	.039	.414
18.	27.89	35.07	1.078	.374	.471	2.546	29.97	3.81	28.0	49.0	1.8	.780	.092	.028	.220
19.	38.71	66.66	1.226	.591	1.018	3.320	38.22	4.46	14.0	50.0	4.4	.714	.257	.142	.715
20.	37.69	45.56	1.382	.292	.353	1.683	33.92	2.8	9.0	52.0	1.5	1.050	.107	.058	.140
21.	33.99	62.12	.959	.570	1.042	4.280	31.88	4.04	19.0	43.0	3.8	.575	.143	.191	.793
22.	26.15	34.96	1.501	.560	.749	4.136	31.10	5.51	8.0	39.0	2.4	.646	.061	.024	.146
23.	25.64	32.30	1.391	.433	.546	2.065	26.54	4.55	7.0	40.0	2.3	.711	.082	.024	.142
24.	10.09	36.21	1.079	.142	.511	1.379	11.75	1.3	35.0	61.0	1.4	.961	.116	.376	.267
25.	21.08	26.93	1.057	.446	.569	2.637	21.76	4.52	11.0	42.0	1.7	.540	.096	.027	.210
26.	26.07	37.44	1.250	.240	.345	2.092	26.77	2.38	15.0	43.0	2.4	.784	.068	.049	.192
27.	19.63	23.39	.964	.500	.596	3.375	21.16	5.26	14.0	46.0	1.8	.544	.078	.021	.194
28.	29.55	38.20	1.459	.640	.828	2.594	32.90	6.47	10.0	51.0	1.3	1.281	.023	.032	.187
29.	34.05	42.85	1.268	.267	.335	1.663	31.67	2.51	11.0	40.0	1.6	.753	.120	.060	.139
30.	38.78	45.29	1.500	.327	.382	2.158	32.78	3.07	9.0	41.0	2.4	.917	.104	.062	.124
31.	31.42	44.56	1.323	.520	.738	5.374	32.70	5.33	14.0	43.0	1.6	.885	.025	.032	.173
32.	7.79	30.84	1.053	.102	.404	1.217	10.19	.89	44.0	51.0	2.0	.612	.137	.581	.393
33.	32.57	43.26	1.422	.235	.313	1.794	29.71	2.20	9.0	31.0	1.8	.543	.062	.016	.141
34.	42.47	42.67	1.603	.449	.451	3.707	31.02	3.11	10.0	47.0	3.0	.974	.082	.039	.103
35.	21.47	26.99	1.223	.482	.606	2.943	21.40	4.73	7.0	40.0	1.7	.533	.081	.030	.121
36.	34.58	42.30	1.238	.311	.380	2.354	31.88	2.38	12.0	54.0	1.4	1.000	.162	.061	.179
37.	34.01	39.60	1.902	.489	.569	1.757	34.14	4.74	8.0	47.0	3.9	.884	.052	.024	.209
38.	37.66	48.69	1.346	.525	.679	3.105	37.70	5.23	11.0	52.0	2.8	.903	.064	.064	.210
39.	36.21	43.07	2.277	.293	.349	2.019	34.35	2.78	8.0	39.0	2.0	.608	.093	.037	.136
40.	28.99	44.77	1.135	.506	.607	2.245	28.46	4.89	18.0	44.0	.9	.660	.112	.032	.143
41.	54.40	80.43	1.627	.554	.819	2.408	67.24	5.15	29.0	69.0	3.8	.948	.076	.075	.237
42.	42.78	50.91	1.324	.339	.403	2.219	36.02	3.24	8.0	46.0	1.3	.961	.075	.049	.140
43.	23.73	29.36	.949	.419	.519	2.379	23.84	4.21	14.0	48.0	1.6	.762	.172	.042	.218
44.	43.11	55.08	1.229	.303	.388	1.803	37.90	2.70	17.0	55.0	.6	1.048	.179	.057	.171
45.	29.50	33.91	1.671	.442	.508	1.944	29.97	4.40	7.0	40.0	1.2	.842	.040	.026	.157
46.	30.91	39.36	1.754	.412	.525	3.348	28.67	3.79	10.0	52.0	3.8	1.050	.069	.095	.128
47.	42.08	64.22	1.698	.397	.606	2.049	49.13	4.09	20.0	64.0	4.6	.981	.057	.106	.186
48.	25.61	32.48	1.344	.337	.427	1.790	26.80	3.58	8.0	40.0	4.5	.744	.073	.035	.162
49.	40.15	67.14	1.215	.524	.877	3.731	29.40	3.61	17.0	47.0	2.2	.679	.127	.123	.796
50.	28.70	39.47	.927	.365	.502	2.916	31.46	3.70	36.0	55.0	2.3	.946	.188	.075	.333

Levels of Technology and the Use of
Hired Labor in the Peasant and Manorial Economy of
European Russia in 1917

N. B. SELUNSKAIA

Introduction

Natal'ia Selunskaia's article synthesizes research, based on the agricultural census of 1917, that she and her colleagues have carried out over more than a decade. The census is an unusually rich and comprehensive source. It has the advantage of recording the components of both the manorial and the peasant economies on the eve of the October Revolution. It has the disadvantage that these components were distorted by wartime circumstances, notably the conscription of men and of draft animals, the state's attempts to control the traffic in grain, and the difficulty of acquiring or replacing equipment. This distortion is worth noting because of the importance Selunskaia ascribes to the balance among components as a measure of the extent of capitalist development in agriculture. The concept of balance becomes more problematical in light of the possibility that optimal balance might exhibit considerable regional variation and because it presents problems of scale: a peasant household could not adjust its draft power to its sown area as finely as a manor could because horses did not come in fractional units.* Finally, it remains to be shown that the balances Selunskaia discloses were specific to capitalism; a precapitalist or even a premonetary economic unit—a beehive, for example—can have an inner logic of its own.

Selunskaia's data have implications, which she is not slow to draw, for the October Revolution, but also for the collectivization of agriculture a decade later. By 1917, owners of estates were making only a minor contribution, compared to peasants, to the nation's cereal production, but they still held vast expanses of land that the peasants coveted. The sway of smallholding peasants helped propel the revolution forward, but, because they consumed most of what they produced, their triumph served to provoke the procurement crisis of the late 1920s. Stalin would use their inability or unwillingness to market more than a small fraction of their output as a major justification for the collectivization campaign. Unfortunately, the census of 1917 provides no direct information on market participation, so this aspect of capitalist development cannot figure in Selunskaia's article.

D.F.

Historians of agrarian relations in Russia at the beginning of the twentieth century must confront both the level and the forms of bourgeois agrarian evolution; they must study them in the context of two basic organizational types, the peasant economy and the manorial economy. The historical significance of this inquiry derives from its relationship to the emergence of the social and economic preconditions of the October Revolution and its linkage to social developments in the villages of the empire around 1900. The literature on the subject has its shortcomings and its controversies, but it does emphasize that to establish the causes of the frustration of Russia's bourgeois agrarian evolution requires a detailed comparison of the social and agro-technical development of

* Except in the statistician's pasture. The census records newborn horses and colts separately,

the peasant and manorial economies. The historian is bound, therefore, to seek out new modes of analysis and to resort, in particular, to systematic quantitative analyis. Methodological questions become crucial on the level of macroanalysis, using aggregate data for the whole country. This essay offers some methodological observations and applications and some results of an analysis of the productive level and the role of hired labor in the peasant and manorial economies of European Russia; the analysis is based on published data from the agricultural census of 1917.[1]

Hired Labor and Productive Level in General

Using the data of the census of 1917, one can establish the significance of manorial landholding and its contribution to agricultural production. In 34 provinces, squires[2] owned one-fifth of all the arable land. The proportion of manorial land was highest in the Volga-Viatka region,[3] Belorussia, and the Ukraine, and lowest in the Southern Ural and Middle Volga regions. Note also the high concentration of forests and hayfields in the hands of the squires; taking European Russia as a whole, they owned two-fifths of all the forest land. In every region their share of forest was higher than their share of arable, and their share of hayfields was higher than their share of plowland everywhere except in the Ukraine. Thus they were well endowed with resources of which the peasants had a painfully inadequate supply.

The following data indicate the role of the manorial economy in the crucial branches of the rural economy, agriculture and cattleraising.[4] In 34 provinces the squires held an average of 7.7% of the total sown area. Their share was highest in the Ukraine (15.4%), Belorussia (13.3%) and the Central Blacksoil region (11.6%) and lowest in the Northern (0.7%), Volga-Viatka (1.2%) and Southern Ural (2.4%) regions. Thus even in regions where the character of the manorial economy was overwhelmingly agricultural, its role in the fundamental branch of the rural economy was insignificant. In the same 34 provinces, only 39.8% of the manorial plowland was under crops, whereas for owners of all categories the corresponding figure was 64.3%. The discrepancy between the intensity of exploitation of plowland by squires and by all landowners was at a maximum in the Middle Volga, Southern Steppe, Volga-Viatka, Southern Ural and Central Blacksoil regions, and at a minimum in Belorussia, the Central

---

and these are usually counted as fractions of a mature animal to determine the distribution of horses.

[1] "Pogubernskie itogi Vserossiiskoi sel'skokhoziaistvennoi i pozemel'noi perepisi 1917 g. po 52 guberniiam i oblastiam," *Trudy Tsentral'nogo statisticheskogo upravleniia*, vol. V, fasc. 1 (Moscow, 1921).

[2] Throughout this essay, *pomeshchiki*, which means "nobles owning estates," is rendered as "squires," and the adjectival form of the noun, *pomeshchich'ii*, as "manorial."

[3] Reckoning from the data for Nizhnii Novgorod and Kostroma Provinces; manorial landholding in Viatka Province was insignificant.

[4] In this article we do not touch upon manorial exploitation of the forests, which relates more to industry than to agriculture, but note that the concentration of forests in the squires' hands served, in many respects, to force the peasantry into dependency upon them.

Industrial and Northwestern regions.

The "underutilized" manorial land was cultivated by peasants, who rented 25% of all the manorial plowland in 34 provinces of European Russia, ranging from 15-17% (Belorussia and the Industrial Center) to 40-50% (the Volga and the Southern Steppe). Comparing the quantity of cattle per 100 *desiatiny*[5] of hayfields reveals that squires exploited this resource less intensively than did landowners in general. On the average, squires in 34 provinces needed only one-fifth of their hayfields to support their cattle; they rented out most of the balance. Peasants, especially in the blacksoil belt, suffered an extreme short-age of hayfields and had to rent them, primarily from squires. Furthermore, only 3.5% of the cattle in 34 provinces belonged to squires, with their share at its highest in Belorussia (8.5%) and at its lowest in the Volga-Viatka (0.6%) and Southern Ural (0.9%) regions.

In short, on the eve of the October Revolution, manorial agriculture did not make a significant contribution to agricultural or livestock production and did not exert much influence on the rural economy, either in the nation as a whole or in most regions, the exceptions being the Baltic region, Lithuania, Western Belorussia, and the Ukraine.

It is important to bear in mind that a considerable part of the land owned by squires was simply not involved in cereal production. In European Russia 24.5% of all manors had no sown area, 20.5% had no draft animals and 16.9% had no cattle. Most of these estates were in the Central Industrial, Northwestern and Volga-Viatka regions, where forests were the squires' principal resource and agricultural lands were rented out. In the Central Blacksoil region, as also in Eastern Belorussia, the proportion of manorial estates without draft animals was higher than the proportion with no sown area, because some estates worked their arable entirely with animals belonging to peasants or other agriculturalists.

The Agrotechnical Basis of the Manorial Economy

The manorial economy was not adequately supplied with draft power. In European Russia as a whole, the squires worked about two-thirds of their sown area with their own animals, relying for the rest on horses and oxen belonging to peasants. On the eve of the October Revolution, then, labor rental of the first type[6] was a notable feature of the manorial economy, especially in Eastern Belorussia, the Ukraine and the Central Blacksoil region.

The data on plows also indicates that the squires could not get along without equipment belonging to peasants, which again points to labor rental of the first type. With 7.7% of the sown area in European Russia, the squires had

---

[5] A *desiatina* = 2.7 acres or about a hectare.

[6] Under labor rental [*otrabotka*] of the first type, a peasant would work a stipulated portion of the squire's demesne, providing draft power and all other equipment; instead of wages, he would get the use of another, usually much smaller, portion of the squire's land. Under labor rental of the second type, the peasant provided only his labor; he was, in essence, an agricultural laborer who was paid in kind rather than in cash.

Table 1: The Level of the Manorial and Peasant Economies of European Russia in 1917 (Units per 100 *desiatiny* of sown area) For each variable, the manorial economy is in column A, the peasant economy in column B.									
Region	**Draft Animals**		**Plows**		**Advanced Equipment**		**Hired Laborers**		**All Cattle**	
	A.	B.	A.	B.	A.	B.	A.	B.	A.	B.
Central Industrial	33.9	42.8	17.6	29.0	16.0	1.7	57.0	1.9	28.0	74.8
Central Non-Blacksoil	27.3	39.3	16.1	24.8	11.6	4.0	24.7	1.5	13.7	62.3
Northwestern	27.0	50.8	13.4	19.8	17.0	5.5	29.8	2.0	30.5	106.3
Volga-Viatka	20.7	28.5	8.9	10.5	8.9	1.7	23.0	1.3	4.1	50.6
Central Blacksoil	16.3	31.2	6.3	9.4	5.5	4.4	12.2	0.6	4.3	44.3
Middle Volga	15.4	32.0	9.2	4.7	9.5	1.5	15.4	0.2	2.4	33.4
Ukraine	16.4	32.6	6.1	16.1	8.2	8.4	11.9	1.4	4.1	39.1
Southern Steppe	19.4	24.0	6.2	8.7	6.1	9.4	8.9	0.8	5.9	25.5
Southern Ural	19.5	50.0	5.1	8.0	5.5	5.1	9.7	1.2	4.2	48.8
European Russia*	18.1	35.0	6.8	13.1	5.6	4.2	14.4	1.1	6.6	55.8

*Data for the Northern Region and Eastern Belorussia are omitted because they are incomplete.

Source: "Pogubernskie itogi Vserossiiskoi sel'skokhoziaistvennoi i pozemel'noi perepisi 1917 g.
po 52 guberniiam i oblastiam," *Trudy Tsentral'nogo statisticheskogo upravleniia,* vol. V, fasc. 1 (Moscow, 1921).

only 5.6% of the plows. As regards machines and advanced equipment,[7] however, which were an important indicator of the technical level of agriculture, the manorial economy was in the lead, having 8.2% of the advanced equipment in European Russia, with a particular concentration in the Central Industrial, Northwéstern, Northern and Middle Volga regions. In the Central Non-Blacksoil and Southern Steppe regions, the squires' share of advanced equipment was less than théir share of sown area; this situation obviously was a function of the insignificant role of agriculture in the former area and of the sway of extensive agriculture in the latter.

It follows that the general agro-technical level was higher in the peasant economy than in the manorial, which tended to fetter the former. The peasant economy was everywhere better provided with draft animals; for European Russia as a whole, peasants had twice as many draft animals per 100 *desiatiny* of sown area than did the squires. Except for the Middle Volga region, the same applies to plows. As regards agricultural machines and advanced equipment, the manorial economy enjoyed a significant edge only in the Central Non-Blacksoil and Middle Volga regions and fell short of the peasant economy in the Ukraine; in any event, given their relative lack of draft animals and of tools, the squires' advantage in this sphere had no great significance.

On the whole, then, the census data show that the peasant economy surpassed the productive potential and technical level of the manorial economy and was the defining force of the nation's agriculture. The squires did not operate on a level of technology so elevated as to influence the general progress of agricultural production.

The role of the manorial economy in bourgeois agrarian evolution can be discerned by examining the use of hired labor. Of the long-term hired laborers counted by the census-takers in 1917, the squires of European Russia employed 63.4%. On this count, capitalism was more developed within the manorial economy than among the peasants, but 42% of the squires—notably in the Northwestern, Volga-Viatka and Southern Ural regions—did not use long-term hired labor. Many of them, like many peasants, got along with day-laborers, and some, apparently, used no hired labor at all.

In sum, the data permit us to say that, on the eve of revolution, truly capitalist methods were intertwined with labor rental in the structure of the manorial economy. Our data do not, however, reveal either the balance between the two or, correspondingly, the extent of the development of capitalist *relations*.

On this count, the most important indicator is the use of hired labor. The census data presented in Table 2 permit us to examine the contribution of hired labor to the peasant economy in three regions. We see that, with respect both to percentage of hired laborers in the working population and to the proportion of peasant households using hired labor, the Steppe region far outstripped the

---

[7] Here and below: steel harrows, reapers, mowing, winnowing and threshing machines, and binders.

Central Non-Blacksoil region and especially the Central Blacksoil region. Note that the Central Blacksoil households were poorly provided both with hired labor and the labor of their members.

Table 2: Hired Labor in the Peasant Economy (Three Regions)						
	Central Non-Blacksoil Region		Central Blacksoil Region		Steppe Region	
	Total	Percent	Total	Percent	Total	Percent
Hired labor in the working population	151,281	1.2	87,688	0.8	125,075	2.4
Households using hired labor	102,041	2.5	60,060	1.4	80,090	5.3
Households with no workers of their own	468,718	11.5	426,437	9.2	149,370	8.6
Workers in total population		62.4		51.8		53.6
Source: As Table 1.						

The demographic data from the agricultural census of 1917 open the way to a characterization of the regional peculiarities of the structure of labor resources in the peasant economy. A full treatment of the social aspects of the organization of the peasant economy would require data on the grouping of peasant households—if we relied on traditional forms of analysis. Data of that kind are largely lacking, however. We have to develop methods to meet this problem.

Correlational Models of the Productive and Social Organization of the Manorial Economy

A macroanalysis of the capitalist evolution of the manorial and peasant economies requires the development of a method of analysis for the aggregate data available. Soviet historians have worked on the quantitative analysis of Russian agriculture in the capitalist period and on the refinement of the massive data concerning it. Now we can construct models based on the correlations in the manorial and peasant economies among variables that reflect the type and the level of economic organization. Accumulated experience opens the way to macroanalysis based on the massive aggregate data of the census of 1917.[8]

Lenin's discussion of the capitalist and labor-rental systems would lead us to suppose that manorial economies of the capitalist type would exhibit close

---

[8] The methods used in the pages that follow do not exhaust the possible approaches to the study of productive level and the role of hired labor. Ongoing research indicates that factor analysis and multiple regression are promising tools, especially in undertaking a regional approach to the peasant and manorial economies in 1917.

correlations between the use of hired labor and the supply of cattle and equipment, by virtue of their resort to capitalist hiring practices. Where labor rental prevailed, the correlation would be weak, since the peasants would provide the cattle and equipment to work the squires' fields. To verify this hypothesis, we turned to the inventories of estates mortgaged to the Nobles' Bank and selected those inventories that, along with statistical data, directly stated how the squire ran his estate, making it possible to group these estates according to whether labor-rental or capitalist relations prevailed.[9] After calculating the linear correlations among the statistical data concerning the two types of estate and analyzing these correlations, we established patterns of correlation characteristic of each type. As a result we obtained quantitative models of the two types, enabling us through correlational analysis to ascertain the socio-economic type of a group of estates without any direct evidence as to their organization or management and so to establish, on a nationwide level, the nature of the manorial economy on the eve of the October Revolution.

The extent of capitalist development in the manorial economy is strikingly expressed in its productive structure: the correlations between the extent of sown area, on the one hand, and the supply of draft animals and equipment and the use of hired labor, on the other, are positive and strong. The significant proportionality between an estate's supply of draft animals and equipment and its use of hired labor attests to the sway of capitalist relations in the manorial economy. On the eve of the revolution, labor rental was outmoded, and the possibilities for its use were very limited.[10]

The extent of capitalist development in the manorial economy was linked to the pattern of landownership. Thus the proportion of estates using hired labor correlates negatively with the overall size of estates (-.48), with the amount of plowland (-.44) and of sown area (-.72) and the proportion of all lands owned by squires (-.40). In areas where manorial landownership most predominated, estates took on the character of latifundia, a form that retarded bourgeois agrarian evolution down to the revolution. Note also that the intensity of the capitalist development in manorial agriculture was closely linked with the extensivity of that development; see the correlations between the use of hired labor and the first four variables in Table 3A.

---

[9] On this source and it use, see N. B. Selunskaia, "Modelirovanie sotsial'noi struktury pomeshchich'ego khoziaistva Evropeiskoi Rossii kontsa XIX—nachala XX v.," in I. D. Koval'chenko et al., eds., *Matematicheskie metody v issledovaniiakh po sotsial'no-ekonomicheskoi nauke* (Moscow, 1975), p. 166, and I. D. Koval'chenko and N. B. Selunskaia, "Labor Rental in the Manorial Economy of European Russia at the End of the Nineteenth and Beginning of the Twentieth Centuries," *Explorations in Economic History*, vol. 18, no. 1 (1981), pp. 10-11.

[10] Of the two types of labor-rental, the second was more enduring; the correlation between the proportion of sown area that squires worked with their own draft animals and the number of hired workers per *desiatina* of sown area was positive, but weak. See I. D. Koval'chenko, N. B. Selunskaia and B. M. Litvakov, *Sotsial'no-ekonomicheskii stroi pomeshchich'ego khoziaistva Evropeiskoi Rossii v epokhu kapitalizma* (Moscow, 1982), Table 30, p. 201.

Table 3: The Internal Structure of the Manorial Economy
of European Russia in 1917

A. Correlation Coefficients, Calculated per *Desiatina* of Sown Area

	1.	2.	3.	4.	5.	6.	7.	8.	9.	10.	11.	12.
1. Hired workers	—											
2. Draft animals	.90	—										
3. Plows	.90	.91	—									
4. Advanced equipment	.89	.91	.89	—								
5. Cattle	.95	.96	.91	.91	—							
6. Percent of estates with hired labor	.80	.80	.88	.88	.73	—						
7. Hired laborers on estates using them	-.71	-.75	-.78	-.78	-.73	-.69	—					
8. Arable per estate	-.67	-.73	-.72	-.73	-.69	-.48	.54	—				
9. Plowland per estate	-.87	-.81	-.77	-.76	-.89	-.44	.69	-.63	—			
10. Sown area per estate	-.81	-.83	-.79	-.78	-.92	-.72	.77	.49	-.74	—		
11. Proportion of plowland sown	.95	.80	.87	.86	.86	.53	-.72	-.53	-.72	.74	—	
12. Manorial percentage of total sown area	-.96	-.89	-.87	-.86	-.87	.55	-.77	-.48	-.76	-.73	-.89	—

B. Correlation Coefficients, Calculated per Estate

	1.	2.	3.	4.	5.
1. Hired workers	—				
2. Draft animals	.89	—			
3. Plows	.86	.80	—		
4. Advanced equipment	.86	.86	.92	—	
5. Cattle	.82	.90	.88	.86	—

Source: I. D. Koval'chenko, N. B. Selunskaia and B. M. Litvakov, *Sotsial' no-ekonomicheskii stroi pomeshchich' ego khoziaistva Evropeiskoi Rossii v epokhu kapitalizma* (Moscow, 1982), p. 200.

It follows that the manorial share of sown area was higher where the reliance of estates on hired laborers was heavier, the proportion of plowland sown was higher, and the average size of estates was smaller; again we see that capitalist development enhanced the squires' share of output and latifundia diminished it.

Nationwide, however, it was clearly impossible for the squires to maintain and enhance their share of agricultural output and raise the productive level of their estates at the same time. The negative correlations between their percentage of the area sown by all producers and measures of socio-economic and productive level (that is, between variables 1-5 and 12 in Table 3A) attest to that. The consequence was a steady decline in their contribution to agricultural output.

Squires could maintain and enhance their position only by shifting their operations onto a capitalist footing; the correlation in Table 3A between the percentage of estates using hired labor (variable 6) and the proportion of plowland sown on estates (variable 11) is characteristic of most regions of European Russia. The laws inherent in capitalist commodity production, however, compelled the squires not only to shift to the exploitation of hired labor but also to elevate the productive and socio-economic levels of their estates. Insofar as their capacity to achieve the latter was limited, their contribution to agricultural output declined.

Let us sum up our observations and findings concerning the manorial economy of European Russia. First of all, at the end of the capitalist era, capitalist relations prevailed everywhere in the internal structure of estates. Labor-rental was completely played out as a viable form of production; squires had to rely on their own productive resources and on hired labor. At the same time, the intertwining of capitalist and semi-feudal modes of organization, as well as the privileges the squires still had and the sheer extent of their landholdings, did have significance. Individual squires could be agrarian capitalists while retaining traits and attitudes associated with serfdom. As a result, the struggle against the vestiges of serfdom (and hence against manorial landholding), the struggle against the squires as semi-feudal lords, was also a struggle against squire-capitalists, against the squire as bourgeois. The proletarian socialist revolution had to finish the tasks of the bourgeois revolution, and the liquidation of manorial landholding and the nationalization of all the land were the prelude to the socialist transformation of agriculture. Furthermore, the manorial economy was objectively incapable, either by virtue of its contribution to agricultural output or by virtue of its role in agro-technical progress, of taking the lead in bourgeois agrarian evolution and determining the course of further development. Towards the end of the capitalist era, the so-called "Prussian"[11] form of

---

[11] The "Prussian" path of capitalist agrarian development, based on the large estate, is conventionally contrasted to the "American path," based on the family farm.

conservative bourgeois agrarian evolution clearly had no objective prospects of success.

Productive Level and Hired Labor in the Peasant Economy

A small peasant holding becomes bourgeois, according to Lenin, under the "domination of a commodity economy" and by virtue of "the conversion of labor power into a commodity and the means of production into capital."[12] Hence, to study the peasant economy one must investigate the process of the development of capitalism within the peasant household as a productive economic unit and then analyze social relations, especially the development of labor hire and its various social and economic consequences.

Table 4: Correlations within the Economic Structure
of the Peasant Economy in 1917

A. Calculated per Household

	1.	2.	3.	4.	5.	6.	7.	8.
1. Plowland	—							
2. Sown area	.79	-						
3. Hired workers	.54	.47	—					
4. Draft animals	.80	.79	.63	—				
5. Cattle	.59	.35	.52	.55	—			
6. Plows	.17	.31	.46	.43	.21	—		
7. Advanced equipment	.65	.84	.63	.74	.18	.43	—	
8. Arable	.42	.11	.40	.47	.58	.02	.09	—

B. Calculated per Person

	1.	2.	3.	4.	5.	6.	7.	8.
1. Plowland	—							
2. Sown area	.76	—						
3. Hired workers	.46	.43	—					
4. Draft animals	.78	.73	.56	—				
5. Cattle	.54	.24	.44	.51	—			
6. Plows	.10	.22	.40	.35	.08	—		
7. Advanced equipment	.62	.84	.57	.69	.08	.37	—	
8. Arable	.40	.08	.30	.50	.60	.05	.04	—

Source: As Table 1.

Soviet historians, drawing on hard primary data,[13] have shown the extent to which the peasant economy became subject to the laws of capitalist commodity production. Although their studies deal with the first stage of this subjection, they are important for a general appraisal of the process as a whole on a nationwide scale, for in developing a method for a macroanalysis of the peasant

[12] V. I. Lenin, *Polnoe sobranie sochinenii* (henceforth *PSS*), 5th ed., 55 vols. (Moscow, 1958-65), vol. I, pp. 450-51.

[13] Notably the rich array of peasant household budget studies. See especially I. D. Koval'chenko, "O burzhuaznom kharaktere krest'ianskogo khoziaistva...," *Istoriia SSSR*, 1983, no. 5.

economy of European Russia, one can take as a point of departure the subjection of all types of peasant households to the objective laws of commodity production. As a result we can use correlational analysis of the indices of the development of the peasant economy to study its economic structure and social organization, drawing on the data of the 1917 census.

We recognize that, when the components of a peasant household economy are well balanced, that balance must be an expression of the capitalist organization of production. Hence it is possible to study the economic structure of the peasant economy on the basis of aggregate data, such as the census of 1917, by analyzing the character and the closeness of the correlation of structural components. This is a tried and tested method. We chose indices of three kinds: mean area of land, of plowland and of sown area to register the size of units and the scale of production; supply of draft animals, of cattle, of tools, and of advanced equipment to register the agro-technical basis of production; and data on the employment of hired labor to register the social aspect of economic organization. We went on to calculate these indices per household, per person, per *desiatina* of sown area and per *desiatina* of plowland. By this means we were able to take into account regional variations in the number of peasant households, in the labor force at their disposal, and in the scale of production. The results are shown in Tables 4 and 5.

Consider first Table 4, which reckons the components of production per person and per household. The correlations among most variables express the fine balance among the components of the peasant economy. Both per person and per household, the correlations among such key elements as the use of hired labor, supply of draft animals, cattle, plows and advanced equipment are rather close. So, too, are those between sown area, which registers the scale of agricultural production, and other variables relating to production. As regards the mean of arable, which simply registers the overall size of the units, it is not so closely associated with the productive base and did not determine the character of the productive structure of the peasant household.

We see the same kind of balance when we turn to variables calculated per *desiatina* of plowland and of sown area, but there are also some significant differences. The correlations in Table 5 are generally closer than those in Table 4, with correlations between advanced equipment and machines and other components of the productive base of the peasant economy representing an exception. This difference derives from different forms of measurement. In Table 5

Table 5: Correlations within the Economic Structure of the Peasant Economy in 1917						
A. Calculated per *Desiatina* of Sown Area						
	1.	2.	3.	4.	5.	6.
1. Workers	—					
2. Hired laborers	.55	—				
3. Draft animals	.71	.50	—			
4. Cattle	.74	.50	.78	—		
5. Plows	.58	.53	.46	.38	—	
6. Advanced equipment	-.50	.06	-.20	-.48	-.07	—
B. Calculated per *Desiatina* of Plowland						
1. Workers	—					
2. Hired laborers	.76	—				
3. Draft animals	.97	.71	—			
4. Cattle	.96	.69	.96	—		
5. Plows	.88	.78	.87	.85	—	
6. Advanced equipment	.12	.32	.22	.04	.21	—

Source: As Table 1.

we have "purer" measures of the extent to which the agricultural production of peasant households was provided with human and animal resources and with equipment, especially equipment for plowing. Our findings are significant, for they make manifest the balance among the components of production and hence the subjection of the peasant economy to the regulative force of capitalist commodity development. We should say a word about advanced equipment and agricultural machinery. We observed above that in this area of agrarian technology, the peasant economy lagged behind the manorial economy—in some regions, the lag was significant; this component of the peasant economy was out of proportion to the others. Note also the negative correlation between fluctuations in advanced equipment and in household labor power per *desiatina* of sown area. As the technological level of peasant agricultural production rose, there was a corresponding decline in the need for labor. Rationalization and intensification claimed their due. The association between advanced equipment and cattle of all kinds is also negative (-.48) because the increase in the size of herds and the improvement of agricultural technology were two manifestations of an emerging specialization of peasant households in either cattle-raising or in cereal cultivation. This conclusion is confirmed by calculation of the same correlation coefficient for various regions.

Our analysis convincingly shows that on the eve of the October Revolution, the productive structure of the peasant economy of European Russia was of a capitalist character. The subjection of the production process to the demands of the commodity market was a general tendency, affecting all types and sizes of peasant households and was manifest in rationalization, intensification and specialization.

<table>
<tr><td colspan="6" align="center">Table 6: Indices of Labor Hiring by Peasant Households
and of Proletarianized Households: Correlation Coefficients</td></tr>
<tr><td></td><td>1.</td><td>2.</td><td>3.</td><td>4.</td><td>5.</td></tr>
<tr><td>1. Proportion of households hiring laborers</td><td>—</td><td></td><td></td><td></td><td></td></tr>
<tr><td>2. Peasants' share of hired laborers</td><td>.86</td><td>—</td><td></td><td></td><td></td></tr>
<tr><td>3. Hired laborers per hiring household</td><td>.86</td><td>.99</td><td>—</td><td></td><td></td></tr>
<tr><td>4. Proportion of households without land</td><td>.61</td><td>.76</td><td>.72</td><td>—</td><td></td></tr>
<tr><td>5. Proportion of households not sowing</td><td>.51</td><td>.61</td><td>.56</td><td>.89</td><td>—</td></tr>
<tr><td colspan="6">Source: As Table 1.</td></tr>
</table>

The impact of the development of capitalism on the social organization of the peasant economy found primary expression in a resort to hired labor. We can measure this impact by modelling that organization and analyzing the correlation of its components with variables registering various aspects of the application of hired labor. Table 6 conveys abundant information about the role of hired labor in the peasant economy on the eve of the October Revolution. The correlations between variables registering the use of hired labor and variables registering households with no land or no sown area reflect the emergence within the peasantry of a body of hired laborers, rural proletarians who had given up cultivation on their own and sold their labor power. The data show that "a necessary condition to the existence of rich peasants was the formation of a contingent of full-time agricultural workers [*batraki*] and day laborers."[14] The increase in the size and importance of this contingent led to an increase in the proportion of peasant households that relied on hired labor—that is, operated on a bourgeois footing.

<table>
<tr><td colspan="2" align="center">Table 7: The Application of Hired Labor in Agriculture:
Correlations with the Proportion of Hired
Laborers among Workers of Both Sexes</td></tr>
<tr><td>1. Proportion of peasant households present using hired labor</td><td>.61</td></tr>
<tr><td>2. Peasants' share of hired laborers</td><td>.52</td></tr>
<tr><td>3. Hired labor per hiring peasant household</td><td>.54</td></tr>
<tr><td>4. Draft animals per household</td><td>.48</td></tr>
<tr><td>5. Plows per household</td><td>.56</td></tr>
<tr><td>6. Advanced equipment per household</td><td>.54</td></tr>
<tr><td>7. Peasants' share of draft animals</td><td>-.55</td></tr>
<tr><td>8. Non-peasant landowners without draft animals</td><td>-.53</td></tr>
<tr><td>9. Hired laborers per non-peasant unit hiring</td><td>.66</td></tr>
<tr><td>10. Draft animals per estate</td><td>.57</td></tr>
<tr><td>11. Non-peasant landowers' share of sown area</td><td>.51</td></tr>
<tr><td colspan="2">Source: As Table 1.</td></tr>
</table>

One aspect of an analysis of the development of hired labor is establishing the interaction between the peasant economy as a whole and the extent of hired labor in the agriculture of European Russia. Table 7 indicates that the overall

---

[14] Lenin, *PSS*, vol. III, p. 100.

development of hired labor was closely linked to its increased use in both the peasant and the manorial economies. The increase in the proportion of hired laborers is closely correlated with the increase in the proportion of peasant households using them, with their weight in the peasant labor force, and other variables reflecting the development of capitalism in the peasant economy, but the increase is also linked with variables registering the use of hired labor in the manorial economy.

The development of capitalist relations in the peasant economy and the increased use of hired labor was associated with the breakdown of the peasantry into strata; the bottom stratum formed a contingent of households with no horses and no cattle.[15] The process of stratification is manifest in the correlations between the proportion of hired workers and the peasants' share of draft animals; the lower the provincial mean, the greater the extent that hired labor was used. The supply of peasant cattle decreased as a result of the increase in households with no cattle, and the consequence was the increased use of hired labor within the peasant economy. The same tendency stimulated squires to acquire their own draft power and to operate with their own equipment and resources, using hired labor. The development of hired labor in agriculture is also linked to the bourgeois character of the productive structure of the peasant economy, which we can see in the fine balance among its basic components, reflected in the coefficients in rows 3-6 of Table 7.

Table 8: The Manorial and the Peasant Economies: Correlation Coefficients					
	1.	2.	3.	4.	5.
1. Proportion of estates without cattle	—				
2. Proportion of estates without draft animals	.91	—			
3. Proportion of peasant households using hired labor	-.52	-.48	—		
4. Peasants' share of hired laborers	-.38	-.38	.86	—	
5. Proportion of estates using hired labor	.09	.13	-.56	-.40	—
Source: As Table 1.					

The relationship between the proportion of hired labor and the characteristics of the non-peasant rural economy deserves special attention. The development of capitalist relations in the manorial economy (expressed through increases in participation in production, in the use of hired labor and in supplies of draft animals) naturally contributed to the overall expansion of hired labor in agriculture; see Table 7, rows 9-11. Correspondingly, in provinces that had a large proportion of estates with no draft animals and were, therefore, bastions of the labor-rental system, hired labor was used on a lesser scale; see Table 7, row 8. Furthermore, these estates impeded the spread of hired labor within the

---

[15] Lenin, *PSS*, vol. III, p. 101.

peasant economy; in the provinces with a greater share of estates with no cattle and with no draft animals, fewer peasant households used hired labor, and the proportion of hired laborers in the peasant labor force was smaller, as we can see from Table 8. Note also the negative correlation in Table 8 between the proportion of estates using hired labor and the proportion of peasant households that did so. Here we can detect regional variations in the Prussian and American paths of bourgeois agrarian development. Where the manorial economy dominated that development, it retarded the development of capitalist social relations in peasant production. Conversely, the sway of the ''American'' variant was accompanied by a more passive approach by the squires to the reconstruction of the socio-economic aspect of their operations. This process is apparent in the correlation between the proportion of peasant households using hired labor and all the other variables in Table 9, including the manorial contribution to cereal production. In other words, the higher the level of the peasant economy, the more peasants resorted to hired labor.

Table 9: Proportion of Peasant Households Using Hired Labor: Correlation Coefficients	
1. Peasants' share of sown area	.59
2. Draft animals per head among peasants	.62
3. Land per person among peasants	.63
4. Hired workers as a proportion of the peasant labor force	.57
5. Manorial share of sown area	-.51
Source: As Table 1.	

These are some results of a correlational analysis of the peasant and manorial economies on the eve of the October Revolution.